Paris After Haussmann

History of the Urban Environment

Martin V. Melosi and Joel A. Tarr, *Editors*

Paris After Haussmann

Living with Infrastructure in the City of Light, 1870–1914

Peter S. Soppelsa

University of Pittsburgh Press

Published by the University of Pittsburgh Press, Pittsburgh, Pa., 15260

Printed on acid-free paper
10 9 8 7 6 5 4 3 2 1

Cataloging-in-Publication data is available from the Library of Congress

Hardcover: 978-0-8229-4882-7
Paperback: 978-0-8229-6793-4

Cover art: Unused J. H. postcard no. 070, "PARIS—Old Montmartre" series, with the caption "Stairway on the Passage Cottin still lighted with oil," ca. 1900s. Author's collection.

Cover design: Alex Wolfe

Publisher: University of Pittsburgh Press, 7500 Thomas Blvd., 4th floor, Pittsburgh, PA 15260, United States, www.upittpress.org

EU Authorized Representative: Easy Access System Europe, Mustamäe tee 50, 10621 Tallinn, Estonia, gpsr.requests@easproject.com

Contents

Acknowledgments

This book reflects twenty years of research, leaving lots of allies to acknowledge. I begin with loving thanks for my parents, Betty and Bob Soppelsa, who taught me to love learning and travel, made French our second language, and introduced me to Paris in 1978. They have provided financial, editorial, and moral support for my entire career, reading drafts, and pounding the Paris pavement for apartments, postcards, and my favorite cologne.

This book began during my studies in the University of Michigan's history department. My faculty mentors deserve special thanks. Gabrielle Hecht taught me science and technology studies (STS) and was my toughest critic, always encouraging excellence in interdisciplinary work. Josh Cole, always kind and generous, sharpened my French history chops and prepared me for work in Paris archives. Dario Gaggio introduced me to urban history and theory and drove with me to my first academic conference. Geoff Eley was an important mentor, who taught me to read broadly and to relish historiography and hired me as a research assistant.

Visiting the University of Michigan (U of M) in 2001 as a prospective student, I met several people who showed how distinctive this institution was and remained strong allies on and off campus through my time there: faculty Kathleen Canning, John Carson, Scott Spector, and Regina Morantz-Sanchez, and classmates Josh First and Matt Ides, who became close friends. Beyond outstanding courses with Professors Canning, Spector, Eley, Gaggio, and Hecht, I was fortunate to study history, sociology, and STS with Laura Downs, Paul Edwards, Joel Howell, Michelle Mitchell, Jeffrey Paige, Jason Owen-Smith, and Ulricke Weckel. U of M's world-class faculty is matched by a large and remarkable graduate student community. I was lucky to share classes, drinks, and much more with fellow students Allison Abra, Sarah Avery, Wojciech Beltkiewicz, Chandra Bhimull, Megan Biddinger, Lucinda Bingham, Ross Bowling, Rebecca Brannon, Katherine Brion, Tamar Carroll, Andrew Cavin, Anna Cichopek, Saskia Coenen, Nathan Connolly, Andrew Conroe, Tyler Cornelius, James Dator, Heloise Finch-Boyer, Sara Babcox First, Tracy Fleming, Jessica Fripp, Erika Gasser, Emma Grama, Andrew Highsmith, Erik Huneke, Andrew Ivaska, Ralph Johnson, Dmitri Krallis, Robert Kruckeberg, Peter Lawless, Ursula Lawrence, Nita Luci, Delphine

Mauger, Will Mackintosh, Clapperton Chakanetsa Mavhunga, Adam McCormack, Minayo Nasiali, Andrew Needham, Jennifer Palmer, Sasha Pfau, Meg Raphoon, Isabelle de Rezende, Andrew Ross, Tanya Saunders, Deborah Solomon, Herbert Sosa, Kim Strozewski, Sean Takats, Sam Temple, Lenny Ureña, Matt Whitman, Jon Wipplinger, and Tara Zahra.

Ken Garner deserves special thanks for truly outsized generosity. In Ann Arbor, in Paris, and at conferences, he chased archival sources for me, introduced me to editors, and once found me a summer job. In 2006, fellow graduate students Roberta Pergher and Anne Berg tapped me and Charles Gentry to organize an interdisciplinary graduate student workshop on visual culture called "Visual Archives." Thanks to Charles for being such an excellent collaborator and friend, and together with Josh First, for sharpening my tools of visual analysis. Years of membership in the Graduate Employees Organization introduced me to a remarkable community who taught me activist practice, leftist thinking, and the power of solidarity. Thanks to: Tom Allison, Andrew Clarno, Jon Curtis, Claire Decoteau, Cedric De Leon, Andrea Dewees, David Dick, Mark Dilley, David Dobie, Ian Fulcher, Liz Goodnick, Federico Helfgott, Julie Jedlicka, Gabe Kirchner, Bella Muntz, Holly Rider, Afia Ofori-Mensa, Peter Morrow, Patty Mullally, Gwynne Osaki, Rodolpho Palma, Alyssa Picard, Kyle Rearick, Tracie Rubeck, Jae-Jae Spoon, Nick Syrett, Urmila Venkatesh, Kevin Wildrick, and André Wilson. Off campus, Rob Haug, Matt Ides, Adam McCormack, and Herbert Sosa were fun and thoughtful roommates. Dear friends kept me sane by sharing activism, hikes, music, parties, bar nights, and TV nights with me: Mark Carlson, Dave Reminick, Matt Rienk, Chris Riggs, Dan Rose, Andrew Tholl, and Mia White.

My research took me to the Bibliothèque Nationale de France, the Bibliothèque Administrative de la Ville de Paris, the Bibliothèque Historique de la Ville de Paris, the Bibliothèque Interuniversaire de Santé, and the Archives Nationales de France, but above all the Archives de la Ville de Paris, whose staff raises the bar for hospitality to foreign researchers. There, I met the kindly genealogist Thierry Caloin, my volunteer French uncle, who hosted a memorable weekend road trip to his brother's Loire Valley home. In 2006, I met Jeff Jackson there while researching the 1910 Flood. Since then, Jeff has edited and cited my work, pointed me to primary sources, invited me to speak, hosted me in his home, and hung out in Memphis, in Paris, and at conferences. At the Bibliothèque Administrative, the librarian Agnès Tartie showed me the library's extensive collection of Métro construction photographs and shared her wisdom on the subject. My time in Paris was sweetened by a lovely and welcoming community of fellow expats and their French allies: Alan Cox and Elsa Lohier, Catalina Alvarez, Jorge Gonzalez, Carmen Arza Hidalgo and Sylvain Piot, and Andres Daniel Sainz. Their outsized hospitality and generosity made this outsider feel at home in Paris. So did visits from Will Blackhurst, Nick Christianson, Tracy Flemming, John Kennedy, Nita Luci, and Suzanne Newcombe. I enjoyed visiting Sasha Pfau and David Barber in the Paris suburbs, Josh Cole's family in Aix-en-Provence, and Kim Strozewski and Josh First in Prague.

During my 2009–2010 postdoc year at Oberlin College, my gracious hosts in the history department included Ari Sammartino, Len Smith, Sam White, Carol Lasser, Gary Kornblith, Heather Hogan, Ellen Wurtzel, and fellow postdoc Brian Yates. I also appreciated conversations with French studies colleagues Grace An, Jack Murphy, and Amy Wartheson. The Cleveland water commissioner J. Christopher Nielson hosted a truly memorable tour of the Baldwin Water Treatment Plant. Oberlin's Office of Sponsored Programs funded undergraduate research assistance from Dennison Goodrich-Schlenker. Oberlin also funded a 2010 trip to Paris, where I was lucky to meet Catherine Clark and Brian Jacobson at a saucy ski-themed H-France bar night. They have remained good friends and colleagues ever since.

Later that year, I joined the Department of the History of Science, Technology, and Medicine at the University of Oklahoma (OU), my professional home ever since. For their unflinching hospitality, mentoring, encouragement, and support, I thank the department chairs Steve Livesay, Hunter Heyck, and Stephen Weldon, and the administrators Clebbie Riddle, Stella Stuart, and Sam Fellows. My *Technology and Culture* mentors and coeditors from 2010 to 2020, Hunter Heyck and Suzane Moon, deserve special thanks. Suzanne taught me the craft of scholarly editing and worked with me in the spirit of deepest collaboration as a coeditor and coteacher, and fellow lover of cats and guitars. Hunter has been a loyal ally on and off campus, from leading the effort to get me a tenure line to sharing childcare and karaoke. JoAnn Palmeri graciously helped me locate illustrations and other sources in the History of Science Collections. I would like to thank all the graduate students in my seminars, "Historical Technology Studies" (2014) and "Infrastructures" (2021) for workshopping key concepts in this book with me. I would especially like to thank Blair Stein and Amy Rodgers, my office assistants at *Technology and Culture* for many years and my co-authors of essays in that journal. Outside my department, Norman life has bonded me with terrific friends and colleagues: Dan Mains and Alise Osis, Erika Robb Larkins and Michael Larkins, Robert Scafe and Lisa Boyd, Dan Brobst, Moira Ozias, Kyle Moore, Jennifer Davis and Rangar Cline, Dylan Herrick and Susan Skapik, Mark Norris, Noah Boone and Kate Vincent, Robert Bailey and Maura McAndrew. Dan Mains and Robert Bailey offer a bottomless well of friendly informal chats about serious academic issues, an indispensable incubator for undercooked ideas.

Robert Pergher invited me to the University of Kansas (KU) in 2012 to talk about Paris's water supply. A 2013 Presidential International Travel Fellowship from OU took me to Paris after the Manchester International Congress for the History of Science, Technology and Medicine, where Fabienne Chevallier, Sabine Barles, Jonathan Strauss, Andrew Aisenberg, and Sun-Young Park joined me to discuss Paris's nineteenth-century hygiene movement. On that same trip, I was lucky to visit Carlos Lopéz Galviz and Liz Hosmer in London, and Dmitri Krallis and Amanda Morley in Oxford. At a 2014 History of Science Colloquium talk about this project at Johns Hopkins University, Lawrence Principe corrected a critical error in my understanding of mercury's role as a conductor in electrical circuits. The 2014–2018 Francophone World Workshop at OU connected me with key allies in French studies: Julia Abramson, Jennifer Davis, Pamela Genova, Sarah Griswold, and Jess Pearson.

In 2015–2016, the OU Humanities Forum, under director Janet Ward, provided a fellowship that funded travel to Paris and postcard hunting. The forum also introduced me to co-fellows Robert Bailey, Laurel Smith, Jennifer Saltzstein, Todd Stewart, and Zev Trachtenberg. Zev's working group on the Anthropocene raised the bar for interdisciplinary dialogue and introduced me to scholars from across the campus who have sharpened my environmental thinking: Amit Baishya, Sarah Hines, Robert Lifset, Bryce Lowery, Asa Randall, Cindy Rosenthal, Jennifer Ross, Ingo Schlupp, Lynn Soreghan, Noah Theriault, Traci Voyles, and Jim Ziegler. Working with Kathy Brosnan and Dave Hambright under the auspices of EarthMonth@OU also improved my grasp of environmental studies. In 2016, my department chair, Hunter Heyck, sponsored a manuscript development workshop for this book. I am grateful to Richard Keller, Miriam Levin, and Rosemary Wakeman for participating and providing transformative feedback. That year, undergraduate research assistance from Lexi Jovanovic and Collin Olson was supported by OU's Honors College Research Assistance Program (HRAP) and its administrator Jeff Cooper.

In 2017–2018, New York University's Tandon School of Engineering organized two summer conferences on urban infrastructure in Paris and in Venice, hosted by Joseph Heathcott, Jonathan Soffer, and Rae Zimmerman. I thank Andrew Needham for inviting me, and all the organizers and participants for stimulating dialogue that helped solidify this book's methodology, articulated in my chapter in the conference's edited volume, *Urban Infrastructure: Historical and Social Dimensions of an Interconnected World.*

In 2019, Abigail Owen invited me to speak about Paris's water shortages at Carnegie Mellon, which improved chapter 7, and reunited me with Emma Gramma, Noah Theriault, and Joel Tarr. In 2020, as the COVID pandemic struck, I found online community with the best group of writing buddies I could ask for, the I-35 French historians Nicole Bauer, Melissa Byrnes, Jennifer Davis, Sarah Griswold, and Jess Pearson. Their criticisms, encouragement, and insight have vastly enriched this book and have reminded me to "write dirty." The OU Arts and Humanities Forum, under director Kim Marshall, provided a teaching buyout in 2022 to help me finish the book, and supported a relaunched Francophone World Workshop in 2023–2025.

Professional associations have nourished this work, too, especially the Society for the History of Technology and the Society for French Historical Studies, but also the American Historical Association, the American Society for Environmental History, the Association for the Study of Literature and Environment, the Urban History Association, and the Western Society for French History. For sharing conference panels and conversations about this research, I warmly thank Chandra Bhimull, Catherine Clark, Andy Denning, Catherine Dunlop, Cindy Ermus, Elizabeth Everton, Brian Jacobson, Laura Kalba, Richard Keller, Scott Knowles, Whitney Laemmli, Miriam Levin, Michael Mulvey, Minayo Nasiali, Andrew Newman, Patrick Luis Sullivan de Oliveira, Davide Orsini, Sun-Young Park, Bob Post, Sara Pritchard, Andrew Ross, Edmund Russell, Rebecca Scales, Barbara Schmucki, Rudi Volti, Alexia Yates, Tom Zeller, and Geoff Zylstra.

I am also grateful to the editors who have encouraged and improved my published work: Elinor Accampo, Masha Belenky, Rob Boddice, Herrick Chapman, Hollis Clayson, Martin Collins, André Dombrowski, Benjamin Fraser, Joseph Heathcott, Jeff Jackson, Gijs Mom, Anne O'Neil-Henry, Donna Ryan, Vanessa Schwartz, April Shelford, Steven Spalding, and Alexander Vari. At the University of Pittsburgh Press, editors Amy Sherman and Kelly Thomas, series editors Joel Tarr and Martin Melosi, and two anonymous referees have helped improve this book in many ways large and small. My deepest thanks go out to all of them for their belief in this project and their efforts to perfect and promote it.

Parts of this book incorporate and rework previous publications. I thank the publishers for permission to reuse material here. These include: "Finding

Fragility in Paris: The Politics of Infrastructure After Haussmann," *Proceedings of the Western Society for French History* 37 (2009): 233–47; "Visualizing Viaducts in 1880s Paris," *History and Technology* 27 no. 3 (September 2011): 371–77; "Urban Railways, Industrial Infrastructure and the Paris Cityscape, 1870–1914," in *Trains, Culture and Mobility: Riding the Rails*, ed. Benjamin Fraser and Steven Spalding, 117–44 (Lexington Books, 2011); "The Instrumentalisation of Horses in Nineteenth-Century Paris," in *Anthropocentrism: Humans, Animals, Environments*, ed. Rob Boddice, 245–64 (Brill, 2011); "Paris's 1900 Universal Exposition and the Politics of Urban Disaster," *French Historical Studies* 36, no. 2 (Spring 2013): 271–98; "Reworking Appropriation: The Language of Paris Railways," *Transfers: Interdisciplinary Journal of Mobility Studies* 4, no. 2 (Summer 2014): 104–23; "How Haussmann Haunted the Early Third Republic," in *Is Paris Still the Capital of the Nineteenth Century? Essays on Art and Modernity, 1850–1900*, ed. Hollis Clayon and Andre Dombrowski, 35–51 (Routledge, 2016); and "The End of Horse Transportation in Belle Epoque Paris," *Interdisciplinary Studies in Literature and Environment* 24, no. 1 (Winter 2017): 113–29.

Finally, I lovingly thank Jen and Felix for tolerating my repeated absence during research trips and reminding me that I do my work so that the three of us can enjoy our best life together. In 2025 our time in Paris with my parents, overlooking the Bièvre Valley, was a joyful celebration of a birthday, a wedding engagement, and a book contract. As I was writing this book, the bridge in front of our house washed out, leading to months of construction that shaped our shared experience of living with infrastructure in Norman, Oklahoma.

Paris After Haussmann

Figure I.1. Unused ELD Postcard, 1914, "The Storm of 15 June in Paris: The piping laid bare on the Rue du Havre." Author's collection.

INTRODUCTION

Living with Infrastructure

On 15 June 1914, a powerful storm in Paris left sinkholes and explosions and fires from broken gas lines in its wake. As often happened during extreme events, photographers walked the battered streets documenting damage. Their photographs became postcards like the one shown in figure I.1, from the Paris printing firm ELD, named for its founder, the photographer Ernest Louis Désiré le Deley (1859–1917), which depicts a large sinkhole with the caption "The piping laid bare on the Rue du Havre." The storm did what scholars call "inverting" or "surfacing" infrastructure, exposing the normally buried and hidden sewer tunnels, as well as water, gas, and electric lines, to air, light, vision, and scrutiny.[1] Inversion also appears here as the contrast between the lamppost standing on the sinkhole's edge at the top left amid the rubble, and the toppled lamppost, trailing its supply line atop the wreckage below. The photograph mimicked the storm in "laying bare" hidden infrastructures, visualizing them for people beyond Paris and for posterity.

By revealing underground infrastructures, the postcard also illustrates two relationships at the heart of this book: the relationship between infrastructure and the environment, which technology studies call "envirotechnical," and the relationship between infrastructure and society, which these studies call "sociotechnical."[2] The environmental aspects of this scene encompass both the extreme weather that opened the sinkhole and the resulting breach of surface and subterranean environments. The social dimension is conveyed by the people depicted or not depicted. One worker (top right) surveys the scene, while another worker (bottom right) descends below ground. They signal the labor of building, operating, maintaining, and repairing infrastructures. Meanwhile, along the image's top edge, dozens of bystanders press against a makeshift barrier for a better view, illustrating the public concern and curiosity that greeted this inversion of infrastructure and normality. These spectators signal popular reception of infrastructure, their emotional and sensory experience of technology, and infrastructure's roles as part of everyday life, a danger, and a spectacle. The people not depicted here include the engineers and planners who designed the infrastructures and the photographer, perhaps Le Deley himself, although attributing postcard photos is difficult. The branded postcard and its anonymous photographer document a popular culture of representing, remembering, and reflecting on experiences of infrastructure. Thus, both the storm and its documentation "laid bare" the entanglement of infrastructure, environment, and society that characterized Paris in 1914 and drew the attention of workers, engineers, bystanders, photographers, and postcard buyers to this moment of disaster.

The sinkhole depicted was located at the edge of Paris's 8th district (*arrondissement*), among its wealthiest, most modernized, and most infrastructurally equipped. By picturing a collapse here, the card subtly indicated the fragility of Paris's most "Haussmannized" neighborhood, thus politicizing Paris's special brand of urban modernization, which had transformed the city over the preceding sixty years, and is explained in the next section, "Paris, Infrastructure, and Haussmannization." This postcard not only inverted infrastructure but also, more precisely, inverted Haussmannization.

Paris, Infrastructure, and Haussmannization

This book provides a history of Paris, which does not allow me to generalize about France. Paris was in many ways an exceptional place in France and has been exceptionally important in shaping the way we understand urban infrastructure, urban modernity, and infrastructural politics far beyond France's borders. Paris has long been recognized for its infrastructures. From the paved streets and fortifications of King Philippe-Auguste (Philippe II) around 1200 to the Pont Neuf (New Bridge), coach service, postal service, and first streetlamps in the 1600s, Paris entered the modern era after the French Revolution with the nickname "City of Light" (Ville Lumière), which metaphorically cast the capital as a beacon of enlightenment, but literally invoked its early streetlights. Thereafter, the enormous growth of gas lamps in the nineteenth century cemented the nickname, bolstering Paris's bid to define the modern city by giving infrastructure a leading role.[3]

Gaslight rode a wave of infrastructure development that swept Paris in the 1800s, creating the transformed roads, houses, subways, tramways, electric systems, sewers, water supply, and rivers discussed in this book. This wave was linked with Haussmannization, Paris's approach to urban planning and building, named for Prefect of the Seine Georges-Eugène "Baron" Haussmann (1809–1891), who led Paris's departmental government from 1853 to 1870, and founded a new Office of Works (Travaux de Paris) to direct an ambitious plan of "Great Works" (Grands Travaux) or "transformations" to modernize the capital.[4] He envisioned and pursued, but never completed, a city served comprehensively with streets, parks, fresh water, sewerage, gas, telegraphs, compressed air, horse-drawn omnibuses, and improved canals and rivers. Influential urban studies by historians, geographers, social scientists, and urbanists have given Haussmann a commanding role in creating an international "modern infrastructural ideal" that envisioned a "networked city," served by ubiquitous, uniform, and universal networks that would supply resources and remove waste, intended to achieve social progress. Haussmann's Paris became an "archetype" and "icon," defining urban modernity—that is, what makes cities modern—in terms of infrastructure. By 1900, James Scott argues, Haussmannizing Paris was "a widely admired public works miracle and shrine for would-be planners from abroad."[5] Thus, beyond its well-documented reputation as "capital" of France, modernity, progress, the nineteenth century, and even the world, Haussmannizing Paris was certainly a capital of urban infrastructure.[6]

Although the City of Light dazzles, we must remember how many places and people remained in darkness while infrastructure networks expanded, and understand infrastructure's unevenness rather than uniformity, its problems alongside its progress. For example, Paris's sparkling gaslights also raised awareness of technological risk and carried costs for environmental quality: odor, light pollution, and carbon emissions.[7] In sum, Paris helps us see infrastructural ideals and realities, distinguishing enlightenment from streetlights with more clarity, because it has not always lived up to its own infrastructural model or ideal. This book therefore limits its geographic scope to Paris to rethink the city's role as an archetype, capital, icon, and model. By documenting how Parisians pushed back against Haussmannization between 1870 and 1914, it offers a case study that will help scholars today to provincialize

Paris. The book counters Paris's claims to universality with a concrete and contingent account of a specific place, contrasting Paris the myth and model with Paris the real living and working city.[8]

This history of urban technologies, the urban environments they saturate, and their consequences for everyday life grew from my desire to find new ways of studying how infrastructures are lived, or how they are integrated into society, used, and experienced every day. Even after studies of technology and infrastructure turned toward social and cultural history and user studies, research has remained largely focused on Thomas Hughes's "system builders," engineering, institutional analyses, and top-down perspectives.[9] Paris's history from 1870 to 1914 offers a unique opportunity to study how people learned to live with increasingly ubiquitous infrastructure in a networking city. Scholars often claim that Paris, like London, was instrumental in inventing the "networked city," and this book interrogates that claim, heeding the call of recent infrastructure studies to "splinter" urbanism and move "beyond the networked city."[10] While French elites promoted Paris as a model city embodying Haussmannization for the world, and it *was* widely imitated, scholars have mostly overlooked how both Parisians and foreigners contested and politicized Haussmannization by inverting infrastructures.[11]

Not all infrastructures debuted in the 1800s; Paris's carriages and streetlights date to the 1600s. But even for older infrastructures, the nineteenth century brought massive changes in physical scale, symbolic meanings, and sociotechnical integration. Not all infrastructures are alike. Those studied here—systems for housing, transportation, sanitation, and water supply—are specifically *physical* and *urban* infrastructures. Many Americans call them "civil" infrastructures, from the field of civil engineering, or "utilities," signaling their accessibility, instrumentality, and usefulness for the public. In nineteenth-century France, and in French archives today, similar infrastructures are "public works" (*travaux publics*). This book only discusses social, information, and knowledge infrastructures indirectly, when they are relevant to, or part of, civil infrastructure.[12] Excellent social and cultural histories of Paris infrastructures exist for the parks, sewers, gas, and omnibuses developed before 1870, so I emphasize these less.[13] Similarly, while touching on electrification, I defer to research on the years after 1870 that analyzes energy-related infrastructures: artificial lighting, compressed air, and electricity.[14]

All physical infrastructure systems blend social, technological, and environmental parts to deliver resources or services to humans in a socially useful way. Water has been especially important in research literature on envirotech because water supplies tap aquifers, lakes, rivers, and springs, while canals and sewers exit into waterways.[15] Transportation and energy histories have highlighted horses as draft animals and engines for omnibuses and trams.[16] Sociologists of technology stress the way such systems join technological, human, and environmental parts to form what John Law has called "heterogeneous networks" and Bruno Latour has called "actor networks."[17] Environmental historians and geographers call them "hybrids" of nature and culture, while Donna Haraway speaks of "naturecultures" and Latour of "nature-culture."[18]

The heterogeneity and hybridity of infrastructures create uncertainty and risk.[19] All network connections are maintained by human ingenuity and work. Human control of nature is impressive, but never complete. The flooding Seine, summer heat, and striking workers sometimes brought Paris to its knees. Weather can snarl traffic. Motors pollute the air. Such are the built-in costs for the benefits, resources, and progress that infrastructures can deliver. How did Parisians come to recognize their society as a risk society? How did they debate decisions about how modern people would live with pervasive networked infrastructures? How did they experience living in a city that was not yet fully networked, but continually and contingently networking? This book argues that they did so by inverting Haussmannization.

I distinguish "primary Haussmannization," directed by Haussmann himself from 1853 to 1870, from "secondary Haussmannization," which spanned 1870 to 1914 under the Third Republic (1870–1940). This distinction supports two historical arguments. First, Haussmannization was far from finished in 1870. Many works commonly associated with primary Haussmannization were products of secondary Haussmannization. Contemporaries knew that Haussmannization peaked after 1870. Second, the terms "Haussmannize" and "Haussmannization," like the term "infrastructure," matured in the republican era.[20] This book therefore questions what Haussmann finished, resonating with recent histories of the early 1800s that contest Haussmann's originality by studying his precursors, shifting our sense of what he started.[21] Primary Haussmannization was both pathbreaking and preliminary compared with the sustained transformation of environments and technologies waged after 1870 by the republican Office of Works, although both phases inspired Parisians to politicize infrastructure.

Shifting focus to the years after 1870 raises questions about the politics of infrastructure. If primary Haussmannization was imperial, Bonapartist, and authoritarian, how was secondary Haussmannization republican, democratic, liberal, and egalitarian? As the liberal-nationalist editor and reformer Jules Simon asked of Haussmannization in 1882, could Paris "achieve through liberty that which was started by despotism?"[22] Given substantial documented continuities in personnel, ideology, and practice between the two regimes, how was what Miriam Levin has termed the republican "riff" on Haussmannization different from its primary version?[23]

One answer, following Levin, leans on the historical context of the Second Industrial Revolution to foreground the new challenges Paris faced amid its signature transformations: electrification and motorization. Powered by a sharp turn toward fossil fuels and the internal combustion engine, this energy transition sent out social and ecological ripples. In Paris and peer cities, related urban transformations included street-use battles, the adoption of urban railways, the decline of animal traction, and a seemingly endless cycle of demolition and construction. Secondary Haussmannization, therefore, witnessed the motorization and electrification of Paris's transportation infrastructures, which filled the streets with motorized trams, with the electric Métro below, as opposed to primary Haussmannization's horse-powered transit regime. Railways therefore play a central role in chapters 3–6. As the twentieth century dawned, the automobile appeared on the horizon.

Another answer, offered by both scholars and contemporaries, is sanitation. Secondary Haussmannization also differed from primary in the depth of its work on water and sanitation, vital public services with biopolitical impacts on death rates from key nineteenth-century killers, especially waterborne fecal-oral diseases like cholera and typhoid. The Office of Works addressed sanitation on many levels, from evacuation of household and industrial waste to cleaning up the river Seine and its tributaries. Sanitation shows the uneven acceptance of the germ theory of disease and hygiene practices in France—even in the nation of Pasteur. Although the biopolitical uses of infrastructure spanned both empire and republic, they intensified under the republic, as concerns about pollution, population, birth rate, and disease came to dominate national affairs, and large infrastructure systems seemed a logical response to population-level problems.[24]

This book shows how the republican agents of secondary Haussmannization attempted to invert the project of urban transformation, but struggled because of historical continuities or baggage, which in infrastructural terms is the problem of retrofit.[25] Retrofitting was especially needed in Paris, where Haussmannization followed centuries of previous infrastructure development. Haussmannization, whether primary or secondary, never reconceived or built the city from scratch; one had to carve a new Paris from the old one. New construction demanded demolitions; David Harvey has called Haussmannization "creative destruction," while James Scott has called its retrofitting "draconian."[26] But even the motives for Haussmannization show continuity from empire to republic, including biopolitics, modernization, and monumentality.

One central piece of baggage carried from primary to secondary Haussmannization was a prominent role for engineering in infrastructure planning. After the Enlightenment creation of a state engineering corps credentialed by the École nationale des ponts et chaussées (National School of Roads and Bridges), French engineers became civil servants, which politicized their work. Ideologically, they were inspired by Saint-Simonian thought, which projected enthusiasm and optimism for technology and technocracy.[27] Scholars often note Haussmann's "engineering" approach to city planning, which promoted technological solutions for social problems.[28] Similarly, scholars and nineteenth-century Parisians alike have described a strategy of "social modernity," which applied science, technology, and medicine to addressing the social challenges of modern life.[29] This strong technocentric orientation of Haussmannization through both its primary and secondary phases inspired many critiques of engineers and engineering, which are documented in the pages ahead.

Closely tied to the Second Industrial Revolution and the French engineering tradition were the universal expositions or world's fairs. Although Paris debuted as a leading host of expositions with the Second Empire's 1855 and 1867 fairs, it was during the Third Republic that the massive and progressively larger 1878, 1889, and 1900 expositions made Paris a capital of expositions embellished with iconic public works such as the Eiffel Tower and the Métro. As Miriam Levin and others have shown, this regular rhythm of expositions every eleven years, which spanned the entire duration of primary and secondary Haussmannization, had special importance for the history of infrastructure. The cycles of demolition and construction that preceded each expo to prepare and upgrade the capital for its global guests ensured that expositions were always hot spots for debate about Haussmannization and infrastructure.[30] The chapters of this book detail how engineers and politicians used approaching expositions to justify infrastructural change, while laypeople used them to criticize, invert, and politicize infrastructures. Expositions thus influenced the tempo and texture of both infrastructural change and infrastructural controversy, especially during secondary Haussmannization.

Joining engineering and expositions, a third "e" word that provides essential context for secondary Haussmannization is "empire." Imperialism was another priority shared by the Second Empire and the Third Republic that accelerated under the younger and more democratic regime, which was as aggressive abroad as it was progressive at home. While this book restricts its scope to Paris, the capital's ties to France's growing global empire provided both institutional and ideological frameworks for infrastructure development. Institutionally, infrastructure was a major tool of imperialism, which often colored the work of major public works contractors that operated both within and beyond France. For example, in 1881 alone Gustave Eiffel's firm spent eight times more on the iron viaducts installed across the Mekong Delta to subdue southern Vietnam than it later would on the Eiffel Tower.[31] The 1893 collapse of the French Panama Canal project and subsequent national scandal further indicates how public works and imperialism were institutionally intertwined. Ideologically, public works were an essential ingredient in French imperialism's claim to civilize, develop, and improve its colonies.[32] The chapters of this book show how infrastructure designers echoed imperialist rhetoric in claiming to civilize Paris's poorest people and neighborhoods, while Haussmannization's critics likened infrastructure problems and failures to the evils of imperialism—corruption, discrimination, exploitation, and violence.

Although Haussmannization famously made Paris an influential model of the modern infrastructural ideal, it also had more surprising consequences as will be documented here. Haussmannization taught Parisians how to invert infrastructure and how to politicize infrastructure, not only modeling how to build, plan, and operate physical infrastructures but also inspiring Parisians to contest, hack, negotiate, politicize, tolerate, and resist them.

This book speaks of "inverting Haussmannization" in two historical senses. First, republican elites during secondary Haussmannization tried to invert primary Haussmannization's politics, to shift its consequences from authoritarian to republican. They wanted to steer Haussmannization toward more egalitarian and utilitarian ends while expanding transportation, sanitation, and water infrastructures. But they were challenged by technological momentum, administrative continuities, and shared Saint-Simonian ideals inherited from the Second Empire, which represent baggage, continuity, and retrofitting.

Second, official inversions were countered by popular responses, through which the public tried to invert Haussmannization, to peer inside, unmask, and politicize it. Their language conveyed this inversion. In the 1880s, disgruntled bohemian tenants threatened to "invert everything" in apartment houses (chapter 2). In the early 1900s, journalists complained that the construction of public works turned Paris "upside down" (chapter 5). In 1910, commentators on all sides witnessed the city inverted by that year's catastrophic flood (chapter 8). Parisians inverted progress narratives, creating counternarratives that reversed the flow of history, and described Paris reverting to previous and worse conditions: the Commune, the Revolution, the Middle Ages—not progress but disaster, not monuments but ruins. To upend history and evaluations of progress like this broke the links between Haussmannization and the modern infrastructural ideal. In Paris, inverting infrastructure recast Haussmannization, questioned whether secondary Haussmannization succeeded in inverting primary Haussmannization, and contested narratives of modernization as progress through public works.

This book also aims to invert our scholarly view of Haussmannization in three historiographic ways. First, the book reframes Haussmannization chronologically, contributing to French history by charting continuities in public works between the Second Empire and the Third Republic, while distinguishing primary from secondary Haussmannization. By shifting emphasis to the years after 1870 for understanding Haussmannization, when the words "Haussmannization" and "infrastructure" solidified, I emphasize a wide-ranging, self-aware discourse on urban technology that became the framework for various acts of inversion and politicization, a diverse series of attempts to steer Haussmannization in new directions with new purposes. Parisians were both critical and creative in contesting what Haussmannization would become long after Haussmann retired and died. I shift emphasis from Haussmann the great man to Haussmannization, a contested historical process that belonged to many in France and beyond.

Second, the book contributes to methodological discussions of inversion in infrastructure studies by focusing on infrastructural experience and studying the biological, social, and affective dimensions of living with infrastructure, detailed in the next section, "Living with Infrastructure." A century before scholars began inverting infrastructure, Paris's Haussmannization inspired its own tradition of inverting and politicizing infrastructure. I build on recent currents in infrastructure studies that move from designers to users of technology, from experts to laypeople, and from engineering to everyday life. I show how infrastructure studies benefit from scholarly study of the sensory and emotional past by engaging visual sources—ephemeral, illustrated, and photographic—that reveal and visualize infrastructure. In Paris, the promise, progress, prices, and problems of infrastructures have long inspired their politization.

Third and finally, the book helps scholars move beyond the "networked city," by shifting focus from the already networked city as a stable artifact to networking as a continuous, contingent, and contested process. By consistently keeping technology in its social and environmental contexts, scholars can see physical infrastructures not as homogeneous, but heterogeneous, not just large technical systems, but sociotechnical and envirotechnical systems. Such a perspective troubles our association of networked physical infrastructures with uniform, universal, and ubiquitous services, social progress, and health and safety so that we can see their limits, problems, and risks, which in turn account for their inversion and politicization.

Living with Infrastructure

Recent research across the humanities and social sciences investigates the relationship of infrastructure to "life," especially social or everyday life. The historian Timothy Mitchell and his collaborators ask "how infrastructures can enable, transform, or inhibit ways of thinking and living collectively."[33] The geographers Stephen Graham and Colin McFarlane have gathered studies of "urban lives as everyday *infrastructural experience*."[34] The geographer Ash Amin implicates "lively infrastructure" in the collective production of sociality and identity.[35] Such scholarship parallels what the historian Martin Melosi calls a "cultural turn" in infrastructure studies, which has inspired research on how people make infrastructures meaningful through practice, representation, and imagination.[36] These studies foreground how infrastructures shape and are shaped by ways of life.

I build on this recent research by investigating Parisian experiences of what I call "living with infrastructure" from 1870 to 1914, as I document and historicize the lived experience of a networking city. By "living with infrastructure," I mean practices that include appropriating, adapting, contesting, domesticating, learning, politicizing, resisting, and above all inverting the city's expanding infrastructural networks and services. This focus provides a distinctly material and technological view of Marshall Berman's "experience of modernity," which shows that what historians of technology have called the "compelling tangle" of technology and modernity is tied to the modern infrastructural ideal. Inverting Haussmannization was a practice of citizenship that expressed what the theorists Henri Lefebvre and David Harvey call "the right to the city," a right for all city dwellers to shape their city's future, which resonates with Berman's account of citizens striving to steer the process of modernization.[37]

As an organizing concept for this book, "living with infrastructure" has three dimensions—biological, social, and affective. By keeping these three dimensions in view, I argue that Paris's Haussmannized networking exposed its residents to new fragilities: dependence on infrastructure, sunk costs and debts, technological momentum, the burdens of maintenance and retrofitting, health and safety risks, resource shortages, and environmental pollution. In turn, these fragilities inspired Parisians to invert Haussmannization by crafting progress counternarratives that politicized infrastructural change, questioning what the liberal daily newspaper *Le temps*, a seasoned critic of Haussmann, called Haussmannization's "supposed benefits" in 1872.

The Biological Dimension

First, "living with" infrastructure means that infrastructures provide resources and services that support ways of life. Physical infrastructures are channels that steer and manage the flow of resources and services that meet human needs such as food, water, energy, mobility, and communication. In 1884, the French science writer Louis Figuier called them "roots of life." In this century, the urban planning scholar Michael Neuman has called them "life supports," while the architecture scholars Andreas and Ilka Ruby have called them "the true lifeblood of the spaces we inhabit."[38] Consequently, scholars across fields have recently studied landscape, nature, and ecosystems "as infrastructure," and infrastructure's relationship to "nonhuman life."[39] The Seine's ancient role as Paris's central water source, sewer, and supply line tidily illustrates these ideas and is detailed in chapter 8. Infrastructures help city dwellers survive and thrive, a fact that binds together technical maintenance, social reproduction, and biological survival. The more technologies mediate our relations with the environment, the more we depend on them to survive, and the more their breakdown threatens our lives directly or indirectly.

One Parisian who knew this was the leading water engineer and professor Georges Bechmann (1848–1927). He explained what he called the city's "artificial life" in 1899:

> When men were spread out in small groups over vast spaces, nature almost always furnished for them all the elements necessary for their health in profusion: the air they breathed was pure, the water they drank contained no harmful substances, the ground they walked on took care of rapidly transforming any perishable organic matter that running water had not carried away.

> But, to the extent that groups became larger and more compact, that the surface occupied by each of them increased, and that greater numbers of human beings found themselves together on the same expanse of terrain, more and more serious causes of insalubrity appeared, in the face of which nature was not long in showing itself powerless. So, it was necessary to come to its aid with more perfected and more complex means, as agglomerations were denser and more extensive.
>
> From this fact artificial life emerged, which is the condition of existence of the inhabitants of cities in general, and without which it would not have been possible to develop the enormous capitals whose rapid growth raises more difficult problems each day and calls without cease for new studies and constant efforts.

Bechmann claimed that urbanization and pollution severed humans from "nature," which could no longer support city life. Like other French engineers, politicians, and scientists who worked on infrastructure, he naturalized urban change, casting the leap to a technologically enhanced and mediated ("artificial") life as an evolutionary necessity.[40] As society outgrew nature, it came to depend on technology. Seeing cities and societies as biological organisms was a hallmark of nineteenth-century European thought.[41] As Bechmann indicated, this view implied that cities could be healthy or sick, following the era's parallel tendency to medicalize cities and societies.[42]

Parisians expressed these views with an idiom of the urban body or organism. This idiom hinted that large technical systems, like everything humans do, are part of nature, not outside of or other than it—a key insight of urban environmental studies.[43] Haussmann helped popularize a language of infrastructures as "organs." The streets were Paris's "face," the central market its "belly." Roads were "arteries," parks were "lungs," and sewers were "bowels." Street lighting and street sweeping were "teats" that nourished street life. Activity was "circulation." Demolition and construction were "surgery."[44] Furthermore, savants studying Paris's infrastructural transformation used "life" as a key concept, from Maxime du Camp's *Paris, Its Organs, Their Functions, and Its Life* (1869–75) to Max Boucard's *The Life of Paris* (1892) and George d'Avenel's *The Mechanism of Modern Life* (1896–1905).[45] Such studies turned the popular urban body idiom into a framework for understanding infrastructure, which is strikingly similar to contemporary scholarship in geography and urban studies that analyzes "urban metabolism."[46] The term is both literal and metaphorical. Cities do resemble living organisms in many ways, including their constant intake, digestion, and excretion of resources and materials. Metaphorically, the term stretches to encompass cities' constant circulation of capital, culture, information, people, and vehicles.

Bechmann's sense of urban metabolism was more literal. As Paris's leading water engineer and a typhoid fighter, he saw that infrastructures had both individual and population-level health consequences. Water and sewer networks, depending on how they worked, could either spread or reduce disease, especially waterborne cholera and typhoid. Tuberculosis haunted Paris's shoddiest housing. Together, housing and water infrastructures were caught up in the era's sanitary elan, which in France combined Pasteur's legacy with a well-established hygiene movement eager for sanitation and republican elites hungry for reform projects. By contrast, transportation and energy infrastructures brought risks of bodily injury and property damage: collisions, electrocutions, and fires. All infrastructures carry risks as well as rewards. Health and safety risks indicate infrastructure's shaping of biological life.

The more Parisians depended on infrastructures to provide basic life needs, the more infrastructural decisions became life-or-death choices. Securing needs with infrastructures created a risk of critical resources shortages. The more nature and technology grew into one another, the less they could be disentangled amid "artificial life." Their entanglement created a risk of cascading infrastructure failures, even disasters. Cities and technology exist within the web of nature that supports all life including ourselves, a fact with political consequences. An emerging literature shows how infrastructure intersects with biopolitics.[47] As Michel Foucault theorized, biopolitics operates where the state maintains the population's life rather than individual lives. Rates of birth, disease, and death are its major concerns.[48] Extending Foucault's "government of things" into bioecological domains, scholars now consider how states pursue control of biological materials including food, water, disease, blood, and waste. Recent research interprets physical infrastructures as instruments of biopolitics, territoriality, and statecraft.[49]

I argue here that, because they support life, public infrastructures serve the state as levers of biopolitics, ways to control population and environments by steering flows of air, light, food, water, disease, and waste. The historical coincidence (in time and place) of Haussmannization and Foucault's biopolitics is no accident. Fittingly, Foucault argued that biopolitics aimed to "regularize" life, a word that Haussmann used to describe his works, and that scholars often use to describe Haussmannization.[50] In the nineteenth century, as the state assumed new powers over the life of both humans and nonhumans, it did so in part through physical infrastructure designed for public health and sanitation purposes. Haussmannization dug these biopolitical uses of infrastructure, especially water and sanitation networks, deep into Paris. One common way of inverting Haussmannization was thus to accuse the government of failing at biopolitics by spreading disease with bad decisions about water and sanitation infrastructure (chapters 7, 8).

The Social Dimension

Second, "living with" means that we live alongside infrastructures as we do our neighbors. The technologies, artifacts, or "things" around us are constituents and agents of urban society, the material equipment or everyday objects that clutter the lifeworld. Historians of technology have built a small literature on "everyday technologies," which usually means small, individual-use, domestic technologies such as the bicycle, sewing machine, zipper, flyswatter, typewriter, or camera. But Paris's history after 1870 shows that even large technical systems or public works can be everyday technologies, as the historian Joy Parr argues regarding dams and nuclear plants.[51] Paris's water taps, gas burners, plumbed toilets, sidewalks, streetlights, and tram stops have been intimate parts of social life that Parisians use daily in or near their homes. Paris's infrastructures have always had social meaning, place, and power.

Recent shifts in the social history of technology change our sense of who makes technological history: users as much as designers, but also workers who operate, maintain, and repair technologies.[52] Social and cultural histories of technology have explored domestication, appropriation, and vernacular technologies.[53] The appropriation of technology now looks more like adaptation than simple adoption. In parallel, histories of Paris urbanism have taken bottom-up and user-centered approaches, while urban historians have observed cities at "street level."[54] Applied to physical infrastructures, these methodological shifts call for more work on informal or vernacular urbanism, "nonnetwork" or "off-grid" technologies, and "inverse infrastructures," defined as user-driven, self-organized, and bottom-up systems.[55] This book documents Paris examples from self-built homes for fighting housing insecurity (chapter 2) to alternative water-gathering practices for those without running water (chapter 7), and improvised walkways and rafts for navigating flooded streets (chapter 8). I often refer to Foucault's conception of "heterotopias," meaning exceptional spaces with their own social norms, to interpret un-Haussmannized places with nonstandard infrastructures.[56]

From Siegfried Giedion's "anonymous history" in 1948 to actor–network theory in the 1980s and thing theory and the new materialisms in the twenty-first century, scholars of technology have shown us how to understand societies by studying their material artifacts.[57] Social and historical studies of technology refer to "technopolitics," or using technology to enact political goals, which echoes the older sociology of technologies as "congealed social interests."[58] Far from neutral tools, infrastructures are frequent objects of conflict. The resources and services they provide benefit some social groups more than others: some are excluded from the network; others suffer outages, irregular service, or differential service by location. Such differences in access and service have real effects on people's health, mobility, social integration, and communication, that is, on their life chances in networking cities. This makes infrastructures levers of social power and resistance that are worth fighting over. Hence, histories of Paris's sewers and subways have often been told as "battle" stories.[59] Another dimension of infrastructural inequality is frequently expressed in terms of necessities or utilities versus luxuries. In Paris, even common resources like water and gas could be defined as luxuries to justify their exclusive character, or as necessities to promote their universal provision, and these definitions were contested and shifted over time.[60]

Although Paris public works were expressions and products of state power, they were constantly appropriated and challenged by citizens. Like the republic, they belonged in part to the people. France's republican political culture inspired wide-ranging attempts to democratize society and make government accountable and responsive to citizens. During secondary Haussmannization, Parisians raised urgent questions about organizing and governing large technical systems. Public or private funding and control? Monopoly or competitive markets? Public-private partnerships were the French norm, through government contracts (*concessions*) for corporations. Amid debates about municipalizing various services, the Paris Métro emerged as a municipal project, outside Office of Works jurisdiction. All other infrastructures discussed in this book were departmental projects, except for housing, legally defined as private property. But all infrastructures—private, municipal, or departmental—were contested and politicized. So, naturally the national government took an interest, too, and sometimes intervened in local decision-making. French histories often rehearse the tripartite municipal-departmental-national structure of the French republic, as well as the dual governance of the Paris region under the Prefect of the Seine and the Prefect of Police.[61] Paris thus condensed and intensified layered jurisdictions within a single city with consequences for the institutional organization of infrastructures.

This book argues that the politics of infrastructure operate well beyond bureaucracy, governance, legislation, and regulation. In Paris, infrastructures entered debates over the "politics of redistribution," which negotiates the distribution of resources, the "politics of recognition," which struggles over which social groups are recognized in media and government, and the "politics of representation," meaning which social groups are represented in government and what are the political consequences of various symbolic representations in word or image.[62] Thus, far outside the corridors of bureaucracies and corporations, Paris infrastructures were consistently caught up in political economy, cultural politics, and citizen struggles for democratic recognition and representation. Broadening the scope of infrastructural politics aligns with the wider sense of human agency in recent histories of technology. Therefore, besides Paris's three governments, this book shows how other organizations—labor unions, reform groups, civil associations, industry groups, the landlords' association, and engineers' associations—also bid to steer Paris's infrastructural development. So did a remarkably diverse range of individual men and women: architects, anarchists, coachmen, engineers, hygienists, illustrators, journalists, novelists, photographers, sewermen, and even a baker and a bartender.

Parisian interest in politicizing infrastructure ranged so broadly in part because both Haussmannization and the modern infrastructural ideal projected futures. Infrastructure studies have shown how infrastructures "promise" to deliver much more than energy, mobility, sanitation, or water. According to the anthropologists Nikhil Anand, Akhil Gupta, and Hannah Appel, "They have long promised modernity, development, progress, and freedom to people all over the world."[63] The modern infrastructural ideal envisioned systems that are centralized, cohesive, efficient, equitable, reliable, standardized, ubiquitous, and unifying.[64] In Paris, Haussmannization promised *embellissement*, meaning both beautification and technical advancement, alongside security, rationalization, regularization, sanitation and its corollary, deodorization. It also promised the state the simplification and legibility of urban territory. Such extensive promises, which the geographers Stephen Graham and Simon Marvin call "lustrous," are not easily realized, making them easily politicized.[65] The chapters of this book show how Parisians often criticized officials for broken infrastructure "promises." The discussion of promises in infrastructure studies dovetails with recent work in technology studies on "sociotechnical imaginaries," visions of the future that braid together social and technological change.[66] By constructing images of the future, promises and imaginaries create expectations that politicize infrastructure when they are not met. Among the challenges facing secondary Haussmannization were rising expectations for infrastructure among officials, experts, and the public.[67]

The Affective Dimension

Third, "living with" infrastructure embraces the experience of technology. As I have theorized elsewhere, infrastructures are powerfully affective; they shape and are shaped by our assumptions and expectations, as well as our embodied, sensory, and emotional experiences. In Haussmannizing Paris, men, women, children, and even horses felt the transformation of infrastructure with their bodies. Social scientists have registered the everydayness of infrastructures as "boring" and "banal," but infrastructures also inspire a range of reactions: love, hate, pride, shame, fear, awe, and even humor.[68] This book draws on established histories of technological ambivalence, enthusiasm, disgust, and the sublime.[69] Literature on artificial lighting has revealed various feelings that attend the technological modulation of light and dark: enchantment, excitement, fear, and arousal.[70] Following Marshall Berman's classic account of modern urban experience, recent research clustering around affect theory, sensory history, and history of emotions has given an experiential cast to studies of cities and technology.[71] Other scholars have shown how infrastructures modulate experiences of time and space, creating the "rhythms" of urban life. Paris's famous expression "Métro, Boulot, Dodo" (Métro, Work, Sleep), which weaves the Métro commute into the daily routine, is a fitting example.[72]

One of the longest-running debates in infrastructure studies builds on the sociologist Susan Leigh Star's famous claim that "the normally invisible quality of working infrastructure becomes visible when it breaks." This claim suggested that infrastructures naturally disappear when they are working well and capture our attention when they break down in accidents, failures, or disasters, which explained how infrastructures became expected, habitual, hegemonic, normalized, or ordinary.[73] Since the late 1990s, scholars have revised the claim in several ways. First, they have shown how some infrastructures are intended as visible, even monumental like Paris's Avenue and Place de la République discussed in chapter 1. Second, they have shown how hiding infrastructure is deliberate and human-made, not simply "natural," and therefore varies widely in different times and places. Finally, they have shown how people contest, debate, and negotiate the visibility of infrastructures. In sum, recent infrastructure studies reveal that normalizing infrastructure's invisibility or hiddenness encourages us to invert infrastructure instructively but also runs the risk of black-boxing infrastructure in fetishizing or mystifying ways.[74]

What these discussions of infrastructural visibility largely overlook, however, is the precisely sensory and experiential dimension of *seeing*, envisioning, and documenting infrastructure, as the postcard in figure I.1 demonstrates. By deploying visual sources throughout this book, I show that seeing infrastructure and inverting its hidden aspects is not only a methodological imperative today but also a historical phenomenon documented in my sources. In Paris, desires to see or not to see infrastructure raised aesthetic, environmental, and political issues that were hotly debated during secondary Haussmannization as Parisians negotiated the experience of living with infrastructure. Because Haussmann began the practice of hiding infrastructure (usually underground), documenting, representing, and visualizing infrastructure became a strategy for inverting and politicizing it.[75] Similarly, debates over Paris's construction sites (chapter 5), water supply (chapter 7), and polluted waterways (chapter 8) politicized infrastructure's olfactory outcomes in a practice I call "making a stink." Sight and smell were particularly central to the way Parisians sensed infrastructure.[76]

Beyond sensory experience, another important aspect of experiencing infrastructures is that we "live with" or tolerate them despite their drawbacks, failures, and risks. As humans become more dependent on infrastructure systems to meet basic biological and social needs, the risks of cascading failures and critical resource shortages increase. Building upon classic sociologies of risk, Charles Perrow's "normal accidents" and Ulrich Beck's "risk society," a burgeoning literature on accidents, failures, and disasters parallels envirotech research in blurring technological and environmental disasters.[77] In Paris, construction snarled the streets, creating dust, noise, and clutter. Familiar itineraries were blocked and then revised. Traffic jams, electrical accidents, and water shortages became common failures in technical systems that otherwise worked smoothly. Minor floods were frequent, major ones disastrous. Depending on and tolerating imperfect infrastructure required skill and creativity. Users learned how to jump trains, hack water meters, navigate detours, and pressure landlords for better amenities. Tramway workers and passengers appropriated trams for protest. When new technologies failed, users dusted off old technologies: candles when electric lights failed, horse- and steam trams when electric trams failed, and public water collecting when taps failed. Living with imperfect infrastructure required patience, hacks, workarounds, and user innovation.

Infrastructures thus elicit strong emotional reactions, giving rise to evocative vocabulary, imagery, and narratives, which technology studies call "scripts," that scholars can find in the archive as the journalistic catchphrases and infrastructure epithets detailed in this book.[78] Chapter 1 documents Paris's most dangerous street crossing, the "intersection of the crushed." Chapter 2 details shoddy houses called "microbe factories," "murderous lodgings" (*logements meurtrièrs*), "haunted houses" (*maisons maudites*), and "funeral homes" (*maisons funèbres*). Chapter 4 shows how subway critics smeared the Métropolitain as a *nécropolitain*, while chapters 3 and 6 record a whole series of tramway epithets involving the terms "murder" and "death." Chapter 5 shows how pervasive construction inspired the general term "Paris-chantiers" (worksite Paris) and the specific term "Opéra hole" for the repeatedly retrofitted Opéra Plaza. Chapter 7 describes recurring water shortages politicized through the phrase "odors of Paris" and chauvinistic references to colonial Africa as the "land of thirst." Finally, chapter 8 discusses how the 1910 flood inspired Parisians to politicize engineering, calling the disaster the "1870 of engineers," which displayed their "powerlessness" against natural forces. These phrases flag popular concerns about traffic, safety, and health, and elite concerns about monumental architecture and civilization as motives for inverting Haussmannization. Close attention to language, text, and discourse allows scholars to document infrastructural experiences and how they were represented and politicized.[79]

Images can play a similar methodological role. An undated photo postcard, published by François Fleury with the common title "Old Montmartre" depicts benches, paths, and a table amid foliage, with the caption: "The last park belonging to the Moulin de la Galette, through which they pierced the Avenue Junot" (fig. I.2). The avenue, built in 1910–1912, roughly dates the card. The word "pierced" deployed Haussmann's surgical idiom to cast the park as a casualty of Haussmannization. The card inverted Haussmannization both visually and chronologically. It documented a time before Haussmannization, preserving traces of the old city destroyed by modernization. It depicted the former green space rather than the future avenue, contrasting old and new, soft and hard, organic and artificial. Above all, it mourned and memorialized a loss, Haussmannization's sacrifice of living space for traffic space, conveying grief and nostalgia. Like the previous postcard (fig. I.1), its location in the city also matters. The park depicted was in Montmartre, a peripheral district with an often radical, avant-garde, even separatist relationship with central Paris (chapter 2). The card thus reversed the order of center and periphery in Paris's geography. Moreover, this park occupied a distinctive space or heterotopia within Montmartre, the Maquis

Figure I.2. Unused F. F. Postcard, ca. 1910, "Old Montmartre: The last park belonging to the Moulin de la Galette, through which the Avenue Junot was pierced." Author's collection.

or "Bush," which blended green space with self-built sheds, shacks, and shanties, and limited access to infrastructure. The postcard, therefore, also inverted Haussmannization by contrasting top-down, centralized, homogenizing planning with the distinctiveness and variety of bottom-up, local, and vernacular urbanism. The bare facts of photographing infrastructural change and printing postcards show that Parisians wanted to document, remember, and reflect on their infrastructural experiences. More specifically, the visual content, composition, and style of each image convey emotions and experiences that Parisians leveraged to politicize infrastructure.

Outline of the Chapters

Each chapter in this book focuses on a particular infrastructure or infrastructural issue. Chapter 1 examines streets to contrast elite and popular ways of politicizing infrastructure. Avenues and boulevards were the centerpieces of primary Haussmannization, which helped politicize them. The chapter contrasts an official republican infrastructure politics, aimed at "moral and material improvement" and celebrated in parades and speeches, with a popular infrastructure politics of concern for traffic safety, which criticized the ways that street planning, building, and maintenance disrupted traditions of pedestrian culture flânerie and the promenade. Emerging from this debate over the politics of streets and traffic safety was Paris's first (failed) traffic light in 1912.

Chapter 2 examines housing, an infrastructure that was privately owned, not public works. Faced with issues including a housing shortage, limited tenant access to infrastructure, and powerful organized landlords, Parisians made the "housing question" a central issue for infrastructural politics. By focusing on rental properties, shoddy houses, and peripheral districts, especially Montmartre, the chapter shows how Parisians inverted housing in different ways. Those inhabiting problem sites publicized their conditions to the world outside, while bourgeois reformers looked in from the outside in diagnosis and judgment. The chapter thus highlights infrastructural inequalities and living with infrastructure versus living without infrastructure, and it explores living with old and domestic infrastructures.

Chapter 3 builds on research that sees streets as defining sites of urban experience to track the "street life" of the tramways. Unlike Métro dreams, tramway experiences offered a much more embodied, finely grained, and experimental perspective on the consequences of motorization, which highlights maintenance, everydayness, and living with failing technologies. Engineers tested new motor systems while road crews tested new ways of clearing snow, laying tram tracks in streets, and paving streets with new materials. Paris elites criticized tramways as foreign to Paris, both suburban and American, using the epithet "barbarians' tramway" to express fears about infrastructure threatening French culture and civilization. A broader public, by contrast, celebrated the accessible "ten-cent" or "democratic" tramways, or used tramway epithets that expressed safety fears: "murderous tramways" and "tramways of death," but above all "crusher tramways," which recalls the traffic safety concerns discussed in chapter 1.

Chapter 4 reviews debates over the Métro from 1872 to 1895, when the municipal council won control, making this the only municipally controlled infrastructure considered in the book. The chapter asks what it means to live with infrastructures that do not exist yet. Métro plans and debates imagined a future transformed by technology, which gave rise to conflicting sociotechnical imaginaries that braided together social and technological futures. Far beyond engineers and politicians, a broad public participated in these debates, as they tried to imagine or predict the consequences of motorization, a process that Parisians depicted as "dreaming."

Chapter 5 shows how secondary Haussmannization blanketed the city in construction sites, which invert infrastructure by dismantling the cityscape to reveal its inner workings and the human labor behind it. These places of physical construction are also places of social construction. How do people experience infrastructures in progress? What are their growing pains? Parisians politicized the experience of a city under construction using two types of inversion. Workers inside construction sites highlighted labor issues including health and safety, publicizing their risks to the world outside. Meanwhile, bourgeois observers looked in in judgment, publicizing the conditions of construction sites as "surgery" that disfigured or injured the urban body.

Chapter 6 examines accidents, many of them electric, to study how Parisians lived with infrastructure's health and safety risks. Like construction, accidents inverted infrastructure, and if they became damaging enough, Parisians cast them as "catastrophes" or "disasters." By narrating minor electrical accidents, fires, and vehicle collisions, mostly falling between two landmark disasters, 1897's Bazar de la Charité fire and the 1903 Métro tunnel fire, the chapter shows how Parisians contested what counts as a disaster to invert infrastructure and politicize its risks.

Chapter 7 shows that, like the Métro and housing questions, there was a long-standing "water question" that reflected the authorities' struggle to provide enough fresh water for Parisians. The water supply demonstrates the biological dimension of living with infrastructure and illustrates how depending on infrastructure for meeting biological needs leaves city dwellers vulnerable to cascading failures and critical resource shortages. Rather than discrete "great stinks" like London's famed 1858 crisis, Paris experienced what I call a "Slow Stink" from 1880 to 1911, during which Parisians politicized water shortages, typhoid risks, and ambient odors in a variety of inversion I call "making a stink."

Chapter 8 considers Paris's rivers as natural infrastructures that offer both resources and risks and shows how disasters invert infrastructure. The first section, "Sanitizing the Seine," examines the question of how the city's sewers should relate to the river, and how to remediate water quality after the 1870s. The second section, "The Flooding Seine," highlights flooding, especially the 1910 "Great Flood." The third section, "Burying the Bièvre," narrates the life and death of the Bièvre River, a heavily polluted Seine tributary that was later entombed as a sewer in 1912. All three sections explore hygiene, enclosure, and environmental quality as reasons for inverting and politicizing infrastructure.

The archival basis for this study comprises documents written by government engineers, inspectors, and politicians who worked in Haussmannization's home, Paris's departmental Office of Works, which are conserved today at the Archives de Paris in the more than three thousand cartons of the V.ONC series. I supplemented this corpus with government documents from the Archives Nationales (AN), series F^{14} from the Ministry of Public Works. I then turned to library sources to survey the daily press, scientific press, and technical literature, from the University of Michigan and Oberlin College to the Bibliothèque National de France, and the Bibliothèque Administrative de la Ville de Paris, the Bibliothèque Historique de la Ville de Paris, and the Bibliothèque Interuniversitaire de Santé. I also combed the University of Oklahoma libraries, especially the History of Science Collections, and digital repositories including Google Books, Archive.org, CNUM, and Gallica.bnf.fr. However, because archival and published sources only reveal so much, to find popular voices, I balanced archival documents with the sources of cultural history: printed, illustrated, literary, and ephemeral works. Finally, my sources include a personal collection of over one hundred photographic postcards of Paris like the two shown in figures I.1 and I.2, many of which appear throughout this book as illustrations. Unless otherwise noted, all translations from French are my own.

By revealing the dark corners of the City of Light, I hope my inversion of Paris's infrastructures in the time of secondary Haussmannization demonstrates the uneven, uncanny, uncertain, and unjust aspects of their history, with the surprising outcome that this famous episode in the history of urban technology teaches lessons quite different from those conventionally understood. Haussmann did not invent the networked city; he invented a city attempting to network—an attempt that was remarkably controversial. His republican successors continued that project, adding electricity, railways, sanitation, and water supply to the city's repertoire, and thus provoked new controversies of their own. Haussmannization invented not only the modern city but also the modern city as an object of conflict.

Today, as we confront our own thorny infrastructural issues—climate change, deferred maintenance, emerging diseases, and smart cities—the case of Paris resonates. It is less the City of Light, the archetype, icon, or model that resonates, and more Paris at street level, the site of a grand experiment of living with infrastructure, inspired by the modern infrastructural ideal and its discontents, which taught that urban technology would be deeply contentious. During the early Third Republic, an era of engineering, expositions, and empire and a period of parallel growth of public works and the public sphere, diverse and sustained public disputation about physical infrastructure opens a window onto the varied lived experiences of a networking city bidding to become an infrastructural beacon for the world.

1

Streets, or the Politics of Infrastructure

All streets are roads, but not all roads are streets. While road researchers stress how states use roads to project power over territory, streets, defined as urban roads, have a less top-down, more complex and contested political history.[1] Streets are clear everyday infrastructures, the stage of daily life. They are workplaces for construction workers, police, performers, vendors, ragpickers, and prostitutes; they shelter the homeless. Paris streets had long nurtured pedestrian practices the promenade and flânerie, but they were also a storied theater of revolution, offering both space and cobblestones to build barricades.[2] The famed anarcho-syndicalist Émile Pouget wrote, "In the street, the people are at home" (*Dans la rue . . . le populo est chez lui*).[3] Walter Benjamin observed that Parisians "inhabit" the streets as an "interior" or "domestic" space, recalling the term "street furniture."[4] Paris's streets were as everyday as house and home.

Histories of urban traffic have shown that around 1900, street use changed sharply in western cities, as multi- or mixed-use streets became increasingly dominated by vehicle traffic. Streets were long commons or public spaces for communication, commerce, and culture, but now bicycles, automobiles, and railways were colonizing the pavement for mobility alone. As two French scholars perceptively noted, "the *rue* became the *route*."[5] Amid the Second Industrial Revolution's energy transition toward electrification, motorization, and fossil fuels, public debates gripped Paris and other cities about the form and purpose of streets, posing contentious questions about aesthetics, maintenance, safety, sanitation, and traffic. By contesting changing streets, these conflicts engaged what Gabrielle Hecht termed "technopolitics": "the strategic practice of designing or using technology to constitute, embody, or enact political goals."[6] This chapter explores the political goals of street planners and street users, which went far beyond vehicle mobility to embrace beautification, flânerie, health, solidarity, and much more. Republican experts and elites explicitly designed streets with multiple goals, while popular voices contested streets' success in achieving these aims and suggested other goals. Rather than being simply citizens who politicized while officials depoliticized, all participants politicized infrastructure in their own ways. While the longer arc of history bends toward the automobile and single-use streets, the period before 1914 was one of unfinished transition and struggle.

As the signature infrastructures of Haussmannization, streets provide a particularly clear example of how Parisians inverted Haussmannization in varied ways after 1870. This chapter thus contrasts popular and elite inversions of Haussmannization. Popular controversies over streetlights, street cleaning, maintenance, and traffic that arose as the republic emerged from war in 1872 show how infrastructural politics shifted from imperial to republican under one of Haussmann's successors, Léon Say. Street politics remained equally practical, material, and symbolic, raising issues of safety, environmental quality, government responsiveness, pedestrian culture, and street naming, which called into question how beneficial changing street design and use would be.

By contrast, republican elites—politicians, doctors, and engineers—tried to invert Haussmannization by appropriating its methods of urban transformation for republican ends. A later prefect, Eugène Poubelle, revealed the republican technopolitics of infrastructure when he promoted the 1891 Avenue de la République as a tool for the republican cause of "moral and material improvement," civilizing social progress for bodies and minds, people and environments alike. Historian Miriam Levin stresses the positivist and Saint-Simonian "culture of change" pervading republican elites, which drew on a much older Western "culture of improvement."[7] This progressive orientation connected the Pasteurians' sanitary elan with France's civilizing mission and the "modern infrastructural ideal." French elites pursued material progress by steering flows of people, microbes, light, air, and water through the urban environment. The Avenue de la République retooled Haussmann's practice of opening new streets to clear slums, as republican public works attempted to invert Haussmannization after 1870. This attempt met with limited success because republican works reproduced the premier negative social consequence of primary Haussmannization, the displacement of working-class homes from the city center.

In the 1870s–1880s, traffic congestion, issues of risk and safety, and traffic research converged on the infamous "Intersection of the Crushed," reputedly the city's busiest and most dangerous. This scene of many accidents inspired debate about whether intensifying vehicle traffic would harm pedestrian practices and pedestrian bodies. It also became a laboratory for traffic fixes, including wood paving, pedestrian bypasses (walkways and tunnels), and railways. As motorization advanced in the 1890s, Paris faced the emergence of early automobiles and the spread of dangerous intersections across the city, which were met with new traffic practices and police interventions, including Paris's short-lived first electric traffic signal in 1912. The failure of authorities to manage booming traffic and its risks constantly inspired popular inversions of Haussmannization that questioned the success of republican "moral and material improvement."

The Supposed Benefits of Haussmannization

The Prefecture of the Seine stumbled after Haussmann's ouster, switching swiftly from Henri Chevreau to Jules Ferry, Léon Say, and Marc Antoine Calmon between 1870 and 1873. In January 1872 amid rebuilding after civil war, Prefect Léon Say, who held office less than a year, oversaw ten new gaslights on the Place de l'Opéra. This plaza, a showpiece of primary Haussmannization, but not completed until 1875 under the republic, was famous for streetlights—both gas and, after 1878, electric. The conservative newspaper *Le Figaro* called the 1872 lamps "brilliant New Year's gifts," but struck a fearful note: "It was only just barely that one could circulate last night in the streets, where gas [light] was almost absolutely absent. We are talking about the center of Paris. How will it be in the farther-out neighborhoods?" Security sparked anxiety in postwar Paris, and streetlights had been surveillance tools since the 1600s. *Le Figaro* worried that while the busy center was lit at night, the periphery, reputedly less safe, was not. If infrastructure grew slowly on the outskirts, how soon could they be secured with lighting? The article staked safety and security on proper lighting and retraced the social and symbolic inequalities of center and periphery.[8]

The next day, the liberal daily *Le temps* queried the Prefecture's priorities. The lamps' glow was "lively, intense, sustained and plentiful," with "a light bluish tint which recalls electric lighting." Illuminating all Paris like this would create "a veritable fairy play," but "why must the splendors of the Place de l'Opéra be like a lamp in a tomb?" This haunting simile suited the postwar moment. Although "exaggerating on purpose," *Le temps* said officials economized, installing "the bare minimum" of gaslights. Ubiquitous lights were unnecessary, but "the major arteries and their offshoots, the streets which put the principal neighborhoods into contact with one another, these should enjoy integral lighting." The Opéra lamps were striking or sparkling (*éclatant*), but not *Le temps*'s priority, which suggested a planning idea: artificial lighting should vary with traffic flow. The editors prioritized traffic and safety over lighting's aesthetic effects.[9]

While *Le temps* was more critical than *Le Figaro*, both papers criticized lighting, linked it with traffic (*circulation*), and imagined more future lights from the Prefecture. *Le Figaro* feared the periphery; *Le temps* stressed the center. *Le Figaro* suggested that lighting let traffic flow, while *Le temps* suggested, conversely, that traffic should steer lighting. Would people follow lighting, or should lights follow people? Should infrastructure serve or steer social practice? Behind their differences, the newspapers agreed that streets and lamps should function to make mobility smooth and safe for pedestrians and vehicles alike.

Postwar reconstruction competed with routine maintenance. Thus, *Le temps* linked lighting with maintenance, complaining that "one no longer sweeps the streets" and noting "puddles of liquid sludge which accumulate along the edges of the sidewalks." This stinky, slippery mess was "a shame for a capital" and hazardous at night: "As the municipal council well knows, we hold them responsible for all the unseemly sludge stains and all the slips and falls taken in poorly lit streets." The historian Sean Weiss has studied citizen complaints and lawsuits that claimed negligent street maintenance by Paris authorities and demanded indemnities, citing two women injured by potholes in 1884 and 1903. A similar 1896 complaint from one Anne-Marie Courcier blamed a cleaning crew hose and lance left on the sidewalk.[10] Infrastructure inspired strong feelings of pride, shame, and vanity, charges of official irresponsibility, and concerns about health, safety, and environmental quality—and all of this fed its politicization. Street maintenance was a common, contentious topic that journalists and citizens used to invert public works.

Le temps's editors addressed the Prefect, showing they valued infrastructure and were watching. "Lighting and sweeping are the two teats of Paris. May Mr. Léon Say pardon us for addressing this maxim to him." Their familiar "urban body" imagery illustrated that the street life of Paris's bustling boulevards required maintenance. Lighting and sweeping fed street life like mothers feed babies; infrastructure nourished practice. Both *Le temps* and *Le Figaro* called gaslights official "gifts" (*étrennes*), which suggested gratitude, but also accountability. Politicizing infrastructure, journalists demanded that the Prefecture maintain pavement, lighting, and cleaning, as essential for urban life as food and water.

In February 1872, a winter storm threatened to glaze the battered streets in ice, illustrating how infrastructure and practice were embedded in the physical environment. *Le temps* criticized the director of works, Jean-Charles Alphand, who "remains passive amid these atmospheric movements: he gave them neither another sweep of the broom nor another pan of asphalt." Alphand countered that sweeping crews who sprayed water awaited temperatures above freezing. Handmade ice sheets would harm pavement and traffic more than doing nothing. He evoked water and weather to naturalize and deflect blame and controversy from public works, but to little avail.[11] *Le temps* maintained pressure for weeks. "I forcefully engage our excellent Prefect to monitor all this. . . . This year we are going to begin, by way of fairly heavy taxes, the liquidation of the Empire: the population cannot ignore to whom we owe these extra charges," meaning to Haussmann. *Le temps* warned Say that he should respect the social contract exchanging taxes for public services as Haussmann had not. Its liberal editors evoked the republican cause of public works in the public interest.[12]

They later criticized "the supposed benefits of the Haussmannization of Paris," arguing, "Many find that this upheaval, excessively extended to all the largest and best constructed streets, touches on monomania and, before even recognizing the price of these high fantasies, they find grave inconveniences in it." Fantastic finances aside, while primary Haussmannization was intended to smooth traffic, it ironically complicated it.[13] *Le temps* rehearsed the refrain of the *culte de l'axe* (cult of the axis), the Prefecture's devoutly rectilinear geometry, which "sometimes produced bizarre results, improbable acute angles or incredible obtuse angles, streets cut up, slashed, where one cannot enjoy more than three minutes of free sidewalk, unformed squares, or the lack of landmarks, [which] forces the pedestrian into the most complicated topographic exploration."[14] Haussmann's streets amplified vehicle traffic, forcing pedestrians onto sidewalks, "so as not to be smashed in the unbridled whirlpool of cars." Pedestrians were left on a "steep island with no coast . . . something like deportation within a fortified area," recalling the Communards exiled months earlier to New Caledonia. Like the Communards repelled from central Paris and later deported, pedestrians confronted Haussmannization with a "revenge of the expelled."[15] Haussmannization made streets harder to cross, by cutting new paths while blocking old ones. Again, quoting *Le temps*: "The perfection of the system has been realized on the Place du Château-d'Eau [later Place de la République]: grand roads vomiting and absorbing torrents of pedestrians and vehicles wildly, with no rule and beyond any direction. If one has the bad fortune to stray into this wasteland in the evening, where the cyclone of vehicles runs wild, one is not certain to make it home in the integrity of one's material person: remember that they count on average two accidents a day, the official figure." Surging horsecar traffic pushed pedestrians onto sidewalks. *Le temps* evoked natural disasters—"whirlpool," "torrents," "wasteland," and "cyclone"—to reflect dangerous streets and to contest Haussmannization's "supposed benefits." Haussmann's roads opened the age of street-use disputes in Paris, which the republic inherited and struggled to settle.

Months later, the nationalist *Le Gaulois* printed André Bechelle's humorous "A Nightmare," in which ghosts haunt Say in his office the way they did Charles Dickens's Ebenezer Scrooge. The ghost of the French revolutionary General Hoche challenges Say's idea to rename the Avenue Haussmann to Avenue Hoche, arguing that Haussmann deserved the honor. Bechelle contended that officials should do "something more important than renaming streets." But the symbolic politics of street names *were* important. Dickens's own Paris chronicles reported that Parisians were still debating removing Haussmann's name from the streets in 1879, which never happened.[16]

Although *Le Gaulois* defended Haussmann and *Le temps* criticized him, both charged Say's Prefecture and Alphand's Office of Works with inaction and skewed priorities. Editors bombarded officials while Alphand deflected criticism by naturalizing public works. In debating the lighting, maintenance, traffic, safety, sweeping, and naming of streets, the popular press drove public debate, contesting what Haussmannization would become under the republic. Paris's street-use debates thus negotiated the technopolitics of public works. What were their purposes and costs? Who should they serve? Were they properly maintained? How did they bring people, technology, and environments into new relationships? Were they glittering gifts or dangerous disasters? Finally, what, other than Haussmann, did they symbolize? As the next section shows, years later, an avenue was named for the republic itself.

Moral and Material Improvement

On the eve of Bastille Day, 13 July 1891, the national celebration started early. President Sadi Carnot paraded from his palace to the Place de la République

Figure 1.1. Neurdein brothers (ND) Postcard 39, postmarked 1905, with the caption "The Place de la République." Author's collection.

with Minister of Education Léon Bourgeois and several army officers. Their carriage rode the eastbound Grands Boulevards, which reach the plaza after crossing the Rue du Montmartre at what became known as the "Intersection of the Crushed." Their route itself embodied the symbolic politics of infrastructure; the Bourbon kings of the 1600s had it built as an official procession route. Thousands attended the parade, bathed in this political symbolism. *Le temps* reported applause and cheering throughout.[17] The convoy first stopped to inaugurate the new Avenue de la République, the final addition to the Place de la République, a long-prominent plaza (fig. 1.1). It housed the northeast gate of Charles V's 1370 city wall and later a water tower, giving it the name Place du Château d'Eau until 1879. That year, the republic appropriated the plaza as a new emblem of the regime and the nation, embellished the space with electric streetlights, and opened a contest for a monumental statue to embody the republic. In 1880, a 14 July celebration inaugurated Bastille Day as the national holiday and unveiled the plaster model of the contest winner Léopold Morice's towering Marianne statue. In 1883, another Bastille Day event inaugurated the final 82-foot bronze maiden, like a smaller Statue of Liberty (fig. 1.2).[18]

Figure 1.2. Unused, undated postcard, *Petit Journal* Collection, with the caption "Paris – Place de la République," ca. early 1900s. Author's collection.

Haussmann's boulevards had linked the plaza with the North and East train stations (Boulevard de Magenta), with the Place de la Nation (Boulevard du Prince-Eugène, today's Boulevard Voltaire), and with the Place de la Bastille (Boulevard Richard Lenoir), making it a crossroads for Paris's most industrial sector. Still highly networked today, it links five Métro lines and nine streets, wedged among the 3rd, 10th, and 11th districts.[19] The French word "boulevard" is a cognate of the English "bulwark," referring to fortifications.[20] Streets smooth and speed traffic by steering motion along an axis, but this also makes them hard to cross, potential barriers for dividing cities (as are other infrastructural channels: canals, fences, pipes, rails, viaducts, and wires). If they do not physically block mobility, they may be social or symbolic barriers. Haussmann's Place du Château d'Eau formed a gateway to Paris's working-class northeast, separated from the bourgeois west with a continuous boulevard chain from Montmartre to the Place de la Bastille. The only street tying the plaza to the 19th district was the medieval Rue du Temple and no road connected it to the 20th. The northeast was ripe for secondary Haussmannization, a sector with a turbulent proletarian past, from the "bloody week" of civil war in May 1871 to homes and workplaces menaced by the slower violence of environmental and health risks.[21] In 1891, the republic celebrated the opening of this major plaza to the northeast.

Members of Paris's three governments received the president's convoy: Municipal Council President Levraud, Director of Works Alphand, Prefect of Police Lozé, Departmental Council President Péan, Prefect of the Seine Poubelle, Senate President Royer, and Chamber of Deputies President Floquet. At the new avenue's head, a table draped in red-and-gold cloth was accented with tricolor flags and fasces, bundles of rods and axes that symbolized state power since ancient Rome and later inspired the word "fascism." The president sat with his guests on either side. An entire company of the republican guard brought their concert band. After speeches by Poubelle and Levraud, and Alphand's presentation of awards to the engineers who designed the avenue, the party rode to its end, passing under two temporary triumphal arches erected by citizens of the 11th and 20th districts. *Le temps* offered this colorful account:

> Poles are decorated along the avenue, holding long *oriflammes* [royal standards] suspended above the street by a chain, from which hang three shiny ornaments. The houses are all decorated with flags. The scaffolds of buildings under construction have been decorated with flags and garlands of yellow lights. The effect is very picturesque. . . .
>
> Yellow lanterns were spread around elsewhere as well, in the branches of trees, on the triumphal arches, on the rooftops. The chestnut trees have transformed into orange trees, with enormous fruit.[22]

The streetscape grew carnivalesque, packed with people and pageantry, and surreally transformed into this new republican avenue's festive grand opening.

The convoy's second stop fêted a renovated high school, the Lycée Voltaire, where the new avenue passed the Père Lachaise cemetery. Minister Bourgeois spoke first, and Floquet recalled that when he was a student the school's walls were "black with mold and humidity." *Le temps* described joyous and unruly excitement around this second ceremony: "Over the entire course the crowd is immense; there are clusters of people on the chimneys of houses. At a certain moment, the security service was overwhelmed, and the crowd, flooding into the street, produced a formidable push. Choral, instrumental, and gymnastic societies penetrated the procession followed by two or three thousand young people. It was not until the rue Oberkampf that they could be pushed back."[23] Such raw popular enthusiasm both boosted republican power and made the authorities "push back." Ceremonious speeches, excited crowds, and ostentatious symbols (arches, fasces, flags, lanterns, standards) continued until the Porte de Romainville. After a brief inspection, the convoy returned to the Place des Pyrénées to dine at the 20th district *mairie* (administrative office).

All this pomp celebrated the new avenue and school, but also the nation, its capital, the republic, and the "universal" progress of civilization, both material and moral. On the eve of Bastille Day, the avenue's unequivocal name and ritualized inauguration show how public works served as what David Van Zanten has called Haussmannization's "political theater."[24] Parisians decorated their neighborhoods, cheered the presidential procession, and broke security lines to rush the parade. Municipal Council President Levraud said, "The Avenue de la République is symbolic by the name it carries and by its function." So, what was its function and symbolism? Which technopolitical goals made streets republican?

The day's speeches by Poubelle and Levraud explained. Poubelle began, "A city like Paris is in a perpetual effort of transformation and growth; the spirit that animates it tends incessantly to renew it." Such "material changes" punctuated Paris's history. How often had the capital "made its city walls crack?"[25] Like Alphand, he naturalized urban change. Poubelle's eight-year term sustained efforts to improve infrastructures, especially for sanitation. Beyond streets, his regime addressed building codes, water distribution, sewerage, and river pollution. His trash collection reform famously led resentful landlords or ragpickers to name garbage bins "Poubelle"—now the everyday French term, and much more common than "Haussmannize" or "Haussmannization."[26] Few people were better poised to express how public works could reflect and serve republican technopolitics.

The "magnificent" four-kilometer avenue, Poubelle said, was over a third of Paris's widest diameter. Almost a decade of construction cost over 40 million francs, 38 million for land alone. He explicitly stated its technopolitical and biopolitical goals. First, the avenue "shortens the distances between the Place de la République, the veritable industrial center of Paris, and the suburban communes." In so doing, "neighborhoods where the working-class population, packed into narrow and unclean little streets, was often decimated by epidemics, are today sanitized." In the 11th district, "between the avenue Parmentier and the boulevard de Ménilmontant," the avenue "made a group of factories as dangerous for workers as they are for the neighborhood disappear." In the 20th district, it removed "the sordid constructions of the rue des Poirriers, of the rue des Oiseaux, of the impasse des Coudriers, of the passage Robineau and the impasse Fanny-Benoit." Beyond mobility, Poubelle's avenue was built for social and environmental engineering. Like Haussmann's streets, it excised perceived blight—Poubelle's "dangerous" and "sordid" buildings. It steered traffic, cleared slums, and reduced pollution: "following the formula, [to] bring air and light into these until now deprived neighborhoods." The familiar "formula" held that wider streets delivered natural resources light and air. Nineteenth-century hygienists regularly prescribed light and air to combat disease, vice, and crime—moral, social, and bodily degeneration. Locals and visitors alike praised Paris's wide Haussmannized streets as channels of light and air.

The new avenue's biopolitics also attempted to steer working-class settlement, and thus displaced people from home. According to Poubelle, the avenue "facilitates access for the population to an important part of the periphery where vast plots of land of lesser value are found, so naturally indicated for the construction of clean and low-cost housing." His administration furthered Haussmann's exile of workers and industry from the center, thinning the population, to further repair, embellish, revalue, and gentrify it. Meanwhile apartment blocks bloomed on the outskirts. Putting a positive spin on displacement, he added, "In the new dwellings which are going to be built for them [workers], they will find more favorable conditions of economy and well-being, and they can enjoy these without being obliged to move too far from their workshops; the steam tramway voted by the general council will bring them yet closer, while a very low special fare will permit them to take advantage of the healthiness of this nice neighborhood, without an appreciable increase in charges." In the future tense, Poubelle claimed the avenue would clear slums, smooth traffic, improve worker homes and commutes, push peripheral building and settlement, beautify a neighborhood, and deliver light and air, serving goals of mobility, ventilation, aesthetics, suburbanization, disease control, and social welfare. Infrastructure enacted the republican elite's social fantasies of a cleaner, healthier, more comfortable working class. Workers would be humanely provisioned, socially uplifted, and politically and biologically neutralized by state gifts like 1872's Opéra lamps or this new avenue. Public works would serve class collaboration, harmony, and "social peace"—the "solidarity" that bound Émile Durkheim's sociology to the dominant political culture of *solidarisme*.[27] Lastly, the avenue embellished the northeast "with a very picturesque square, from which one discovers the most beautiful panorama of Paris."

Poubelle saw public works as patriotic, progressive, and civilizing. He praised President Carnot's "solicitude for anything that relates to improving the lot of workers," continuing that "the most noble way to celebrate the national anniversary—is it not to take an encouraging retrospective look at ourselves and our works, because it will lead us to recognize that our efforts are not sterile and that each day we come closer to that future which should assure for all the children of the common mother a more equitable participation in the benefits of civilization, to the very fatherland of those destined and worthy of France's constant love for liberty and justice?" Politicians and engineers celebrated the new avenue near the national holiday because it served the core republican values of progress, universalism, equality, and civilization. The "fatherland," "children," and "common mother" evoked a national family's affective ties. Poubelle beamed at "the ensemble of progress accomplished by our dear Paris, under the government of the republic," as the inauguration elevated infrastructure and built environments, the materials of civilization, to striking heights. Infrastructure reflected France's national honor and glory, republican politics, and "universal" human progress. As Poubelle said, the avenue "is but one more artery in the network of streets that, for twenty years, have renewed the physiognomy of the capital. The ravages of war slowed this movement of transformation for some time, but, from the moment that France began to recover, Paris put itself to work. Works of general utility proceeded rapidly." He could not cover all the works done in "this useful and arduous period," but his twenty-year timeline, 1871–1891, reviewed works *after* Haussmann, and thereby ostensibly republican.

Compared with primary Haussmannization, the mark of republican works under Alphand (director of works, 1871–1891) and Poubelle (prefect of the Seine, 1883–1896) was prioritizing equality, hygiene, and utility. The US social scientist Albert Shaw observed that "services of education, of cleanliness and of health, on a vast and varied scale," occupied the authorities "once so engrossed with boulevards and architecture." Shaw found republican works less decorous, pompous, and visible than imperial "Great Works," but they "more closely touched the lives of the people," and made everyday life "more perfectly" fit the ideals of "modern civilization." Poubelle argued that the republic's most urgent tasks had been "the addition of potable water, the evacuation and use of contaminated water." Such sanitary work, echoing Shaw, revealed what Poubelle, inverting hidden infrastructure, called the "less apparent but no less appreciable benefits of underground piping."[28] His regime tapped distant new sources, more than doubling the water supply. They built aqueducts for delivery (one over 100 km), storage reservoirs, and pump stations to draw water from area waterways and pressurize distribution. They installed smaller pipes (*petite canalisation*) along each street to distribute water. Poubelle decreed in 1894 that every apartment must have a shared or private toilet, draining directly to the sewer, called the *tout-à-l'égout* (all to the sewer), which dashed Haussmann's dream of keeping human feces from the sewers (discussed in chapter 8).[29] Poubelle hyped the water system with technical details, citing liters of water and lengths of pipe.

Contrasting with the new avenue's multiple purposes, Poubelle demanded only that water infrastructure prevent disease. He said, "Typhoid fever has diminished, and cholera knocked at the gates of Paris without being able to implant itself there"—a reference to 1884's cholera outbreak (see chapter 7). His survey of material progress stressed sanitation, a hybrid effort to control buildings, people, pipes, water, and microbes. Politicians and engineers saw the city not only as an organism but also as a human habitat, sharing a widespread environmental determinism (Alain Cottereau's "ecological slippage") drawn from the biologistic human sciences. If human health depended on environmental quality, then environmental engineering could influence human behavior and social progress. Improved freshwater and wastewater infrastructures could increase water quality, and reduce cholera and typhoid, weakening disease and strengthening the ailing population. Planners intended for infrastructure to "fix" social and environmental problems by managing disease, pollution, and other risks. French republican social fantasies cast infrastructure as the material arm of France's civilizing mission and an important tool of biopolitics.[30]

Poubelle revealed city planning ideas in arguing that the avenue and its tramway would increase mobility and "shorten distances." Paradoxically, in practice, they *lengthened* distances, integrating center and periphery, distancing home and work, and suburbanizing workers. These spatial shifts suggest that Poubelle valued the functional sorting of space later called "zoning." It also reflected the older taboo on urban industry, which so defined contemporary railway controversies (see chapters 3–6).[31] He also prefigured the suburbanization and greenbelts of the Garden City movement, which reached France after 1900.[32] His speech disclosed how republican planners promised to improve life by shaping material environments, a project framed by hygiene, sanitation, and biopolitics. Infrastructures for housing, transport, and sanitation came to define notions like "standard of living" and "quality of life," as planners cleared slums, cleaned and greened working-class environments, and separated the city's functional zones.

After Poubelle spoke, Levraud addressed sanitation's "double character," both "moral and physical." For Levraud, the republic was inverting Haussmannization by attenuating modernity's social inequalities. Promoting *solidarisme*'s class harmony and social peace, he argued that primary Haussmannization's class segregation discouraged social mixing, "fusion," "contact," "attractions" and "sympathies." But the new avenue would combat social division, "attenuating this fatal consequence"—a phrase that naturalized urban change. He described infrastructure less than Poubelle did but similarly linked it with social equality. He also patriotically praised Carnot: "You accentuate the character of social peace, union, and republican solidarity which emanates from this popular celebration." Public works should soften capitalism's frictions, replacing class conflict with collaboration. On the eve of the republican national celebration, engineers and politicians celebrated infrastructures as sociotechnical and biopolitical resources, nourishing the social body with public health and social welfare that managed urban modernization's evolving benefits and costs.[33]

Poubelle's often-quoted speech was excerpted in a glowing 1892 patriotic review, *The Social Works of the Parisian Municipality, 1871–1891*. These two decades paralleled both Poubelle's timeline and Alphand's term as director of works (from his appointment to his death), heavily feted in the review. The editor Le Mansois Duprey, of the Municipal Council Secretariat, wrote:

> All the works enumerated by the Prefect of the Seine have assuredly contributed to the moral and material improvement of the conditions of life in Paris. But some of them, such as the piercing of new streets, the construction of barracks, markets, etc., constitute only the development of a city assured by a municipality vigilant and concerned for the interests it represents, in view of the incessant growth of the population. Others, on the contrary, have as an immediate goal either assuring the material well-being of its [the city's] inhabitants, the poorest in particular, or contributing to the elevation of their spirit. Among those of first importance are public assistance, sanitation, hygiene, education, fine arts and numerous special creations that do not exactly fit into any of these major divisions. It is to these that this work is consecrated.

Duprey stressed two points. First, he separated population management from works that raised the "material well-being" or "spirit" of the poor. The former sated growing appetites, meeting basic infrastructure standards, while the latter, uplifting "social works" targeted the lower classes. He promoted progressive, republican public works that *raised* living standards to "civilize" the poor and script expectations for modernity.[34] Second, he further divided progressive works into those securing "material well-being" and those that "elevated the spirit," structuring his entire book around "moral and material improvement." Part 1 concerned "moral and intellectual development," and part 2 the "amelioration of the material conditions of life: public health." He admitted the division was "a bit arbitrary," because "the more educated man knows his real needs better, observes the laws of hygiene with more care, in such a way that his health and well-being increase. From another side, guaranteed by the extent of what is possible against sickness and poverty, more vigorous and healthier; he has fewer bad thoughts and becomes better. There is thus a sort of repercussion of the spirit on the body and vice versa." He easily equated "material conditions of life" with "public health." Like Poubelle and Levraud, Duprey made public health and public works republican and biopolitical missions. Infrastructures that steered flows of light, air, water, microbes, and settlement would shape healthy citizens.[35] The formula "moral and material improvement" (*amélioration morale et matérielle*) codified one of the Third Republic's deepest commitments, applied both across France's empire and in the capital, where it fell mostly on the poor. By "improving" Paris pipes, roads, and schools, the Office of Works was civilizing citizens and city alike.[36]

The formula had blossomed around 1830, when the historian-politician François Guizot surveyed the moral and material progress of civilization globally, and the physiocrat doctor P. J. G. Cabanis theorized and medicalized their "relations."[37] As the social question engulfed the 1840s, Victor Hugo wrote, "To improve material life is to improve moral life. Make men happy and you make them better." Such materialism could and did justify public works.[38] The archbishop of Paris, Cardinal Morlot, claimed that Haussmann's works were the material mirror of the church's moral efforts.[39] In 1882, the poverty researcher Armand Baron discussed employer paternalism, in which bosses "freely imposed" sacrifices on themselves for the "moral and material improvement" of their workers. E. de la Hautière's 1895 political study *The Constitution and Institutions* argued that taxes were more than an insurance premium (*prime d'assurance*), a fee for individual insurance or protection, because some taxes always went "to public works, to education, to the improvement of the material and moral conditions of the nation." Taxes fund the state's "civilizing and moral mission." One 1900 writer called the new Métro an "admirable agent of moral and material progress."[40]

This familiar formula illuminates the two ceremonies of July 1891: the avenue represented "material improvement," while the school promised "moral improvement." Ideas of hygiene and sanitation guided both projects as measures of civilization. The avenue promised to invert primary Haussmannization. Rather than neglecting the working-class east, Poubelle's regime intended to uplift worker neighborhoods. Because Haussmann's works increased inequality and segregation, republican works would attenuate modernity's social ills, fostering moral and material improvement for all, fortifying the social body and making social peace, steering Haussmannization toward republican, egalitarian, and modernizing ends. Public works in the Third Republic were loaded with the lustrous, but often broken, infrastructural promises of the modern infrastructural ideal.

Crossing the exterior boulevards, a limit for primary Haussmannization, the avenue further integrated center, periphery, and suburbs. In time, it transcended Haussmann's horsecar mobility by supporting motorized transport. Poubelle's dream of a steam tramway was realized in an 1896 electric tramway, inaugurated with another festive celebration of transportation infrastructure. Again, a parade of experts toured the new works and lunched at the Romainville *mairie*, but this time took another meal at the Lilas terminus.[41] The party included the city engineers Bruman, Huet, Leroux, and Menant, and the chief departmental engineer Hétier; alongside General Omnibus Company President Cuvinot, Municipal Council President Landrin, and the mayors of the suburbs Lilas and Romainville. Their tram, sporting tricolor flags, was filled with engineers invited by Claret, the designer of the tramway's surface-contact power-delivery system, whose official contract to operate the line required that he provide electric street lighting along its course inside the city limits.[42] The avenue served the northeast's gradual electrification as a utility corridor.

Le temps cast the tramline as progressive by detailing its technical, especially electric, features. This was Paris's first tram powered by "underground conductors," more commonly called "surface contacts" (see chapter 6). "Metallic paving-stones" set between the rails every 2.5 meters transmitted charge, which trams received by a first dragging "cursor" that fed current into two 15-horsepower receptors. A second, smaller cursor triggered an alarm if the current was interrupted. Three Corliss steam engines driving Hillariet dynamos in a power plant near the Porte de Romainville terminus generated current. Engineering magazines including *Le génie civil* offered detailed technical descriptions for specialist readers, calling Claret's system a "very interesting innovation."[43] Politicians, engineers, journalists, and citizens celebrated or contested trams, as they did roads and pipes, as embodiments of republican ideals.

Republican elites designed infrastructures to embody and enact their politics. Public works were justified by their "public utility," a legal authorization required for all projects. Republican technopolitics, in attempting to invert imperial works, projected a public-minded ethos, but also linked public works with public health, or infrastructure with biopolitics. Decorous public works celebrations masked the tensions between public and private that ran deep through republican politics. Although ostensibly based in the utilitarian ethic of the greatest good for the greatest number, "public utility" often served elite power and private profit as much as public welfare. The displacement of poor Parisians to build this avenue is a fitting example.

"Intersection of the Crushed"

Living with intensifying traffic made the "e" words "encumbrance" (*encombrement*) and "congestion" (*embarrass*) Parisian keywords and drove social scientists to study traffic more closely.[44] In 1882, France conducted its first national traffic survey. The Ministry of Public Works Special Commission on National Roads included the Ponts et Chaussées engineers Alfred Durand-Claye, known for sewerage, and Émile Cheysson, celebrated for contributing to cartography, economics, and statistics. The commission found that Paris was a national exception. By daily horse collars (*colliers*) pulling vehicles, Paris traffic was nearly 20 times the national average (4,700 vs. 220 collars/day). By annual tonnage, Paris was similarly 15–21 times the average. Finally, Paris was the only place where passengers made up 57 percent, thus outweighing freight.[45]

Just as Paris stood out nationally, one Paris intersection was distinct. Centrally located, crossed by over 100,000 vehicles annually and hosting many accidents, it was called the *Carrefour des écrasés,* "Intersection of the Crushed" (fig. 1.3). Likely coined by the editor Émile de Girardin, the phrase also graces Jules Verne's 1882 play *La voyage à travers l'impossible.* The intersection stood where the roaring Grand Boulevards, at the Boulevard Montmartre, crossed the smaller Rue de Montmartre. The boulevard sloped down before bisecting the rue, making it hard for drivers to stop amid cross traffic. Congestion clouded visibility and increased the risk that pedestrians would be crushed between or under vehicles. This infamous intersection became a laboratory and microcosm of Paris's traffic troubles, which the historian Mathieu Flonneau aptly named the "heart of the urban maelstrom."[46]

Traffic studies were only one way to address the dangers of streets. The Paris historian Edmond Beaurepaire argued that the arcades, gathered near the Grand Boulevards, had been early shelter for pedestrians against traffic and weather.[47] In 1879, a tense Chamber of Deputies debate about rebuilding the central Post Office (Hôtel des Postes), with Haussmann attending, weighed

Figure 1.3. Unbranded postcard, postmarked 1902, with the caption "288—Paris—Boulevard Montmartre—The Intersection of the Crushed." Author's Collection.

renovating the Intersection of the Crushed via the long-disputed extension of the Rue aux Ours, which finally opened in 1910.[48] In 1881, the municipality contracted a British company to repave the intersection with tar-treated wooden blocks in a concrete bed. Engineers found wood paving cheap, durable, and clean, thus reducing dust and mud and absorbing noise and vibration. But the statistician Jacques Bertillon found that accidents during equivalent periods before and after repaving grew from 25 vehicle and 6 pedestrian collisions to 31 vehicle and 19 pedestrian incidents. Hygienists worried that the smoothness of wood paving made it hard for carriages to stop and its quietness made it difficult for pedestrians to hear the vehicles.[49] In 1883, the illustrated satirical *Le charivari* noted that elevated walkways (*passerelles*) over the boulevards had been debated "since the time of Haussmannization" as a "means to escape crushing," but without any results. They hoped the 1900s might bring some change, asking, "What say you, routine?" Other sources suggested tunneling under the boulevards.[50] Similarly, Parisians proposed both elevated and underground railways (see chapter 4). Additionally, the weather still battered the pavement. The years 1882 and 1883 were "almost constantly rainy." Combined with increasing vehicles, tonnage, and passengers, this amplified wear and tear, convincing engineers that wood should replace stone wherever possible. Yet, by 1887, of 8,635,990 square meters of paved surface, only 428,600 or 5 percent was wood.[51] Engineers, hygienists, and even humorists debated how to stop the crushing but with little resolution.

In 1886, the science writer Louis Figuier promoted the long-awaited Métro to relieve traffic, invoking the Intersection of the Crushed and rendering the experience of traffic in affective detail:

> Thus, circulation becomes more difficult each day in the streets of the capital. Vehicles, whose number increases without end, occasion permanent congestion, amid which the pedestrian cannot engage without risking his life. The omnibuses and tramways are always full when everyone needs them, and the irritation of the voyager is even more vivid, as distances have become enormous, following the extension of the city.
>
> How we regret the absence of an interior railway, with fixed stops, its cars always sufficient in number, its speed, and its safety, while, feet in the mud, with the wind and the rain, we await, in the uncertain shelter of a porte cochère, a private coach, which never passes, or whose coachman refuses, sneering, to respond to your desperate call. If instead one decides to take to the road on foot, in the adventure, omnibuses, trams, and carriages that crisscross in all directions, hurl the mire of their wheels at your face, and you graze heavily laden carts, which threaten to grind you under their dreadful mass.

For Figuier, living with traffic was deeply emotive, inspiring feelings of annoyance, desperation, discomfort, fear, and impatience. The word "circulation" embodied several concerns: bad weather, bodily harm, dirty pavement, inadequate transit service, surly coachmen, and wasted time.[52] It also suggested that traffic flowed through the city's arteries or avenues like blood. While experts found a quantifiable growth of traffic in terms of repair cost, tonnage, passengers, and vehicles, everyday passengers and pedestrians felt this intensification with their bodies. Figuier's verbs and adjectives vividly capture a street user's perspective; throughout, he addressed "we" (*on/nous*) and "you" (*vous*), parroting public opinion, but his text also reads very personally. Paris's circulation and congestion were nothing to celebrate if crossing the street meant risking one's life, but the republic needed things to celebrate. Celebrating infrastructure was another of Haussmann's hallmarks that republicans inherited, inverted, and turned to new uses, as the section "Moral and Material Improvement" indicated.

In the 1890s, the automobile's early experimental days, Paris was a capital of car culture. As traffic intensified, problems previously localized at the Intersection of the Crushed overtook the city.[53] In June 1898, the conservative critic Jules Claretie devoted his long-standing *Le temps* column "Life in Paris" to cars, writing, "The coachman is an adversary, the chauffeur is an enemy. There used to be one intersection of the crushed, now there are ten, now there are 100. We crush everywhere." He scorned the coming automobile exposition for promoting "these machines of crushing, these moving accident factories." He argued that fear of being run over had ruined flânerie, the storied Parisian practice of walking and writing urban experience.[54] Meanwhile, the humor magazine *Le pêle-mêle* announced a poetry contest, "Future Paris: The Last Coachman."[55] A 1901 history of nineteenth-century Paris argued that there was only one Intersection of the Crushed in 1870, but by 1900 each district had "one of its own, or rather three." Echoing Figuier's earlier language, the author described the streets' "craziness" (*affolement*), "bewilderment" (*ahurissement*), "disorientation" (*désorientation*), and "lack of peace" (*manque de placidité*).[56] Parisians used this name for other crossings, too: the Carrefour de Croix Rouge (today Place Michel Debré), and the Rue Saint-Lazare.[57] Some sources described intersections of the crushed in Marseille, Algiers, and New York.[58] The architect Eugène Hénard, who studied traffic for the Office of Works, is often credited with inventing the roundabout or traffic circle (*carrefour à giration*). His 1904 study on intersections noted that Parisians used this name for many crossings, and "not without reason," citing 3,125 automobile accidents in 1903.[59] In 1905, the editor-satirist Pierre Veron blared, "Paris is not a city . . . it's a crusher" (*écrasoir*).[60] In the coming years, other writers summed up, "The too-famous Intersection of the Crushed now exists a bit everywhere" and "more than ever, we crush."[61] In the fin-de-siècle era, horsecars had created the first Intersection of the Crushed; in the Belle Époque, tramways and automobiles helped revivify safety concerns as more and more intersections proved dangerous.[62]

The police tried to control traffic flow. In 1892, officers (*guardiens de la paix*) began standing in busy intersections to direct traffic, adopting a white baton in 1896 and a whistle in 1900.[63] The government created the Permanent Committee on Circulation, an advisory board, in 1898. It soon advocated modern drivers' licenses to replace the previous police practice of issuing *coupe-files*, permissions to pass, which created a "culture of exception." France's first speed limit of 20 kilometers per hour came in 1899, but it was weakly enforced because most cars lacked speedometers, and many Paris traffic police rode bicycles. Police also renewed enforcement against "marauding" (*la maraude*), in which coachmen trolled for customers.[64]

The period 1907–1909 brought a flurry of activity, including tests of Paris's first one-way streets and roundabouts, and the rule of "priority on the right" (*priorité à droit*), which made the right lane the passing lane. Today, mobility scholars distinguish "geometric" approaches that physically steer traffic into built channels from "symbolic" approaches that encourage drivers to steer traffic with signs and signals.[65] Paris tested both approaches simultaneously, but with little relief. Intensifying traffic and its dangers suggested a sick urban body. The historian André Guillerme showed that the traffic term "congestion" had medical roots. In 1905, Octave Mirabeau's famous criticism of automobiles called the need for speed "a disease, a sickness," while Hénard compared stalled traffic to heart failure.[66] In 1907, Prefect of Police Louis Lépine acknowledged that "Paris suffers from the ill of a too-intense circulation. It's a bother and a danger." Émile Massard's often-cited 1910 traffic report said, "Each stop in circulation equals a stop in the life of the city, which suffers thus from its very excess of vitality, and risks dying of congestion."[67] Another author called Paris "a monstrous congested brain, where blood pools. The flow of it is thick enough that it coagulates."[68] If traffic was the urban body's lifeblood, traffic problems were organ failure.

There was one brief attempt at a technological fix or pacemaker for this congested heart: Paris's first traffic signal, *le kiosque-signal*, at the original Intersection of the Crushed (fig. 1.4). In 1910, the city council authorized Doctor Edmond-Alfred Goupil to construct, entirely self-funded, a small iron and glass cabin on an island no more than 2.5 meters across. Councilors specified certain technical details: white paving would indicate where vehicles should stop, and the signal would pair vision and sound. The entire effort was a strict trial to be removed at the first injunction.[69] Before the trial started, the newspaper *Comoedia* interviewed an optimistic Goupil, who explained that a red lit-up disk meant stop, while a white one meant go. The police handled seven to ten cases per day of "disobedience of agents with the white baton," and he hoped the signal kiosk would solve the problem.[70] On the trial's first day, 12 April 1912, Goupil observed the intersection from a café terrace, while newspapers reported worse traffic congestion than before. The police scrambled to steer traffic while technicians managed electrical problems amid general confusion.[71] *Gil Blas*'s press review of papers from across the editorial spectrum found a remarkable "consensus" (*entente*) about the kiosk given their usual bitter debates: it was ugly, dangerous, and ineffective. *Le matin* said

Figure 1.4. Unused Neurdein brothers (ND) Postcard 3622, with the caption "The 'Intersection of the Crushed,' Seen from the Rue du Faubourg-Montmartre." Although undated, the presence of the signal kiosk in the center of the photograph dates it to spring 1912. This image gives a clear sense of the intersection's chaos and crowding. Author's collection.

onlookers felt "ironic curiosity," while *Le rappel* morbidly predicted that an autobus would knock the kiosk over with the agent inside.[72] Demolition began on 22 May. Parisians tarred the *kiosque* "goupillon" ("bottlebrush"), a word derived from the liturgical wand for sprinkling holy water, which also echoed Goupil's name. *La petite presse* called it the "object of gentle mockery" and *Le petit journal*'s editorialist Jean Lecoq thought it spoiled "the beauty of Paris," the conservative rallying cry against modern technology.[73] The short-lived trial, although futuristic, was singularly insignificant to contemporaries, not infrastructure to celebrate, but infrastructure to question, mock, dismiss, and ultimately demolish.

Over forty years, nothing worked to calm the Intersection of the Crushed, neither 1880s wooden pavement, nor new 1890s police techniques, nor 1912's electric audio-visual signals. At the same time, dangerous intersections multiplied, revealing problems so pervasive that any purely local or technical fix might seem quaint. No existing technology could mitigate the outsized power of automobiles. Paris's *kiosque signale* was a casualty of the growing pains of gradual motorization and shifting street use.

In this chapter, conflicts over streets have illustrated Paris's politics of infrastructure. Major debated issues included street finance, maintenance, safety, and environmental quality. Streets were as congested as they were contested. The street lighting, sweeping, and maintenance controversy of winter 1872 revealed a popular politics of infrastructure, often articulated through the press, which staged conflicts with the authorities, and contested public works as matters of public concern. The intensification of traffic, and the Intersection of the Crushed as a focus for managing traffic, continued this controversy about streets by highlighting risk and safety. By contrast, the 1891 Avenue de la République shows how republican elites scripted public works to serve "moral and material improvement," civilizing progress intended to improve traffic and health—measures of modernity—by shaping the material environment. The long liberal intellectual pedigree of "moral and material improvement" helps explain why republican public works congealed and embodied the political tensions emanating from liberal values. Chief among them were tensions between civilization, discipline, and expertise on the one hand and social liberty and citizen input into urban planning on the other. Elites envisioned socially engineering working-class settlement and environmentally or biologically engineering away cholera and typhoid. Citizens asked that infrastructure nourish everyday social practice, securing safe streets for pedestrians and practices like flânerie. All parties in the debate agreed that infrastructure had *social* purposes, that physical infrastructure was also social infrastructure that could help or harm specific social outcomes: health, safety, and equality. But their goals differed: republican elites intended infrastructures to breed patriotism, pride, progress, and power, while the citizens, journalists, doctors, and hygienists who contested these aims had other priorities: convenience, efficiency, safety, security, and the right to the city. Different parties in the debate also operated on different scales: elites prioritized the entire population, bourgeois reformers and professionals scrutinized social practice, and citizens prioritized individual users. All sides tried to invert Haussmannization in their own ways; as republican elites attempted to reform it from within, popular voices criticized it from without.

2

At Home with Infrastructure

Houses are more than places to live. Spatial studies argue that buildings are no mere containers; they actively shape us as we shape them.[1] Shelter, a basic human need, is an infrastructure of daily life and biological survival. Like streets, houses are everyday things. But unlike the streets, sewers, and railways discussed elsewhere in this book, French housing was legally private property, not public works. Without the state's securing of shelter, housing issues migrated into the private, domestic, or civil sphere. Free market ideas and powerful landlords, who organized to resist change, kept public housing from emerging sooner, and hence the protagonists of existing studies have been bourgeois reformers and their organizations, including the reformist think tank the Musée Sociale (founded in 1894) and the French Society of Urbanists (founded in 1911). The Society for Affordable Housing promoted reforms from 1894 to 1912, thus creating financial incentives and a design contest for building affordable housing, as well as financial aid to help poorer people own property, sometimes collectively. Meanwhile, the 1902–1903 public health law expanded the state's public health jurisdiction into private buildings. Yet, until the 1920s, housing was not "public works," and this pushed domestic and private issues forward.[2] Paris homes therefore illustrate living with infrastructure beyond public space and public works.

Houses are infrastructures that provide shelter, but they meet this need unevenly and unequally. Some houses are better built and maintained, or more connected to shared infrastructures, than others. In Paris's increasingly financialized housing market, known for "cramped, overpopulated, unclean and expensive dwellings," living space and domestic access to infrastructures were commodities and privileges.[3] Supply and quality of housing were among modern Paris's most intractable infrastructural issues, reflecting an "urban crisis."[4]

The nagging "housing question" inspired politicized inversions of domestic space. Stories of furnished rooms, shoddy houses, off-grid homes, and peripheral districts illustrate the gap between universalizing expectations for infrastructure and the infrastructural realities of Paris homes. Paris housing clearly conveys all three dimensions—biological, social, and affective—of living with infrastructure. Elite inversions of domestic space centered on exposing and transforming the stigmatized spaces of cheap, furnished rooms for rent (*garnis*). Popular inversions, by contrast, are illustrated by the northern peripheral district of Montmartre, home to many low-cost lodging houses, vernacular architecture, and anarchist and avant-garde subcultures, all of which inspired ambitious and hotly politicized inversions of Haussmannization.

The Quantity and Quality of Housing

Histories of architecture, urbanism, social reform, and public health richly document Paris's chronic housing insecurity and the bourgeois reformers who fought it.[5] Since Friedrich Engels's 1872 essay "The Housing Question," Haussmannization has been criticized for real estate speculation and rising prices. Political economies show that rising land values brought declining access and supply, because land and buildings were prone to speculation that priced out workers.[6] Haussmannization sharpened existing housing inequalities, creating a surplus of costly bourgeois apartments in the center and the west, sometimes lying vacant, while the periphery lacked affordable housing. Private development under the republic maintained these patterns.[7] Popular encyclopedias tracked rents from *Le magasin pittoresque* (1882) to Jacques Bertillon's *Je sais tout* (1912), as did savants from the Orleanist politician Othenin d'Haussonville (1883) to the lawyer Henri Biget (1913).[8] Political discourse assembled these problems into a compound crisis—"crisis of rents," "crisis of lodging," and "crisis of the cost of living"—which resounded from the Commune through World War I in policy debates and the social sciences.[9]

A related problem was population density, crowding, or overpopulation. Across the 1800s, booming migration and population made Paris a "sick city" that "grew too fast."[10] By 1913 Paris's 30 square miles housed almost 3 million people, an average 95,000 per square mile, compared to Berlin's 80,000 and London's 37,500 per square mile. From its roots on the Seine islands, Paris has been a "fortress" enclosed by moats, walls, and today by a belt highway. Such physical limits were compounded by practices and scripts for using space.[11] One powerful script was the centripetal, homeward pull of neighborhoods (*quartiers*)—hence the popular image of the "true Parisian of Paris" (*vrai Parisien de Paris*), like Jules Verne's Jean Passepartout, a "peasant" rooted in this "city of villages."[12] Studies of Paris apartments show how crowding inspired scripts like the (contentious or cordial) interaction and intimacy of neighbors and the vis-à-vis—their views into each other's windows. High-density living pushed the limits of public and private spaces, threatening delicate social values like privacy, morality, and decency.[13] Contemporaries used words like "bursting," "cracking," "spilling," "stifling," and "strangling" to capture the experience of physical crowding and emotional tension.[14]

Crowding compounded wear and tear, degrading environments and health. Urban studies have linked Paris density with ill health from the 1832 cholera to the 2003 heat wave.[15] In *The Odors of Paris* (1867), the conservative writer Louis Veuillot called Paris "musty and dense," fit to ripen stink. "Odors of Paris" became a catchphrase used through the 1920s when debating sewers, garbage, industry, and pollution (discussed in chapter 7).[16] Many Parisians reported humid, rotten, sagging walls, and mold.[17] Paris was both literally and figuratively sick. Republican "moral and material improvement" drew urgency from Paris's local crisis, nested within the national context of degeneracy theory, the medicalizing discourse of national decline, a sagging birthrate, and aspirations toward "regeneration."[18] French experts pathologized urban space, discovering the "harmful city" or "hostile environment" and making the worker's "home" (*foyer*) and "unhealthy lodgings" (*logements insalubres*) objects of intense concern.[19]

In response, the state took more control over health and homes: routine inspection and disinfection, building codes, construction permits, disease reporting, slum clearance, and vaccination.[20] These trends gradually accelerated in the century after 1830. After 1871, republican elites practiced what scholars and contemporaries alike called "social modernity"—approaching social problems with reason, science, technology, and medicine.[21] The Third Republic was a distinctly medico-scientific regime.[22] George Weisz's "medical mandarins" and Jack Ellis's "physician-legislators" took power in France by both professionalizing medicine and holding public office.[23] Many Paris studies stress doctors' social and political power, and their collaboration with other experts: architects, engineers, hygienists, politicians, statisticians, and reformers.[24] The medicalization of social problems helped define French modernity, feeding the growth of biopolitics, urbanism, and the welfare state. In the civil sphere, sickly spaces preoccupied two leading and overlapping social movements: hygiene and housing reform.[25]

Michel Foucault's biopolitics originated at this intersection of state formation, public health, social sciences, and social welfare. Intellectually, biopolitics followed the biological and ecological strains of nineteenth-century social thought, which in France foregrounded relations between organism and environment captured in the terms "milieu," "habitat," and "living conditions."[26] Although Foucault traced biopolitics to Enlightenment rationality, and well beyond France, the experience of nineteenth-century Paris housing was instrumental in reifying biopolitics as normal governance under the Second Empire and Third Republic.[27] France's perceived biological decline, sagging birth rate, and unhealthy lodgings inspired the state's intervention into life and death, making Paris houses one birthplace of biopolitics.

In sum, scholars know a lot about Paris housing governance, policy, reform, and research. Scholarship says less about material conditions and physical infrastructure—and even less about the embodied experience of inhabiting unhealthy houses. Following closely on the quantity or supply of housing were questions about environmental quality, health, and safety, often framed as questions about access to infrastructural networks and services.

In the 1860s, Paris landlords began posting enameled metal plaques on building facades boasting "Water and Gas on All Floors" (*Eau et gaz à tous les étages*), selling points because access was rare. Similar plaques adorned Marseilles, Strasbourg, and Brussels, or read "healthy house; all-to-the-sewer [drainage]" (*maison salubre; tout-à-l'égout*). French hotel and apartment ads called water, gas, and sewer "modern comforts," setting standards for the material equipment of modern domestic life.[28] In 1884, the science writer Louis Figuier called underground water, gas, and telegraph conduits "the roots of life in a modern city." Yet an 1893 history text noted that few Paris houses enjoyed "the classic 'water and gas on all floors,' being absolutely primitive sanitary conveniences [*commodités*] and, most of the time, common to all tenants." The travel writer Jules Huret wrote in 1906, "The Germans amiably mock us, who put on our buildings—even in Paris!—the inscription 'Water and Gas on All

Floors' as if these were extremely rare commodities [*denrées*]."[29] These modern "comforts" or "conveniences" were no mere physical facts, but normatively charged, inspiring mocking jokes and jealous comparisons among nations, disrupting France's modernity and national identity. Figuier's "roots of life" also met human biological needs. Cities depend on infrastructures, which Michael Neuman calls "life supports," for securing urban metabolism, the circulation of resources and waste. Behind infrastructural norms and standards lurked life-or-death questions of resilience, sustainability, and survival.[30]

Modernization's boosters projected progressive infrastructural evolution. Figuier and the engineering journal *Le génie civil* predicted that elevators would follow water and gas. Figuier's 1885 telephone history argued that after gas came other infrastructures: "water, gas, elevator, heat, and time," time being the Victor Popp Company's Paris domestic compressed-air clock service after 1879. Reflecting on early entertainment telephones and distance listening like the Théâtrophone, Figuier imagined that "opera" would soon reach "all floors," anticipating radio. Henry de Graffigny's bright 1905 forecast for domestic electrification, *Electricity for All*, predicted, "The day is near, without doubt, when all houses will be electrically equipped, and where, instead of the notice 'water and gas on all floors,' one will have added, in well-kept buildings, the signage 'heat, telephone, and electric lighting in all apartments.'"[31] Such forecasts, like early science fiction, performed the discourse of modern "wonders" (Figuier's "marvels") of science and technology, amply documented by cultural histories of technology.[32] Imaging the future, these men scripted expectations for ubiquitous, universal, and updated infrastructures as in the modern infrastructural ideal—all floors, all apartments, all tenants, all services.

At the 1889 Universal Exposition, the Municipal Sanitation Service compared "clean" and "unclean" houses, staking hygiene solely on "modern comforts" light, heat, water, and sewerage. One reviewer found the exhibit unpersuasive, classist, and technically naive: "It is not necessary to be a sanitary engineer (which today is a consecrated expression that is often abused) to perceive that all of this is nothing but an affair of the purse and of taste," a question of privilege and preference, not necessity but luxury, less sanitation than chic bourgeois gadgetry.[33] For another reviewer ("Dr. P. L."), the exhibit essentially contrasted rich and poor homes. The doctor stressed drainage, detailing exterior drainpipes emptying onto the street, stinking sewer pipes without syphons, and above all, "those frightful toilets" called "turning toilets" (*cuvettes tournantes*), whose bowls rotated to deposit urine and feces in separate receptacles. He described "dirty, stinking, oxidized" toilets, "veritable sources [*foyers*] of infection." One could detect frequent "leaks and spills" (*des fuites et des débordements*) in swivels rusted in place or touched-up "patches of paint" (*taches peintes*) on nearby walls.[34] Shoddy houses were the social question materialized, and social inequality embodied, which inspired debate about living conditions, modern comforts, social standards, and rights. Houses and the infrastructures threaded through them raised consistently moral, social, and political questions, like the streets discussed in chapter 1, only in an intimate, indoor, domestic register.

Inverting Furnished Rooms

Bourgeois novelists, hygienists, social scientists, and reformers took special interest in cheap, furnished rooms (*garnis*), looking in on lodging-house life from the outside. The century preceding 1920s public housing saw expanding housing regulation, often justified by public health. The Prefect of Police's May 1878 furnished rooms ordinance began the early republic's building code reforms. Though older architectural histories argue that these reforms barely changed building form and style (art nouveau emerged later in the 1890s), lawmakers expanded Eugène Poubelle's "less apparent" infrastructures under streets and behind walls, gradually pushing landlords to maintain and upgrade houses for disease control and biopolitics.[35] Lawmakers regulated infrastructure to steer practices, people, and environments toward republican moral and material improvement. The 1878 ordinance tracked each dwelling's volume of air and access to sunlight. Furnished rooms for two or more people demanded direct light, thus limiting the size, shape, and placement of courtyards, rooms, and windows. Lawmakers also regulated toilets, arguing that tenants ignored shared toilets located far from apartments and dumped waste in courtyards or streets. Later laws in 1882–1884 limited building height by street width to ensure that light and air could reach streets. Other new standards included 14 square meters of space (*air*) per resident, direct daylight (*jour*), water, toilets, and garbage disposal, which pushed landlords to renovate houses, though with limited enforcement and jurisdiction in the private sector.[36]

From 1877 to 1883, the Commission on Unclean Dwellings under the hygienist Octave du Mesnil studied the *garni*, or furnished rental room, "concerned for the dangers which could result from the encumbrance . . . produced during the 1878 Exposition." Du Mesnil used striking words: "encumbrance" for congested streets, exposition traffic, and visitors packed into rooms; and "danger" for critical health risks. The *garni* might embarrass Paris if foreign guests got sick or spread disease from shabby hotels. Exposition construction brought migrant laborers, mostly pick-and-shovel earthmovers (*terrassiers*). The population grew an average of 5 percent annually from 1850 to 1914 but rose 14 percent from 1876 to 1881 with the arrival of 280,000 migrants. More single young men in short-term rentals expanded the long-stigmatized "floating population" of migrants, vagrants, laborers, ragpickers, bohemians, and homeless. These demographic shifts affected infrastructure by compounding wear and tear or by gendering facilities: heavily male-occupied lodging houses with shared bathrooms made urinals common long before toilets. The 1878 police ordinance flagged urinals as unhealthy and required a water flush. Crowded houses were contagious hotspots, intensified by the exposition's migrant workers and tourists.[37]

Commissioners flagged shared spaces (*dépendances*) as health risks: "courtyards, stairways, landings, hallways, restrooms, utility sinks, etc." These heavily used, poorly maintained spaces spread disease. Landlords delegated cleaning to domestic workers or concierges, but with common spaces falling between public streets and private apartments, and little oversight, both tenants and concierges could neglect them. Thus, building codes shaped not only room size, ventilation, lighting, and plumbing but also tenant practices that included gathering water, disposing waste, using toilets, and cleaning house. Commissioners tried to shape physical spaces and infrastructures to affect daily habits. They connected infrastructure with practice, regulating design and use, as they braided social, moral, architectural, and sanitary concerns. Like the sociologist Michel Callon's "engineer-sociologists," Paris engineers, hygienists, and reformers both designed ideal equipment and scripted ideal users.[38]

The novelist Émile Zola starkly depicted lodging-house disrepair and filth. Sharon Marcus argued that the apartments in his *Pot-bouille* (1883) were "restless," suffering a "strained, simmering containment," and like the title's boiling pot, given to "explosion, discharge, excess, escape, and overflow." Similar terms filled Zola's *L'assommoir* (1877): "The lodgings, all too small for their occupants, seemed to be bursting, showing scraps of their misery in every crack." Zola depicted the infrastructure: sinks were "open" and "rusty"; windows were "bare panes" "without shutters." Loudest of all, he wrote: "The grey walls, partly eaten away by a kind of yellow leprosy, were streaked by the drippings from the roof."[39] The protagonist of his 1883 novel *Au bonheur des dames* (often translated *The Ladies' Paradise*) was Denise, a young provincial woman seeking department store work in Paris. Her small, dark rental room left her "uneasy." Its window faced a narrow courtyard, which, "sodden and filthy, was like the bottom of a well; a circle of sinister light fell into it." Through Denise, Zola echoed experts, arguing that the provinces were cleaner and healthier than cities, thanks to open and green space. Later, though she feels more at home, the same eerie, sickly scenery disorients her as she scrutinizes infrastructure in another courtyard that was like "a stagnant pond": wall plaster was "old, damp" and "black with filth"; stairs were "warped," doors were "cracked and peeling," and "the tenants left their dustbins on the stairs." A third "dark" and "narrow" courtyard's stink "poisoned the hotel."[40] Zola's deft depictions of dreary dwellings were meant to inspire disgust, pathos, outrage, and action.

Others joined Zola's obsessive detailing of unclean dwellings with an eye for abjection. Experts and journalists alike granted buildings the ability to spread disease, calling them "microbe factories," "murderous lodgings" (*logements meurtrièrs*), "haunted houses" (*maisons maudites*), and "funeral homes" (*maisons funèbres*). The built environment could spread sickness, signaling urban modernity's risks and the limits of biopolitics in private space. The Pasteurian bacteriologist Alexandre Yersin tracked diphtheria by conducting home visits across working-class Paris by bicycle. *Le petit Parisien* reported parquet that was so "used-up, spongy and cracked" it resisted disinfection, adding that "old, worn wooden floors harbor murderous microbes in their cracks."[41] The hygienists Léon Colin and du Mesnil criticized Poubelle's plan to suburbanize workers for public health. Opening and cleaning old central neighborhoods did not itself improve new peripheral neighborhoods. As workers flooded the 13th, 19th, and 20th districts from 1876 to 1886, living conditions stagnated.[42] Du Mesnil inverted the dark corners of Paris housing for bourgeois readers to see. He noted one rental's "viscous" walls, "black" ceilings, and "windows stripped of their wooden slats." Another with "pounded earth" floors was "covered with tar paper"; its windows were poorly sealed with cloth scraps. "Its occupant sleeps on a bed of straw. He pays 12 francs a month in advance for this space, and in addition he is charged with caring for dogs and chickens of the tenant [from whom he sublets]." Du Mesnil used charged words—"slum" or "hovel" (*taudis, bouge*), the adjective *infect* (literally infected or figuratively foul), and the verb "entasser" (to stuff or cram)—highlighting unequal and unsafe living conditions.[43]

In the 1890s, social scientists diagnosed overcrowding or overpopulation (*entassement, encombrement, surpeuplement, cantonnement*) at various levels: apartments, buildings, city blocks, and neighborhoods. In 1891, Bertillon showed that 14 percent of Parisians suffered "excessive overcrowding" of more than two per room, worst in the 19th and 20th districts, which census data confirm through 1901. Despite declining mortality across Europe from typhoid, smallpox, measles, scarlet fever, whooping cough, and diphtheria, Bertillon indicated that this decline missed Paris's periphery and suburbs, which had higher rates of morbidity and mortality, especially during the 1882 typhoid and 1884 and 1892 cholera outbreaks (see chapter 7). Even modern germ theory, demography, and epidemiology led Bertillon, like du Mesnil, to the familiar view that density, disease, poverty, and deficient infrastructure clustered together.[44]

Tuberculosis, the disease of unhealthy houses, did not decline. After 1894, the zoologist Paul Juillerat directed the Prefecture of the Seine's new "Sanitary houses cell" (Casier sanitaire des maisons) and argued that crowding caused tuberculosis by increasing filth and blocking light and air. The Casier staff compiled disease statistics and mapped them to produce medical geography but also inspected houses and chemically disinfected 7,000–11,000 homes each year.[45] Disinfection was already booming; the Prefecture counted 78 disinfections in 1889 and over 34,000 in 1893.[46]

Hygienists and reformers across France joined "the struggle against tuberculosis." City Councilor Ambroise Rendu brought the "struggle" to a 1902 Musée Sociale meeting.[47] Doctor Paul Brouardel's book *The Struggle Against Tuberculosis* called the disease "contagious, preventable, curable." With the politicians Casimir Perier and Léon Bourgeois, and the doctors Louis Landouzy and Jacques-Joseph Grancher, Brouardel founded the 1902 Anti-Tuberculosis Federation (Fédération antituberculeuse), later renamed the French Anti-Tuberculosis Association (Association antituberculeuse française). He also joined the 1905 Alliance for Social Hygiene (Alliance d'hygiène sociale). The "struggle" spread globally, labeling health projects in Europe, Asia, and the Americas. London (1901), Paris (1905), and Washington (1908) hosted International Tuberculosis Congresses.[48] Brouardel and Landouzy, with the politician Paul Strauss, organized Paris's 1895 congress.

The October 2–7 conference featured the engineer Émile Cheysson, the Pasteurian Dr. Paul Roux, the Casier director Paul Juillerat, and the architect Louis Bonnier, before a three-week public exhibition starting the 8 October. Organizers counted 184,762 visitors but estimated 200,000.[49] The exhibits featured furnished rooms. The public exhibition began when the French president, Émile Loubet, toured two similar exhibits: pairs of model rooms juxtaposed as in 1889. L'assistance publique, the city welfare office, hosted the first exhibit: "A room where one could see a cell from the Prison at Fresnes-les-Rungis and the room of a domestic [servant] in a nice house on the avenue des Champs-Elysées side by side. The contrast between this cell, which is perfectly hygienic, and the room, which is obviously unclean, is striking!"[50] The exhibit raised a moral question: how could prisoners live better than honest domestics? Both populations inhabited small, cheap, sparely furnished, epidemic hotspots with high turnover, "murderous" lodging like the *garni*.

Loubet later visited two more rooms: "Two bedrooms were set up looking at one another, so that the visitor would easily establish their comparison right away." Organizers deliberately paired these separate exhibitors to amplify familiar expert and reformer calls for light, air, space, and improved infrastructures. The first, "modern" and "hygienic" room, furnished by the Touring Club of France, had light curtains, sleek iron-framed furniture, hard floors, and smooth walls covered with brightly colored, washable paint rather than wallpaper. The club wanted "to clean up old hotels and to this effect, introduce some brightness and cleanliness. . . ." Sunlight "combats humidity and makes more cheerful [*gai*]." Open space drew in air, "the biggest dimensions possible. To be clean, it should be large." The second, "unhygienic," room was designed by the Society for Preservation Against Tuberculosis by Popular Education: "With its curtains, its fabric wall coverings, its furniture, its rugs, its traps for dust and for microbes, its lack of air and of light, [it] reminds one of the better part of current bedrooms and certain hotel rooms." The unhygienic room exaggerated the claustrophobic Victorian interior's heavy drapery and upholstery, Walter Benjamin's "plush as dust collector." Event organizers added a sign urging "go sleep across the way" in the Touring Club's clean, modern room.[51]

Hospital rooms shared similar problems. At the 1900 Universal Exposition, L'assistance publique contrasted two hospital rooms representing the dark past and the bright future. A reviewer called the past room "the sinister 'retrospective section,'" evoking graphic stereotypes of medieval medicine without modern pharmacy and with patients sharing beds. The modern room, "irreproachably clean," boasted a "soft and comfortable" bed. In its "incessantly renewed atmosphere," surfaces and bedding were cleaned often, and patients enjoyed the "latest materialized benefits of science."[52] These hospital rooms, like previous hotel and prison rooms, were spatial models that politicized furnishings, articulating the inequalities of and standards for infrastructure. In major and minor expositions, model rooms medicalized space, staging an inversion of domestic workers' quarters, prisons, and hospitals by contrasting them with clean, modern living spaces.

The crusading enthusiasm around tuberculosis prevention faded after 1905. Local government took up the struggle around 1906, as tuberculosis control turned toward sanitation. Under Juillerat, the Casier's dozen-year study mapped six neighborhoods that had the highest tuberculosis mortality. The 1902–1903 public health law required departmental authorities to intervene if a municipality's mortality rate exceeded the national average. Paris mortality did not, but Juillerat pushed intervention, anyway. Like its traffic, Paris's mortality was exceptional. Its average mortality rate was below the national average and steadily declining, but certain neighborhoods, blocks, and buildings did not fit the pattern. The municipality was too broad a unit of analysis to capture Paris's diversity and inequality. Paris's six densest neighborhoods suffered heavily from tuberculosis, mostly on lower floors. Juillerat argued, typically, that tuberculosis thrived without light and air in the crowded depths.[53]

Between 1906 and 1909, citing Casier studies, the municipal council designated the six tubercular areas as "unhealthy blocks" (*îlots insalubres*). Until 1914, the departmental Office for Architecture, Sidewalks, and Gardens made plans to demolish buildings and widen streets.[54] These familiarly Haussmannian plans widened roads to clear perceived blight and slums. Planners overlaid fainter maps of existing constructions with darker lines detailing desired changes to roads and buildings. Partial demolition plans removed facades to widen streets, while complete demolition plans slated entire blocks of houses for conversion to streets. These projects stalled until the 1920s, slowed by haunting Haussmannian problems—expropriation and finance. But planning documents represent dreams of urban change that sought to control poverty (society), crowding (space), and tuberculosis (biology). Experts mapped social and biological danger, locating slums as heterotopic kinks in the urban fabric for improving or removing. Their plans neglect the fact that even while housing was scarce, its demolition would have displaced people from home as had happened with the Avenue de la République.

From narrating and researching unhealthy furnished rooms to displaying model healthy ones, various bourgeois actors—novelists, hygienists, journalists, and reformers—made cheap, short-term, badly maintained rooms the infrastructural embodiment of Paris's housing crisis. For a decade after 1909, the municipal government began working to remove the capital's six most tubercular blocks of houses but made little progress. These bourgeois responses to the housing crisis observed unhealthy housing from outside, inverting houses, peering inside and describing what they saw to express social concerns. But what was it like to live inside, and how did popular inversions of rental lodging differ from these bourgeois examples?

Montmartre Boils Over

Montmartre is both Paris's 18th district and its bohemian Other. The historian Richard Sonn called it a "refuge," "liminal realm," "borderland," and "counterweight," where avant-garde and anarchist subcultures shared turf. This home to radical politics and modernist culture was thick with lodging houses and slums. From 1877 to 1883 it was the Commission on Unclean Dwellings's most visited district.[55] In 1891, around 90 percent of its apartments (66,532 of 74,038) had the lowest-tier annual rents under 500 francs. In 1906, the district's 809 lodging houses held 17,884 people, an average 22 per house, and more cases of contagious disease than any other district. These figures indicate poverty and overcrowding.[56] Neglected by Haussmann and bloodied by bitter combat in the Commune, the former suburb (*faubourg*) resisted merging with Paris. In 1884, the founder of the Chat Noir (Black Cat) cabaret, Rodolphe Salis, lost his run for Montmartre municipal councilor on a platform to secede from Paris. Salis was protecting his outsider status and promoting local rule like the Commune. An 1898 visitor found Montmartre oddly rural, seeming "at first view to be one-half country village and one-half provincial town . . . one would believe himself more than two hundred miles from the metropolis." In 1904, the British critic Frank Emanuel placed it among "the so-called eccentric quarters of Paris—that is to say on the soiled fringe of nondescript outlying districts of the *Ville Lumière*, which is separated from the city proper by the shabby-gentile exterior boulevards." Cementing its eccentricity, Montmartre, like Montparnasse and Ménilmontant, housed a large cemetery. Like the Bièvre Valley it hosted windmills. The owners of the Lapin Agile (Agile Rabbit) cabaret mounted another failed secession in 1920.[57]

Montmartre also hosted Paris's most noted area for informal or vernacular urbanism: the Maquis, literally "bush," like a bramble or warren. This neighborhood of sheds, shacks, and shanties tucked among gardens and greenery was near the city limits and infrastructurally off-grid. It was home to crooks, drunks, ragpickers, and others toiling in the informal economy—street vendors and dealers of drugs, junk, or scrap metal.[58] It was also the "heart" of Paris's artistic bohemia, an "ignition point" for culture, home to the illustrators Francisque Poulbot and Théophile Steinlen.[59] Houses here were "rickety shelters" (*abris branlants*) or "shacks pieced together" (*cabanes rapiécées*) with cloth, tar, and "sardine cans."[60] The historian Octave Charpentier described "plaster tile constructions, perfunctory workshops, [and] huts of planks."[61] In the Haussmannian tradition, the Maquis was demolished after 1910 to make

Figure 2.1. Postcard, K. F. series 1089, no. 6, postmarked 1906 (on card back) with the caption "The Old Montmartre—The last vestiges of the Maquis. 1904." Author's collection.

Figure 2.2. Unused, undated Neurdein brothers (ND) postcard 502, with the caption "PARIS (Old Montmartre).—Scene in the Maquis." Author's collection.

way for the Avenue Junot. Early 1900s postcards document its "last vestiges" (figs. 2.1 and 2.2; see also fig. I.2 in the Introduction). Montmartreans held long-standing local, leftist, communard, and anti-national grievances, reflexively recognizing their district's un-Haussmannized, heterotopic allure.

Montmartre's "bohemian" cultural modernists struggled to secure shelter. Charles Baudelaire had fourteen addresses in sixteen years, some in Montmartre.[62] Jules Lévy founded "incoherent art" at the first Salon des Incohérents (1882) in his studio, consisting of "my bedroom," also "my living room, dining room, kitchen, and bathroom." Lévy embodied what Philip Dennis Cate called the avant-garde "Spirit of Montmartre," which inspired the famous modernists Henri Toulouse-Lautrec, Alfred Jarry, Erik Satie, and Guillaume Apollinaire.[63] Other notable incoherents known as "Hydropathes" and "fumistes" ran the Chat Noir cabaret: Eugène "Sapeck" Bataille, Jules Jouy, and Alphonse Allais. The illustrators Poulbot, Steinlen, Caran d'Ache, and Adolphe Willette worked with the Chat Noir, the Incoherents, and the periodical *L'assiette au beurre*. These radical, populist freethinkers were absurd, biting, comic, cynical, irreverent, parodic, profane, sarcastic, and satirical. They staged shocking cabaret acts, and printed short-lived, cutting-edge 'zine-like papers that mixed poetry, prose, and illustrations in cartoons, rebuses, and lettrist word-pictures. Their minor literature clung to the "eccentric" periphery, tapping outsider ideas from anarchism, symbolism, and primitivism.[64]

When Apollinaire visited Jarry in 1897, the concierge directed him, oddly, to the "third floor and a half," which set up a joke. Jarry's building owner profited from its high ceilings, dividing each floor horizontally to make twice as many stories. The house had fifteen floors, but it was still standard height and therefore "merely the reduction of a skyscraper." His studio was also "a reduction." Jarry was shorter and able to stand under the demi-ceiling, but Apollinaire had to stoop. Jarry's bed was "a mere pallet" but also his table, for he "wrote flat on his stomach on the floor." Hence, in a turn of Montmartre wit, "the furniture was the reduction of furniture—there was only the bed." The room was sparely appointed with a partial painting and a "reduced library." Most curious was the "large stone phallus" on the mantel: "Jarry kept this member, which was considerably larger than life size, always covered with a violet skullcap of velvet, ever since the day the erotic monolith had frightened a certain literary lady who was all out of breath from climbing three and a half floors and at a loss how to act in this unfurnished cell. 'Is that a cast?' the lady asked. 'No,' said Jarry, 'it's a reduction.'"[65] Apollinaire enjoyed a phallic joke at the lady's expense, but his crude cultural combat pursued serious politics. Jarry's "reduced" studio was funny, shocking, and a bohemian bona fide, but Apollinaire also depicted a greedy landlord squeezing tenants into half lodgings and he mocked prudish literary ladies winded after only three and a half flights. A later political cartoon visualized this "half apartment" narrative; after paying half his rent, a man came home to find half his room rented to another man. Apollinaire's blend of housing politics and dirty jokes exemplifies the experience of housing in Montmartre. Steeped in Zola's boiling pot, the shoddy infrastructures and strained social relations of lodging houses became staple themes for Paris's burgeoning avant-garde, its humor, vulgarity, and populism.[66]

Several prominent Montmartre periodicals illustrated the experience of infrastructure in peripheral neighborhoods: *L'anti-concièrge* (the Anti-Concierge, 1881–1883), "the official organ for the defense of tenants"; *Le père peinard* (Father Cobbler, 1889–1902), printed by the leading anarcho-syndicalist Émile Pouget, and *L'assiette au beurre* (the Plate with Butter, first series 1901–1912), an iconic large-format caricature rag edited by Samuel-Sigismond Schwarz (1901–1904) and André de Joncières (1904–1912). Their writers and artists valued independent printing as a medium, and rights and freedoms of speech and press. Their rhetoric blended humor, parody, and satire with vulgarity and worker vernacular to mock the powerful, especially bourgeois republicans, socialists, and imperialists. They projected militant anticlerical, anti-militarist, and anti-colonial values. They consistently stressed the political economy of housing, controversially advocated violence as a lever of social change, and gave a strong sense of the right to the city.

Printed during the early 1880s housing debate, *L'anti-concièrge* styled itself "the official organ for the defense of tenants."[67] More topical and timely than "incoherent," its raucous humor vented tenant stress while its mock pomposity masked a litany of resentments. Foremost was a misogynist rant at concierges—often older, poor, or single women—whom it lambasted as cannibals and phrenological oddities, or dehumanized as guard dogs, even Cerberus at the gates of hell. Concierges anchored a broader critique of apartment house micropolitics involving male-dominated roles such as doormen (*gardiens*), property managers (*gerants*), and "Mr. Vulture" (M. Vautour), a generic landlord-speculator evoked by leftists from the cartoonist Honoré Daumier (1852) to the Marxist Paul Lafargue (1909). Vautour often endured antisemitic attacks; Sapeck disparaged Shakespeare's "usurer of Venice."[68] *L'anti-concièrge* oozed ennui and spleen as if, like Zola, to recruit disgruntled tenants or bourgeois reformers.

The *fumistes* saw concierges as prying, spying micromanagers, whose middle rank between landlord and tenant let them "[climb] the wall of your private life." Faced with poverty and the social fear of falling, and, for women, sexism against "old maids," concierges jealously hoarded scarce social power. Relations with tenants were strained, awkward, and conflictual. Concierges opened, read, and lost tenants' mail, or left tenants waiting in the rain. Jouy admitted that *fumistes* were bad tenants—hence the rhyme "surging from all corners of the universe / [come] the fumistes, to your house, turning everything inverse." But they sympathized little with concierges' social position. Jouy quipped that the homeless have no shelter, heat, or privacy, but also no concierge![69]

The regular column "Echoes in the Stairway" exposed obnoxious concierges posting their suggestions for minimum holiday gifts or printing business cards as "The Concierge of Victor Hugo." Mock ads on each back page fantasized futuristic consumer gadgets to annoy, prank, sedate, or even murder concierges. Articles told transgressive tales of subverting everyday hierarchy by tricking and cheating concierges. As Alphonse Allais suggested, "annoy your concierge while amusing yourself." After bribing the concierge to open the door after hours, pretend to leave something outside; when he offers to get it, lock him out and he will have to return the bribe to get back in. Allais exclaimed: "Oh, the magical power of metal!" The piece ends coyly, "Allais doesn't have mold in his place," because skillful tenants can work lodging-house relations to win better infrastructure. *L'anti-concièrge* was a radical tenant manual, offering what Michel de Certeau called "spatial stories" or "operations on places," scripts that subvert power relations by rescripting and empowering users as redesigners.[70] The magazine highlighted the deprivations, fragilities, and inequalities of lodging-house living, and imagined alternate futures.

Sapeck, like the Marxist Lafargue, feared the 1881 Landlord's Union, which demanded information about "bad tenants" (*mauvaise locataires*). Concierges, "transformed into spies," would deepen surveillance, sending information "to the central police station, which would be located in the office of the Paris Landlords' Union." He wailed, "But this, O bourgeois liberalism, is universal espionage taken to the tenth power!" Sapeck bared the class limits of republican equality, claiming a conspiracy of landlords and police. Further, he argued, tenants in conflict with landlords, simply too poor, or judged "in an incorrect family situation," might also be evicted, unlikely to get credit from other landlords, ending "in the impossibility of finding housing," and "forced to take refuge in some nasty hovel for which they pay double or triple." The bourgeois reformer Jules Simon echoed this scenario, but Pouget later seethed in *Le père peinard* that Simon was his landlord. Sapeck sneered that the law was already "made for" landlords, and now their association would confront the tenant with "a band of man-eaters, hell-bent on his complete ruin and organized for it." A caricature of the poet-politician Clovis Hughes, later the deputy of Paris's 19th district, who had recently proposed rent reduction laws, adorned Sapeck's editorial: "Bravo, dear poet!" This was serious politics, engaging with class conflict, civil associations, policymaking, and surveillance.[71]

Lafargue demanded a Tenants' Union to plan rent strikes, while the socialist city councilor Jules Joffrin urged British-style public housing, which Lafargue opposed because it would strengthen the state, postponing revolution.[72] *L'anti-concièrge* leaned into this debate. But their populist boosting of social underdogs and playful class warfare revealed deep antisemitism and misogyny. Their "serious" tenants' rights magazine was a send-up, mocking the arrogance of a self-styled "official organ" for anything, including tenants. Contributors parodied the earnest *prudhommie* (male virtue) of social reformers and civil associations, distancing themselves from the "bourgeois liberalism" of Bertillon, du Mesnil, and Simon. *L'anti-concièrge* depicted everyday lodging-house micropolitics among tenants and concierges, the common power struggles of Montmartre lodging houses. Dropping flowerpots on granny's head, locking her out, or withholding the rent payment occupied the contributors more than organized social change.[73]

But bohemian circles intersected anarchist circles, and Paris's strongest tenant organizers after the 1880s came from Montmartre's far left, inspired by rhetoric like that of *L'anti-concièrge.* In 1889, the anarchist Casimir Pennelier founded Paris's first Tenants' Union (*Syndicat des locataires*), which aped expert concern about "unhealthy dwellings," but wanted *landlords* to clean houses, demanding new laws against seizing tenant property, new taxes on rents received, and no more hidden fees or holiday gifts (*étrennes*) for concierges. The union also organized clandestine move outs, fleeing debt or eviction with family and possessions, usually by night, known as "moving out by the wooden bell" (*déménagement à la cloche de bois*). It may be related to the Parisian word "clochard," slang for a homeless or transient person.[74] The Tenants' Union faltered in 1906, before regrouping in nearby Clichy in 1909.

Émile Pouget, the later labor leader who coined the term "sabotage," printed the radical rag *Le père peinard* in Montmartre from 1889 until the republic closed it during 1894's anarchist crackdown. He fled to London, printing the

1894–1895 issues, before regaining Paris to print 1896–1902.[75] *Le père peinard* was equally scatological and scathing, steadily unmasking the bourgeois hypocrisy of the "cake-eating socialists" (like today's "limousine liberals") who "lick Carnot's ass."[76] The radicals were "radishes"—red (leftist) on the outside, but white (monarchist) inside and served with butter—while "we remain Communards."[77] The regular columns "Death to Landlords" (*Mort aux proprios*) and "The Wooden Bell" (*La cloche de bois*) promoted housing concerns. The conservative *Le Figaro* scoffed at its "colorful accounts of clandestine move outs." Pouget reprinted Jouy's song "By the Wooden Bell" in 1898, and dozens of drawings by Montmartre illustrators across his print run.[78]

Pouget inverted housing to display the inequalities of infrastructure and disease. He documented stinking houses with cesspits ("shitters" *chiottes*) instead of sewerage.[79] Noting how slowly all-to-sewer (*tout-a-l'egout*) drainage was replacing cesspits, he advised the people to send "charitable Christians and fraternal republicans" all to the sewer![80] He penned scandalous stories about police harassing homeless people, or landlords ("vautours") murdering indebted tenants and chaining shared water taps shut except for one hour a day.[81] Regarding influenza, he argued that "the people" (*populo*) suffering in "hospice," "miserable sheds" (*cambuses miserables*), or on the "street corner" (*au coin des bornes*) "have been its only and true victim."[82] During cholera's last bout in Paris, killing 1,700 in 1892, Pouget recycled a sixty-year old conspiracy theory that cholera was a bourgeois germ weapon.[83] Cholera was a running image: later illustrations caricatured Jules "Ferry-Choléra" and depicted a giant king rat on a throne, declaring the bourgeoisie "the real cholera."[84] Pouget tracked the costs of cesspits, disease, and capitalism for poor homes.

Housing horrors continued in the large-format illustrated satire magazine *L'assiette au beurre* (The Plate with Butter, 1901–1911). The illustrators took up Montmartre's housing scripts in special issues on vagabonds and ragpickers (1902), concierges (1905); cholera (1908); and rent (1908). The latter was called "Three Months' Rent [*La terme*]: issue drawn by Bernard, Ricardo Flores and Poulbot—so they can pay theirs." The cover image showed an evicted family trudging through snow with no horse as the father pulled the cart holding all their possessions.[85] A 1901 Steinlen cartoon, "Winter," portrayed a homeless man and woman huddled on a park bench; she turns and asks, "Prisons—are they heated?" An Adolphe Willette cartoon later that year depicted a male concierge "his majesty Pipelet," using his slang nickname, ranking him among the "annoyers" (*emmerdeurs*, literally "enshitters" like cholera) cataloged in Willette's special issue. Many drawings told sad stories: a mother who killed herself and her children the night before eviction, or a concierge feeling "ennui" when she could no longer read tenants' mail.[86] An ironic 1907 cartoon pictured two scruffy men talking in a suburban shantytown: one reported moving from Saint-Ouen to Ivry, "more distinguished" because of its gas streetlights—though with very little infrastructure in homes.[87] A 1909 special issue spotlighted vagrants and police along the Seine's banks (*berges*), another urban margin.[88] The magazine's biting humor, striking original artwork, and multicolor printing raised the visibility of housing issues in peripheral neighborhoods, inverting lodging houses for all to see inside.

Another Montmartre master of visibility was George Cochon who, like Pouget, merged housing advocacy and syndicalist militancy. A weaver, and the author of *39 Ways to Make Your Concierge Groan*, in 1911 he became the secretary of the heir to Pennelier's union, the Syndical Union of Workers, Employees, and Tenants (Union sydnicale des locataires ouvriers et employés). Cochon's landlord disliked his activism and evicted him in January 1912. Cochon caused a media circus, barricading himself in his apartment for five days with large stones, and flying a red flag and a banner on his balcony blaring "Under the Third Republic the law is broken by the Police." With union carpenters, he later sneaked a small, prefabricated house into the Tuileries Garden and assembled it for a large homeless family. A mock sale sign read, "House with garden offered by the Syndical Tenants' Union and the Building Union for a family of ten persons without lodging, abandoned by L'assistance publique." This spectacular prank appropriated public space, offered it for sale, sheltered the homeless, stressed assistance for large families, and suggested that unions worked better than welfare, embodying Cochon's ambitious and creative leftist activism.[89]

Unlike smaller-scale anarchist tactics, Cochon's large peaceful collective actions raised visibility for housing issues. Cécile Péchu analyzed his "methods of spectacularization," his tactical publicity and media baiting. He helped squatters occupy buildings and planned mass move outs for families fleeing eviction. Squats both housed the homeless and appropriated and occupied abandoned, empty, public, shared, or wasted spaces. Cochon rescripted clandestine move outs as noisy and visible with his band, the "Racket" (Raffut). With an ever-changing lineup and instrumentation, they made terrible anti-music using everyday objects, sounding the alarm for fleeing tenants while disorienting concierges, landlords, and police. In 1913, the Racket crashed the Moulin Rouge with a recently evicted family and disrupted a show. Such actions mimicked the cabarets' cheekiness and prefigured Dada: mocking the powerful and producing outrage with humor, lies, and shocking, subversive spectacles.[90]

Cochon embarrassed and infuriated the authorities with practical, tactical *fumiste* pranks that sent serious messages. Like the Chat Noir's Salis, he ran for office, for *maire* of the 20th district in 1912. He lost the election and lost credibility among some anarchists. Thousands of members left his Tenants' Union, which reformed as the new National and International Federation of Tenants (Fédération nationale et internationale des locataires), with posters illustrated by Steinlen. Amid these ups and downs, Cochon remained a popular hero, the subject of a postcard series and several songs.[91] Perhaps most significantly, Cochon formulated and popularized the "right to lodging," tying housing to the French political discourse of rights and thereby contributing to France's eventual turn to public housing.[92]

From *L'anti-concièrge* to Cochon, Montmartre avant-gardists and anarchists confronted the housing crisis with a carnivalesque explosion of symbolic resentment, satire, and cruel jokes, but also with serious community organizing and direct actions. The close quarters, cost, low quality, and strained sociality of lodging houses made them Zola's "boiling pots." In this neighborhood packed with cheap rentals, the pressures of lodging houses nurtured some raucous and radical responses, which rescripted rental apartments, abandoned buildings, shantytowns, and other spaces, to appropriate them for disgruntled tenants, the homeless, and anyone with precarious or insecure housing. The words used to describe Montmartre—"eccentric," "shabby," and "soiled"—convey its peripheral geography, in both the physical and symbolic senses, but also subtly capture the texture and tensions of living with infrastructure in relatively off-grid spaces, from lodging houses to shantytowns, which inspired inversions of domestic space.

Living with infrastructure at home was an experience fractured by class and geography. Some had homes; others did not. Some homes had infrastructure; others did not. Different levels of access, types of infrastructure, and relations with neighbors determined whether houses enjoyed modern comfort, convenience, hygiene, and luxury, or became "boiling pots" suffering stink, sickness, and seething social conflict. Between these extremes was a spectrum of living conditions, a patchwork of partial access to housing and infrastructure. Infrastructure both embodied and delivered social inequality—nagging social inequality that embarrassed the republic's egalitarian aims and took material and environmental form as the houses of Paris from glitter to gutter. Stories of Paris housing underscore that living with infrastructure is always shadowed by living without infrastructure. For all these reasons, housing remained a privileged site for debate, activism, reform, and conflict, and a subject for studies, novels, illustrations, and songs throughout secondary Haussmannization.

Paris's inadequate housing was a nagging socioeconomic and public health problem. But it signaled much more: anarchists, bohemians, and migrants; margins and peripheries; cholera and tuberculosis. In infrastructural terms it meant Zola's dark, dripping, destitute courtyards, half apartments, broken toilets, stinking cesspits, and living at home with inaccessible, unreliable, worn, or badly maintained infrastructures. Bad housing raised urgent questions: Which infrastructures were luxuries, necessities, or rights? Whose houses would enjoy "modern comforts" or become "murderous lodgings"? This chapter has charted a counter-discourse to republican progress narratives, which inverted infrastructure to make its problems visible and defended the people's right to the city. The novelist Zola, the reformer du Mesnil, the firebrand Pouget, and the activist Cochon carried this counter-discourse, alongside Montmartre's rowdy anarchists, activists, and artists. The lived experience of domestic infrastructures sharply diverged by class and place in ways that reflected people's chances to find shelter, water, comfort, or disease. Through varied acts of inversion that publicized the faults and failings of private spaces, Parisians questioned whether republican moral and material improvement and the modern infrastructural ideal would improve life at home for all.

3

The Street Life of the Tramways

Amid the transformation of streets introduced in chapter 1, among the most visible shifts as mixed-use streets yielded to vehicle traffic was the expansion of Paris's tramways. Horse-drawn trams in the 1870s, motorized trams in the1880s, and electrified trams in the 1890s were vehicles for more than mobility; they overturned Haussmann's horse-powered transit regime by bringing mass transit, light rail, motorization, and electrification to Paris. While tramway promoters and designers sought to invert primary Haussmannization by claiming the social and technical superiority of motorized light rail over horse-drawn alternatives, tramways were both criticized and cheered on by working-class Parisians and suburbanites clamoring for expanded transportation. Meanwhile, bourgeois and conservative Parisians protested trams as threats to urban culture, order, and tradition. All kinds of Parisians politicized the everyday experience of trams that were crowded, uncomfortable, inconvenient, inefficient, and dangerous, and were operated by companies struggling with finance, corruption, and competition from omnibuses and later subways. The tramways thus offer a clear case of varied and competing efforts to invert and politicize Haussmannization by scrutinizing infrastructure amid motorization.

Three scholarly threads illuminate the lived experiences of trams. First are studies of everyday life that interpret large structures, institutions, and processes "from below" by studying ordinary people's experiences of them.[1] Second are social histories of urban railways and of everyday technology, which promote the user's perspective in technological history.[2] Third are recent "street level" urban histories of Paris and elsewhere, which see streets as defining sites of urban experience.[3] Applied to tramways, these approaches emphasize drivers and riders. As large technical systems, urban railways were also social infrastructures for new groups of workers and riders, users who often saw infrastructure with less certainty and optimism than designers did. Thus, while Paris designers emphasized the economy and efficiency of motorized trams, users stressed their dangers, difficulties, and democracy.

Efforts to retrofit streets and tramways in the 1870s–1880s faced social, spatial, safety, technological, and weather issues. Street railways were a significant sociotechnical novelty, which officials hoped could invert primary Haussmannization, while laypeople contested this process. In the 1890s, economic and institutional issues impacted the expanding mass market for motorized transport. As horse-drawn omnibuses and tramways struggled to satisfy ballooning demand for mobility, transport remained contentious and politicized. Paris's rail development was contingent, layered, and uneven; 1850s rails predated 1870s motorization, 1890s mass transit and electrification, and the 1900 Métro. Historian Arnaud Passalacqua has shown that Paris tramways "met with only reluctant acceptance." They embody Edward Jones-Imhotep's deft distinction between *failed* technologies, which become disused, and *failing* technologies like the trams, used despite their faults. The trams teach that entire infrastructure networks can be weakly appropriated and integrated into everyday life even after operating for decades.[4] Parisians used a broad range of epithets to criticize and politicize tramways. This rich language of tramway critique illustrates my methodological claim that linguistic analysis helps reconstruct past emotions and experiences.

The tramways contrast instructively with the Métro (chapter 4). While Métro dreams illustrate sociotechnical imaginaries of the future, the tramways embody day-to-day, hands-on, routine use and maintenance.[5] Tramways and subways grew in tandem from the Prefecture of the Seine's 1872 urban railway commission. The General Omnibus Company (CGO) had operated a few horse trams since the 1850s, making them familiar, though never above reproach. Hence, 1870s officials quickly approved new tramways before debating the controversial, futuristic Métro for another quarter century. While imagining subways to articulate sociotechnical futures, Parisians experienced urban railways aboard the existing tramways, which offered a testing ground for new motors, rail designs, paving materials, social practices, and cultural meanings. Tramway troubles sometimes cast shadows over Métro debates, but testing Métro dreams on the tramways improved the future subway, while the tramways persisted through recurrent crisis. Métro dreams both dark and bright had their waking mirror on the trams. Thus, while the Métro burrowed deeper into Paris over time, the tramways came and went in 1855 and 1929. No matter how routine trams were by the interwar years, twentieth-century daily life never came to revolve around trams as it did around the Métro. Even after trams returned to Paris in 1992, they have never spawned an expression like Métro, Boulot, Dodo.[6]

In sum, this chapter highlights how embodied experiences of tramways varied with class, place, and technology, reflecting the sociotechnical and affective dimensions of living with infrastructure. Living with tramways was often a rough ride.

Retrofitting Roads and Rails

In January 1872, amid postwar reconstruction, Prefect of the Seine Léon Say created a special departmental engineering commission on "local interest railways and tramways." Having escaped war by traveling, he returned in 1871 convinced that Paris, otherwise a beacon of modern infrastructures, had fallen behind in transport. London's subway, New York's elevated trains, and horse trams in many cities now set the pace. Soon Constantinople and San Francisco installed cable cars. Say's commissioners thought Paris should remain a modern infrastructural model but struggled to imagine living with railways.[7]

Financial worries loomed as Paris was still paying dearly for primary Haussmannization. Social and spatial questions abounded. Some commentators worried that the public would reject railways as clashing with French "customs." Yet another problem was power or traction. While Say's commissioners knew of pneumatic (compressed air) and funicular (cable) traction, they assumed that Paris's railways would employ locomotives, whose sound, steam, smoke, sparks, and speed would be environmental nuisances on busy streets. To protect traffic and "the beauty of the capital," they rejected surface and elevated rails. Standard-gauge steam railways did not fit the Paris streets they knew, but subways could bypass them. Haussmann's ambivalent appropriation of technology, which expanded but also hid infrastructure (usually underground), remained hegemonic. These concerns and London's example led the commission toward a mixed network, with a standard-gauge central

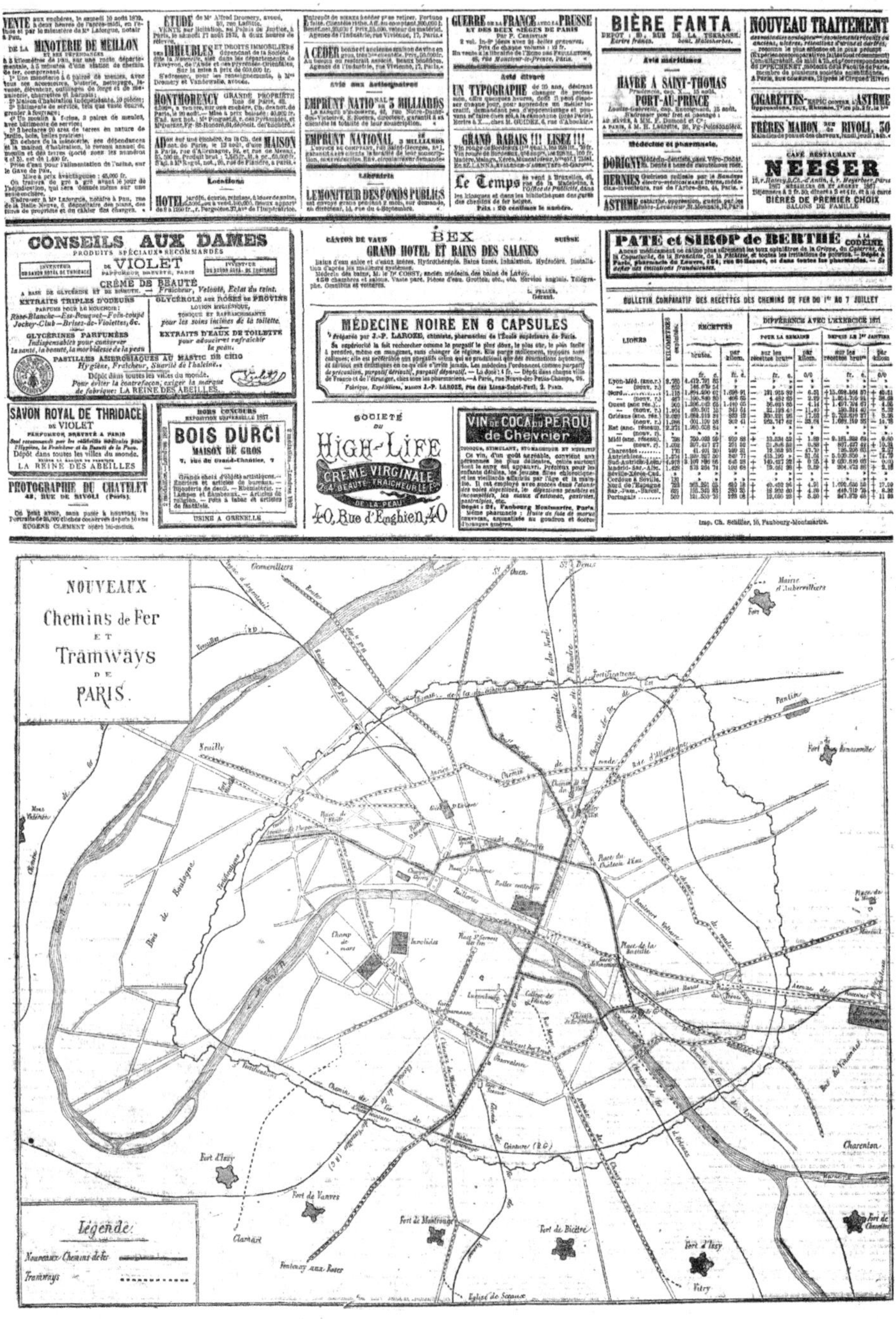

Figure 3.1 Map of subway and tramway lines proposed by the Prefect of the Seine's 1872 Special commission from *Le temps*, illustrated supplement, 29 July 1872. Bibliothèque Nationale de France. Note that dark black lines represent projected railways and dotted lines represent tramways.

subway (for which the source of traction was pending), and peripheral horse tramways. Their deliberations laid the groundwork for Paris's tramways.[8]

Say's commissioners expected friction between railways and city life. Their modernizing dreams and war-torn patriotism cut against their railway doubts, as they faced a tough modern problem, Wolfgang Schivelbusch's "tracks in the city," which the French called "interior" railways. The earliest railways (ca. 1830–1870) moved between, not within, cities. Born of British industrialization, they were loud, dirty, heavy machines that menaced environmental quality and were unfit for civilized and monumental Western capitals. As the civil engineer Félix Sérafon noted, "Railways interior to large cities" were "a contemporary question."[9]

In July 1872, *Le temps* publicized the initial plans from Say's commission (fig. 3.1). A limited central locomotive subway would meet the Ceinture railway in three peripheral places (Montmartre, Boulogne, and Porte d'Orléans), connect the major train stations (excluding Saint-Lazare and Montparnasse), and serve a central north–south axis, the Grands Boulevards, and Les Halles. Horse trams would serve the rest of the city: on an east–west line along the Seine's quays, across the Left Bank and the exterior boulevards, and on radial lines to the suburbs. Motorized subways would bypass the center, dense with traffic, monuments, and population, while horse trams would serve passengers in the sparser periphery.[10] Different ideas of aesthetics and traffic governed core and periphery, which needed railways for different reasons. The historical and Haussmannizing center was an outdoor trophy case of architectural patrimony, where proud patriots rejected surface railways. The periphery's lesser traffic, fewer monuments, and lower property values inspired planners to project street-level radial lines from the center to the suburbs.

While the proposed subways awaited a bidder, hungry investors grabbed three concessions for new tramways: a "north network" serving the Right Bank periphery and suburbs; a "south network" on the Left Bank; and a line along the Right Bank exterior boulevards granted to the General Omnibus Company (CGO) (fig. 3.2).[11] All three contracts specified horsecars, called "American railways" (*chemins de fer américains*) since the French flat-rail innovator Émile Loubat had imported them from New York in the 1850s. While Parisian subway dreams modeled London, its tramway plans modeled US cities. France's jealousy of the anglophone world reflected having its administrative, military, and technological pride hurt by Prussia in 1870.[12] Many nationalists suffered an inferiority complex. As *Le temps* printed in 1872, "Our great city, which has so often taken the initiative of perfecting and embellishing [itself],

Figure 3.2. Neurdein brothers (ND) Postcard 1120 CGO (General Omnibus Compny) Horse Tram with roof deck on the Rue Monge, ca. early 1900s. Wikimedia Commons.

has remained 25 years behind in adopting American railways."[13] From the start, therefore, Paris's tramway debates articulated a sense of delay, of "waiting" for the transportation that Parisians deserved.

Tramway opponents, however, saw these "American" imports as unfit for French life despite their French importer. Some critics pushed this "alien" reputation, suggesting that tramways threatened Americanization. Others called them "impossible" because "the public would never accept this mode of locomotion." Confined to rails, critics argued, trams were not mobile enough. Their "rigidity" clashed with "our mores," which better matched the horse-drawn omnibus, whose "mobility permits it to choose all directions instantly."[14] Omnibuses handled any pavement and stopped on demand. Appropriation, which absorbs and integrates novel or foreign artifacts into cultures, is particularly evident in the tramway's "alien" status, linked with American innovation, foreign investors, and suburban commuters. Tramway critics thus appropriated tramways as aliens to keep Paris French, omnibuses and all. While future passengers appropriated tramways more overtly by riding them, nonusers were appropriating tramways more subtly by assigning them scripted social roles.

As the tramways opened in 1873, debate emerged in popular periodicals such as *Les causeries scientifiques*, *La revue des deux mondes*, *La nature*, and *Le magasin pittoresque*. The encyclopedia *Le magasin pittoresque* tried to disarm tramway critics, writing, "the drawbacks cannot be compared with the advantages."[15] But the conservative economist-politician Raphaël-Georges Lévy's passionate 1878 editorial called tramways "a nuisance [*embarras*] and an obstacle to the free movements of merchant carriages and pedestrians." He called the tram a "monster" that would "mutilate" Paris and he rejected rumbling, smoky subways. Tramways were an "attack on good taste" and on the "right to promenade," especially the new La Chapelle–Alma line, which crossed the Avenue des Champs-Élysées, a planning showpiece since the 1700s. Lévy defended Paris's pedestrian culture and beautiful cityscape, while progressives responded that his "imaginary" worries could "immobilize many thousands." He favored aesthetics and nationalism; they valued accessible mobility. Conservatives envisioned citizens moving freely, slowly, and reverently through a stable city, while progressives hoped that technologically enhanced movement would transform daily life for all.[16]

Tramway plans survived this controversy: construction would start in 1873, and service in 1874. Paris's first tramway boom lasted until 1879, fitting the familiar pattern of development that accelerated around universal expositions. Elites feared being caught by foreign guests with outmoded or inadequate infrastructure at the 1878 exposition, which also inspired new Métro plans.[17] French national tramway development spans 1855 to 1929, and concentrated from 1872 to 1914 in three waves: horse trams (1872–1884), mechanical trams (1875–1889), and finally an explosion of electric trams including the Métro (1890–1919). Minister of Public Works Yves Guyot collected national tramway statistics in the 1890s. In the Seine department, he counted 70 new tramlines from 1874 to 1894: 59 horse, 6 steam, 3 pneumatic, 1 electric, and 1 funicular. These numbers predate the surge in electric trams before the 1900 exposition (discussed in chapter 6), but they show that the technical problem of traction was vital and shifting during the energy transition of the Second Industrial Revolution, while horses long remained dominant.[18] Politicians and engineers struggled to invert the horse-powered transport regime inherited from Haussmann, which finally ended in 1913.

Motorization promised savings over costly horse traction. The science editor Émile Gauthier wrote, "Animal traction is the ruin of the tramways. Thus, all cities are preoccupied with replacing it with various methods of mechanical traction." Each tram required several horses, which were worked hard and fell

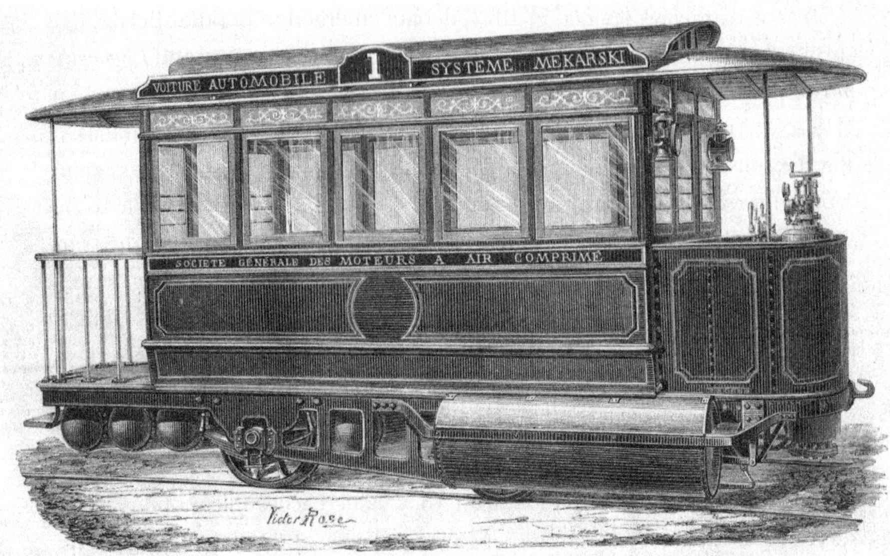

ASPECT EXTÉRIEUR DE LA VOITURE AUTOMOBILE À AIR COMPRIMÉ, SYSTÈME DE M. MÉKARSKI, EXPÉRIMENTÉ SUR LA LIGNE DES TRAMWAYS-NORD.

Figure 3.3. Illustration, Mékarski compressed air tram from *l'Illustration* no.1708 (20 November 1875), 332. Wikimedia Commons.

prey to disease during short careers. Operators therefore owned many horses; by 1900, Paris counted 98,000. Horses needed stables (i.e., land), training, veterinary care, food, and workers to move hay and manure. Horsecars and manure were ubiquitous reminders of hybrid environments and the impact of vehicles on environmental quality. These costs were colored by the common notion of technology mastering nature as humans master horses; replacing animals with machines seemed clean, efficient, and modern.[19]

In 1875 the North Tramway Company asked Minister of Public Works Eugène Caillaux to allow trams running the British engineer G. Palmer Harding's new steam engine. The minister established a special commission to test three mechanical traction systems. First, they tested Louis Mékarski's pneumatic engine, powered by metal cylinders filled with compressed air in a central plant and hung beneath the cars (fig. 3.3). Mékarski argued that motorized trams demanded "very great flexibility," modulating power usage for frequent stops and starts and climbing or descending hills. The commissioners liked his vehicles' power and handling, but feared the cylinders might explode, and set up a three-year trial. Pneumatic trams were only marginally more successful than atmospheric subways. By 1888, they served only four of over forty Paris tramlines.[20]

Figure 3.4. Unbranded postcard, postmarked 1908, depicting a Compagnie des tramways de Paris et du département de la Seine Francq tramway engine and its driver. Wikimedia Commons.

The second system tested was Harding's "locomotive avec foyer," a traditional steam furnace locomotive. The engineer Auguste Graeff raved to the minister of public works that it could cost half as much as horses, which might have worked without further safety concerns. The ministry approved Harding engines on the South Tramway Company's Montparnasse–Austerlitz line, but, fearing fire, required two agents per engine car—one operating equipment and another monitoring safety. The cost of hiring two people per car largely undid the savings of steam, and by 1881, Harding retired his locomotives under financial strain.[21]

Finally, the commissioners studied Léon Francq's "locomotive sans foyer," a steam engine with no furnace, powered by metal tanks of pressurized hot water (fig. 3.4). This system also tested smoothly and went into trial on several lines. Francq touted his system's safety: "As the furnace [*foyer*] does not exist on this locomotive, the drawbacks of smoke, sparks, soot, odor, the glow, the grill, the noise of friction, or of exhaust from the smokestack, are completely and absolutely removed. The metal no longer has reasons for alteration; incrustations are no longer possible; there is no longer a rigorous surveillance to exercise over the apparatuses of power or of security in general. Finally, the dangers of explosion are definitively removed, and security becomes complete, absolute." The commissioners worried about passenger safety on Harding and Mékarski trams, but trams also had to safely share streets with other vehicles. Francq wondered whether motorized trams would blind or frighten horses with noise, smoke, steam, and sparks, causing collisions with vehicles or pedestrians.[22] The problem was neither animal nor mechanical power, but mixing the two. Could animals and machines share streets, or would motor vehicles disrupt traffic in a city still so dependent on horses? In 1876, the engineers Mékarski, Harding, and Francq worked on adapting locomotives and replacing horses but found few definitive solutions.

Engines were only one safety risk; collisions were another. A city averaging what *Le temps* called "two accidents a day" needed ways to reduce harm. In 1876 the prefect of police began reviewing designs for a "body catcher" (*chasse-corps*), a metal skirt to cover tram wheels and prevent objects or people from slipping underneath. Periodicals printed lurid stories as reminders of modern risks and leverage in demanding improved equipment. Hence, according to *La lanterne* in 1879: "We have already demanded body catchers for the tramways. Here is yet another demonstration of their utility." The paper reported a tram on the Place du Châtelet hitting a nurse carrying a little girl. "Without the body catcher . . . the unlucky woman was literally cut in half." The luckier child was thrown from her nurse's arms away from the wheels and survived.[23]

Fitting trams with body catchers progressed unevenly. The Ministry of Public Works received citizen complaints and new design submissions through 1880. In 1877, one Melon, an elderly man from the wealthy west suburb of Neuilly wrote the ministry: "Since the tramways have existed . . . accidents happen quite often, either by the fault of the agents of this administration, or by the inattention of the public." Drivers "sit on their benches and read the newspaper," he claimed, ignoring approaching passengers. Melon cited a crucial fact of transport practice: until 1896, the tramways and omnibuses lacked fixed stops or stations—passengers could hail them anywhere to board. Drivers were supposed to stop for boarding, but in practice often only slowed down. Running to catch trams became a recurring image in the press.[24] People boarding moving trams demanded close attention from drivers, who were supposed to sound the horn as people approached the tracks. Mistrusting drivers, Melon suggested furnishing trams with body catchers and horses' collars with "little bells" (*petit grelots*). He neglected to note that some trams already had body catchers since 1876, thus claiming the innovation as his own. He wrote again in 1878 complaining that trams west of Paris were still unequipped, and in 1879, confirming that they had been outfitted, but also requesting payment for his "invention" to supplement his pension, which drew the ministry's curt dismissal.[25]

In 1878, a tram driver named Vallet served eight days in jail for assaulting a passenger. M. Dubard, a Neuilly businessman, threatened to report that Vallet "made me run" by not stopping for boarding passengers. Vallet rejected Dubard's threat, insulted his intelligence and masculinity, and punched him in the eye. The director of roads and navigation at the Ministry of Public Works, the engineer Armand Rousseau, wrote to the prefect of police: "As facts of this kind, just like the fairly numerous accidents which have happened on various tramway lines, are generally being attributed, by the public, to a lack of surveillance, I beg you, Monsieur Prefect, to examine well . . . an effective control of operation." He suggested a tramway control authority. Attached to his letter was an unidentified news clipping about Dubard and Vallet, dated 9 October 1878: "In this affair, we need to know if a tramway or omnibus conductor can honor his duties and help people get on and off, and pay their fares; then again, maybe he should, when he is on his stand, pay attention to those who give him the sign to stop and thus spare the elderly, women, and children the ennui and even the danger of running after the vehicle. One sometimes sees people following one of these large vehicles, at a run, for several minutes, making desperate gestures at a conductor whose attention is wholly elsewhere."[26] Dubard and Vallet reveal two persistent issues in public debate: coachmen's misbehavior and the need for stronger tramway governance or "control."

Common criticisms of tram drivers are reflected in the bohemian Félicien Champsaur's surreal and satirical 1885 short story "Ballad of the Tramway." It opens with a scruffy, drunken coachman snoozing at the reins of the La Villette–Étoile horsecar. Struck by love at first sight of a blonde bourgeoise in her private coach, he takes his passengers hostage, deliberately derailing the tram to chase her. Hurtling through space and time, his tram "evolves" in explicitly Darwinian terms—its wheels become flippers to cross oceans, and later wings to cross skies. Centuries later, it finally reaches the terminus. Paradoxically, the tramway's unreliable mobility (arriving centuries late) justifies its cosmic voyage. In the 1880s, Paris's omnibuses and tramways were often criticized as inefficient and unreliable, driven by drunken, distracted, rude drivers, whose capricious detours wasted time. Champsaur's story self-critically parodies stock critiques of Paris transport, while its fantastic imagery illustrates how tramways piqued his imagination.[27]

The stories of Melon, Dubard and Vallet, and Champsaur show that tramway practice remained as malleable as tramway technology in the 1870s–1880s. Engineers, drivers, and riders remained uncertain about the design and use of light rail systems. One observer thought certainty would come with time: "In effect horses are frightened, but the education of horses happens in time just like that of humans."[28] But both people and horses slowly learned to live with tramways, whose hybridity was as challenging as their novelty. Horses were not the only nonhumans involved in tramway function; because tramways operate outdoors, the weather also contributed to their risk and uncertainty.

The winter of 1879–1880 was, according to one scholar, "the most rigorous ever observed in France," and it stranded railways nationwide.[29] A colossal December 1879 snowfall in Paris stifled traffic and provoked public debate about snow removal.[30] On the eleventh, Director of Works Alphand informed the municipal council that its extra 500,000 francs for snow removal funded 13,940 men, 3,900 horses, and 2,400 carts (*tombereaux*), moving over 2.22 million cubic meters of snow.[31] Even so, traffic remained paralyzed as officials sought new snow-removal tools. The departmental engineer Edmond Humblot deiced the Paris roads with salt for the first time. Other engineers later designed a sheet-metal snowplow (*chasse-neige*) in 1881 and a pneumatic snowblower in 1883.[32] Besides rain and snow, fog also endangered trams.[33] Because tramways were embedded in streets, operators maintained cars, rails, and pavement but also watched the weather. Multiple infrastructures, species, and environmental conditions met at the road–rail nexus, creating complexity and uncertainty in tramway practice.

Uncertainty exacerbated costly horse traction, spelling trouble for the tramway companies. Paris's first tramway boom went bust after 1878. Like all railways, tramways demand heavy initial investment to be (ideally) recovered through tickets or tolls. Hence, big financial interests like the Société Financière de Paris, the Franco-Italian Bank, and several Belgian firms invested in Paris tramways, amid ongoing talk of mergers and acquisitions.[34] The first tramway boom was a classic railway investment craze, and when growth stalled in 1878, the gamble was often lost. The crisis first struck the South Tramway Company, which lost 3 million francs in 1879–1881 by hiring two men for each Harding tram.[35] Soon the North Tramway Company followed. The CGO's 1860–1910 mass transit "monopoly," a dreary piece of Haussmannian baggage, stoked the financial crisis. Because the North and South tramways crossed the city limits, their 1873 charters charged the

CGO with building and repairing rails within the city, which the tramway companies would pay to use, plus "an indemnity for privation of traffic and the partial abandoning of the rights conceded [to] the Omnibus Company by the city of Paris."[36] As public demand grew in the periphery and suburbs, tramway companies that expanded ties to Paris therefore increased what they owed the CGO.

These expenses increased because the CGO's "derailable" trams rode rails without *entretoises*, wood or metal bars like railroad ties, which keep rails rigid and evenly spaced. Within Paris, the CGO built tracks to its own standards: Loubat's "American" rails were laid directly in the pavement. These rails did not fit the North and South Companies' "underailable" cars, whose nonpivoting axles were vulnerable to smaller variations in the rails. So, the tramway companies retrofitted CGO tracks to their cars, which engineer Sérafon calculated to cost 2.25 francs per meter. Handily for the CGO, its trams easily rode the stiffer modified tracks, though some critics complained that its vehicles damaged both rails and pavement.[37]

Despite growing demand, ticket sales could not offset these mounting costs. In lean times, expanding the network and updating equipment were doomed. In 1881 the crisis overtook the tramway companies, and the CGO bid to acquire them. This sparked controversy and a flurry of pamphlets with the keyword "verité" (truth), conveying a sense of Haussmannesque corruption. Sérafon described shady corporate mergers, writing, "Each day sees another new combination, whose authors, little known in the financial world, conceal speculators interested in staying in the shadows." Foreign tramway investors inspired some minor nationalist uproar, sustaining tramways' "alien" reputation. But officials disliked the merger as much as the pamphleteers did and denied the CGO buyout. The North and South Tramway Companies went bankrupt in 1884, replaced in 1887 by Tramways de Paris et du Département de la Seine (north) and the Compagnie générale Parisienne de tramways (CGPT, south).[38] Haussmann haunted tramway finance. The CGO monopoly stacked the competition between the CGO and smaller transit companies, while it and similar government contracts made it harder to repay Haussmann's massive debt, thereby slowing infrastructure development.

Minister of Public Works Severiano de Heredia admitted in 1887 that railway finance laws were often "lost from view or poorly interpreted." Official bungling or corruption amplified trouble in the tramway companies: missed deadlines, broken contracts, and roads and rails battered by wear and badly maintained. As struggling companies trimmed the fat, the numbers of trams and employees declined, maintenance slowed, passengers waited longer for trams, and companies demanded government bailouts.[39] In 1889, Heredia's successor, Yves Guyot, tried to bolster the tramways with a new "Tramway Control Service" (Service du Contrôle des Tramways) to make regular inspections in the Seine department, realizing Armand Rousseau's dream. The service comprised three engineers: two Ponts et Chaussées men to monitor construction and operation, and one from the École des Mines to inspect motors.

The Ponts et Chaussées inspectors concentrated on the rail–street interface: how to set rails in streets, with what paving materials, and at what risk to roads. New designs for integrating streets and rails continued through the 1880s. Even this most basic light rail infrastructure was not fully stabilized or standardized after a decade of development. Previously incompatible, rails and streets were slowly interwoven by technically changing both rails (their shape and the kind of "lock" between wheels and tracks—the issue of *entretoises*) and pavement (new materials). This interweaving helped steer Paris's long, slow shift from wooden and stone cobbles (*pavés*) to asphalt and cement pavement. Rails laid between cobbles rattled, causing the stones to separate over time and damaging both rails and roads, but asphalt and cement hugged the rails. The challenges of building streets and railways together were compounded by questions about traction systems, safety, social relations aboard trams, finance and administration, and the impact of the weather. Though engineers of the Tramway Control Service could not effectively address this manifold uncertainty, ballooning public demand for mechanized mobility kept the troubled tramways running.

Ten-Cent Tramways

As the economist-editor Paul Leroy-Beaulieu noted, the 1890s brought a social shift because tramways and bicycles increased mobility for average Parisians.[40] The transportation market became an economy of scale, giving most Parisians access to mechanized and motorized mobility. By 1891, Paris's forty tramlines were established, if still controversial, features of daily life. Journalism and slang soon called them "ten-cent tramways," capturing the popular importance of their price and accessibility. Since the 1870s, Paris's population had grown from about 2 million to 2.5 million people, horse traction had struggled to manage expanding demand while a discourse of public demand solidified, motorized alternatives proliferated, and the dreams of officials and engineers scaled up—more lines, cars, and riders meant more money saved and earned. Combining motorization and cheaper fares would increase the feasible scale of transport networks. Experts from the railway engineer Léon Francq to the political economist Paul Vibert argued in the 1890s that Paris could only enjoy the awaited means of transport and the predicted savings of motorization by cultivating mass ridership. The tramways' financial and technical problems had the same solution: a larger scale.[41]

Paris's mass market for motorized transport emerged around 1880, following other emblematic mass markets including department stores and daily newspapers.[42] From 1879 to 1883, ridership grew faster for larger vehicles, in which each seat cost less, thus making each conveyance more profitable. Older 28-seat omnibuses grew 1 percent, while newer 40-seat omnibuses increased 50 percent, but tramways expanded even faster. Annual tramway rides increased almost 300 percent between 1879 and 1895, and around 400 percent from 1895 to 1928, when tramway ridership peaked at around 700 million rides. No network grew faster than the Métro, designed for mass transit. From 1901 to 1910, first-class ticket sales increased from just over 6.5 million to almost 27 million (over 400 percent); second class, from 34.5 million to 164 million (almost 500 percent). By 1914, both liberal Métro dreams of booming traffic and socialist dreams of mass mobility came true (see chapter 4).[43]

Such growth would have been challenging for politicians, entrepreneurs, and engineers even under ideal historical circumstances. Demand not only grew with population but also with expanding availability and changing expectations. More people across the department wanted to use motorized transit as it became more accepted and accessible. Seats were open to a broad public, if not enough for everyone. Politicians and engineers stoked demand for motorization before they could satisfy it, scripting future expectations. Chapters 2 and 7 describe the same pattern with piped water and other domestic infrastructures. Thus, the effort in the 1890s to add new tramlines soon shifted toward overhauling departmental tramways. As the Ponts et Chaussées inspector general René Lefebvre wrote in 1898, "It is not merely necessary . . . to complete the current organization of mass transit in Paris and its suburbs, but more to completely transform it."[44]

Public pressure to expand and modify tramway service persisted. The winter of 1892 stoked debate over tramcar heating. As the newspaper *Le rappel* reported, "The tramways and omnibuses are a never-ending subject. What improvements doesn't the public demand, whether in the timetable of trains, or in the renovation of cars?" Far beyond heaters, users were reacting to the chaos creeping into daily operation as the tramway companies struggled.

Parisians learning to live with trams created a litany of common complaints: agents were rude and broke rules, there were not enough tramlines or tramcars, existing lines neglected important places, transfers were difficult, cars were uncomfortable, and their wheels damaged roads. Stories of cold or wet feet were common; rain worsened the tedious wait for trams. In hindsight, these criticisms look well-founded. They inspired the Tramway Control Service back in 1878 and were now inspiring the socialist Caumeau's campaign for heated trams in the departmental council.[45]

As they shifted from luxury toward necessity, the tramways were increasingly politicized. Anatole France's 1901 novel *Monsieur Bergeret à Paris* introduced municipal councilor Raimondin, a radical republican who "lost the confidence of the electors," and would lose reelection because he "neglected the interests of the neighborhood. He didn't even get a tramway, demanded for 12 years, and they say he has sympathy for the Dreyfusards." While a failed tramway could end a radical politician's term, some who were farther left dismissed tramways as tools of the elite. The profane anarchist wordsmith Émile Pouget derided "simple-mindedly" (*bonassement*) socialist politicians as "little shits" (*merdaillons*) for providing themselves with tramways to commute to City Hall or the National Assembly. He thought striking workers made better use of trams as "attractive" (*galbeux*) incendiary devices.[46]

The tramways (unlike the municipal Métro and private housing) were departmentally controlled. Like their "American" origins and foreign investors, this made them exotic to Paris. They both embodied and perpetuated the historically tense relations of city and suburbs. In 1894, negotiations for a new "penetration network" (*réseau de pénétration*) or "complementary network" (*réseau complémentaire*) began, a departmental project to better connect Paris with the suburbs. Planners envisioned thirty-three new lines, half of them radials between the suburbs and Paris, the other half an "exterior beltway" (*ceinture extérieure*) linking the suburbs. From 4 June to 4 July 1896, these plans were subjected to a "public utility survey" (*enquête d'utilité publique*), a procedure for gathering opinion about public works required for obtaining government approval.[47] The *enquête* dossier included three important details. First, fares would be cheaper than they had ever been: 5 centimes for trips outside Paris and 10 centimes for trips into and out of Paris. In 1878, seats cost 15–50 centimes, priced by length of trip, and whether one rode inside the tram or on the roof deck (*impériale*), which was always open air but sometimes shaded, and opened to women after 1879. Second, the older omnibus-based practice of hailing vehicles anywhere to board would be replaced with "fixed stops."

Third, existing horse-drawn lines and all new lines would be motorized, probably electric.[48] *Enquêtes* provided important occasions for competing attempts to invert Haussmannization. While officials deployed them as proof that secondary Haussmannization would be more transparent and responsive to citizens than primary Haussmannization, citizen participants used them to articulate criticisms of republican works.

The 1896 *enquête* produced positive reviews. Citizens across the department generally approved of motorization and fixed stops but made various suggestions: the *impériales* should be covered, cars should be staffed by two agents or electrically lit, fares should be lower, and the system of transfers (*correspondances*) should be reformed or ended. Finally, citizens demanded special routes for morning and evening rush hours, or discounted fares for workers and everyday commuters. *Enquête* participants were invested enough to tinker with details of technology, organization, and pricing. *Enquêtes* and their participants inverted infrastructure by bringing the technical details of engineering practice to the attention of the lay public.

Parisian *enquête* participants echoed their suburban neighbors; each district produced five to twenty observations, no more than one of them negative. Participation rose with social class: more from the Right Bank compared to the Left, and more from the west than the east. Working-class east Paris (in the 10th, 11th, 19th, and 20th districts) registered no comments or suggestions. The themes of fare reduction, the transfer system, and passenger safety recurred, as did the divided geography of center and periphery. One observer in the 18th district suggested "electric traction by underground conductor for the center of Paris, and by overhead wires for the exterior boulevards." Several comments about safety specified that using rails to return current posed an electrocution risk, expressing worries about the dangers of living with electricity (see chapter 6). The Water Company (Compagnie générale des eaux) warned that its pipes could conflict with underground wiring. Third rail traction was also unpopular; although it had just been chosen for the Métro, it could be more dangerous outdoors. The southern CGPT, meanwhile, wanted to electrify some trams by underground conduit (*canniveau*), though its real ambition was the controversial trolley: "the use . . . of the electric system with overhead wires wherever it can be admitted, or with underground conduit in those parts of the interior of Paris where overhead wires will not be accepted."[49]

As Parisians and suburbanites constantly clamored for expanded transport, *enquêtes* were perhaps drolly formal. The few hundred commenters in

1896 constituted a tiny fraction of Paris's nearly three million people (excluding the suburbs), which suggests limited public access to, or interest in, the process. *Enquêtes* were legally required, but their ever-positive results only confirmed what many already knew: the newspaper *L'éclair* referred to the tramways as "so impatiently anticipated." Suburban papers alleged "hundreds of thousands" of petition signatures favoring new trams.[50] As the engineer Gauthier wrote in an 1897 report, "From all sides, they demand the expansion and improvement of our ways and means of transport, as well as a reduction of fares." Local officials, he said, should work "to give prompt and legitimate satisfaction to the voices of the populations in the department."[51] The whole department waited hungrily for mobility.

Dashing Gauthier's hopes, tramway deliberations outlasted the 1900 Universal Exposition and dragged on. Ministry of Public Works documents show that, of the twenty new tramway lines conceded from 1899 to 1902, seventeen were granted in 1899–1900, rushing to serve the exposition, but half were still not operating in 1903. After the 1896 *enquête*, it took three years to concede and seven years to build some new tramways. In 1902, Deputy of the Seine Coutant brought the "scandal" of the "nonexecution" of tramway contracts before the Senate. His argument that private companies could not be trusted with such contracts prompted heated discussion about municipalizing Paris transport. Coutant joined a small but growing number of leftists who advocated public ownership of infrastructure (the gas company was another popular target). The deeper question—whether private companies could reliably provide public services—was widely debated, and even some liberals, who generally preferred private to public control, began to doubt that private companies could manage Paris transport. The concession system itself carried weighty Haussmannian baggage. Transportation was eventually municipalized in 1921.[52]

Departmental councilors, radical-socialists, and journalists feared delinquency from the tramway companies around 1900. The CGO competed with, obstructed, and set a bad example for other transit companies; its sluggishness chafed against the era's steady technological change. Paris's largest transit company and operator of horse trams and omnibuses was timid concerning the use of new equipment, still yoked to animal power and milking its 1860 charter and "monopoly" until 1910. The political economist Paul Vibert campaigned against the CGO before the 1896 senatorial elections. Article 7 of the CGO charter specified that "if the adoption of a new system would have as a result a notable increase in the net profits of operation, the company will be obliged to include the public and the city of Paris in this advantage by means of a lowering of fares." But the CGO stalled for decades, dabbling with steam, gas, and electric motors, while maintaining high fares and blaming expensive horse traction. If the CGO tried to update equipment, Vibert charged, it was "fragmentary, I would say almost infinitesimal, absolutely insufficient and under conditions of bad faith." He criticized the "disastrous" and "defective" cars on the Saint-Augustin–Cours de Vincennes line, "where it [the CGO] placed only enormous vehicles that have the air of war machines, even though it knows perfectly well that it was easy enough to establish tramways with covered impériales, light, simple, graceful and able to pass down all streets." *Le Figaro* found them "frightful" (*affreuse*); Albert Robida called them "monstrous." The CGO used clumsy, dated, and expensive steam and pneumatic motors like those tested in the 1876 special commission. In 1907, the engineer and former city councilor Jules Armegnaud called CGO omnibuses "horrible mastodons," ancient beasts unfit for modern streets.[53]

The CGO oozed corruption. In 1897, the suburban journalist L. de Laere sarcastically claimed he could not distinguish Inspector of Public Works Lorieux from Senator and CGO President Cuvinot, to highlight corrupt coziness between the public and private sectors. De Laere also evoked the socialist Armand Grébauval, another prominent suburban journalist and a local politician. To prevent dirty deals between the state and the CGO, Grébauval proposed to the departmental council that the prefect of the Seine, Justin de Selves, should be forbidden to fraternize with Cuvinot "even to drink a bock at the café of the National Guard," until he declared the public utility of the new tramways. The next year, Stanislas Ferrand, an architect and engineer, a deputy of the Seine, and the editor of *Le bâtiment* (Building) took up the torch of critique. Suburban leaders maintained pressure on officials, inverting Haussmannization to politicize the contracting system and to connect city and suburbs.[54]

Such popular, leftist, and suburban complaints about the CGO and the tramway companies echoed broader critiques of public works aimed at the municipal, departmental, and national governments. Parisians waited until the CGO charter lapsed and was rewritten in 1910 to see the company move toward motorization, which had been legally required since 1896. The CGO effectively waited out the period of Paris's initial electrification before adopting gasoline autobuses after 1910. Despite the rush and controversy of the 1890s and the spur of the 1900 exposition, tramway "reorganization" continued for decades; Ministry of Public Works documents show that it was an ongoing administrative worksite through the 1920s. The CGO disappeared in 1921, bought out by the municipal Société des transports en commun de la

Figure 3.5. L. J. & Co. Postcard, "Everyday Paris" (Paris Vécu) series, postmarked 1904, with the caption "Wait for the tramway." Author's collection.

région Parisienne, as the municipalists finally had their moment. Tramway "reorganization" from the 1880s to the 1920s reflects the rigors of motorization and suburbanization in the CGO's shadow, a prominent example of Haussmannian baggage ripe for inversion.[55]

A postcard from the Everyday Paris series mailed in 1904 depicted a familiar street scene, the "wait for the tramway" (fig. 3.5). The card visualized passenger complaints about inadequate service and long waits, but in the context of ongoing corporate restructuring and government reorganization, "waiting" had a double meaning. While impatient passengers waited for specific trams, an impatient population waited for expanded tramway networks and service. What at first glance seems an "ethnographic" image of a typical street scene therefore reveals a subtle politicizing of infrastructure when read in historical context.

Tramway Epithets

Cultural challenges compounded institutional and technical challenges. Once the masses rode the trams, a series of striking tramway epithets emerged: anti-tramway elites called trams "barbarian," "criminal," and "odious," while a wider swath of commentators highlighted safety by calling tramways "crushers," "murderous," and even tramways "of death." Leisure travelers such as automobilists, cyclists, pedestrians, and tourists coined the phrase "tramway of death" in 1899 for the line from Paris to the western suburb of Maisons-Laffitte. Protest began when the cycling magazine *Le vélo* (The Bike) and the conservative papers *Le correspondant* and *Le Figaro* criticized the tram for leaving little room on the bridge over the Seine for leisure travelers, who risked being crushed or thrown into the river. The cycling group French Velocipedic Union snidely joked that officials were adding ambulance cars so trams could collect victims en route. The popular paper *Le petit Parisien* coined the phrase "murderous tramways" to describe several collisions in 1900 between horse and motor trams that shared rails (see chapter 6). While the more popular tramway epithets highlighted safety, the more elite epithets enacted class conflict and cultural negotiation, evoking the discourse of civilization and barbarism.[56]

Unusual accidents inspired custom epithets. In 1901, when a crowded tram filled with toxic fumes caused by lead reacting with sulfuric acid in a failing accumulator, *Le radical* described the "poisoner tramways" (*les tramways empoisonneurs*).[57] In 1905, after a woman lost her pregnancy in a tramway crash, the phrases "aborter tramways" (*tramways avorteurs*) and "nauseating tramways" (*tramways nauseabondes*) appeared.[58] But the use of these case-specific terms was dwarfed by the most widespread safety epithet, "crusher tramways" (*tramways écraseurs*), which echoed the older phrase "intersection of the crushed" (chapter 1) and circulated from the 1890s to the 1920s.[59] "Crusher tramways" became its own journalistic "special rubric," as *Le Figaro*'s Paul Meyan wrote, usually a headline for brief, detailed, and gory accident reports in the news items (*faits divers*), most used from 1901 to 1908.[60] The phrase echoed the auto critics' term "crusher automobiles" (*automobiles écraseurs*), and was sometimes joined by the word "everyday" (*toujours*), flagging accidents as frequent and familiar. The term "crusher" was widespread, used by journalists from the far left to the far right, indicating that safety was a far more common concern than civilization.[61]

Perhaps the most colorful epithet was "barbarians' tramway" (*tramway des barbares*), coined for a new 1897 horse tram from the southern working-class suburb of Montrouge to the Saint-Philippe-du-Roule church in the ritzy 8th district. Moderate and conservative editorials criticized the new tramline, and residents organized opposition. Like Lévy had done in 1878, critics guarded the Champs-Élysées. *Le Figaro* cried "mass protest" and Jules Claretie demanded "outcry" and "fury" against the "odious tramway," but it never came. Recalling decades of controversy over Haussmannization, Claretie feared that tramways would wake Haussmann's "ghost."[62] Conservative municipal and national politicians conspired to block the line with a police declaration of unsafety and a government buyout. But at this high tide of the Radical Party, they were too small a minority; the Chamber of Deputies approved the tram 337 votes to 114. Socialists cheered the "democratic tramway," arguing that it would "not injure the appearance of the Champs-Élysées at all." Conservative journalists responded with snide quips: "The democratic tramway . . . will be pulled by democratic horses." For them, the tramway and its leftist boosters were barbaric threats to the civilized and monumental New Rome.[63]

One journalist called working-class commuters barbarians, writing, "Every day the barbarians make their descent into the heart of Paris"—implying that trams let marauding commuters "sack" the New Rome. Three of the four "barbarian" trams were part of the new complementary network connecting Paris with the suburbs. Radial tramways were creating Paris's "streetcar suburbs," to evoke a classic of urban history that resonates with Paris's "dangerous classes," urban "frontier" or "margins," and the "red belt"—fixtures of nineteenth-century history that convey deep city–suburb conflict. Beyond aesthetics and safety, therefore, new radial tramways unsettled conservative social geography by making outsiders too mobile, lowering barriers between the city and the suburbs, opening Paris to "alien" movement.[64]

The new 1897 tram operated quietly for two weeks, until organized working-class Montrougeans appropriated it for protest. On May 28, they filled the controversial tram to capacity and hung a banner from the upper deck reading, "The Barbarians of Montrouge to the Recalcitrant Press." En route, protesters "shouted ultra-revolutionary slogans," and reaching the ritzy Avenue d'Antin, they rushed local bars, already drunk on protest, and denounced "the savages" of the 8th district. This spirited reappropriation of conservative language answered one tramway-based protest with another. Ironically embracing the term "barbarian" themselves, they insulted the 8th district's "savages," who tried to block them from northwest Paris. The Catholic newspaper *La croix* quoted one protester, "We are civilized people. And en route to Saint-Philippe-du-Roule!" to stress that suburban worshipers rode the line to Sunday mass. The protesters threw civilization and Catholicism back at conservatives.[65] Protesters also bragged that "the Montmartrois, who have a reputation for wit, could have done no better," but that summer, Montmartre's irreverent cabarets produced several satirical sketches titled "The Barbarians' Tramway." One sketch featured a ridiculous cardboard tram pulling two sheep through a crowd of riotous protestors overwhelming the police. Foreign observers also saw humor here; Britain's celebrated *Punch*, a magazine of humor and satire, printed a sing-song poem demanding, "Why this sudden *melée* / is a mad dog astray?"[66]

Dozens more tramways were proposed for the 1900 exposition, while conservatives kept complaining.[67] In 1900 construction of the new Bourse–Opéra line was stopped by angry neighbors protesting the "barbarians' tramway." *Le temps*, their self-appointed ally and "echo," said the tramway would spoil the neighborhood's beauty, encumbering the Rue du Quatre-Septembre and the iconic Place de l'Opéra.[68] Historians have shown that these protests waged elitist exclusion, fearing that new trams would let "barbaric" workers invade upscale neighborhoods. Several distinct lines were called "barbarians' tramway."[69] This epithet reflects the view of those who used, or refused, tramways—evidence that, unlike popular demand and enthusiasm, elite Parisians still feared motorized mass transport. They saw a darker side of modern infrastructure, which sometimes seemed more barbaric than civilized, especially for conservatives, elites, or Luddites, such as Lévy and Claretie, who feared modernization, democratization, sociotechnical change, and foreign influence.

The Catholic magazine *Le correspondant* orientalized the new trams, describing "such mutilations, profanations, deformations, which will fatally end by making of the former aesthetic capital of the world the gigantic and vulgar caravan-station of the future democracy." The conservative *Le Figaro* offered this striking 1907 use of the epithet: "Oh universal Americanism! Oh the vandalism of governance! . . . Oh barbarians' tramway! Oh the plague of democracy!"[70] That same year, Ernest Levallois, the mayor of Paris's 2nd district, founded "The Parisian Association for Art in the Street" to protect his neighborhood from the new Rue Réaumur line, which he called "the criminal tramway." Decrying accidents, construction, noise, stalled traffic, and neighbors who "lose their clientele, lose air, [and] lose sleep," Levallois published photographs of vacant storefronts and construction gutting the roads. He praised local posters depicting the tramline as a sword piercing Paris's heart—the Place de l'Opéra—with the liturgical legend "deliver us from evil." Decades after the debut of tramways, some Paris bourgeoises and conservatives still found them fearsome and foreign, which one author dismissed as "aristocratic pretensions."[71] Their dramatic laments about declining French civilization linked tramways with common markers of modernity: new infrastructure, foreign influence, and republican democracy.

By contrast, Montrouge's protesters and their republican and socialist allies celebrated tramways as tools of mobility, modernity, democracy, and equality. While elitists called tramways "barbaric," more egalitarian voices called them "democratic" or "ten-cent" tramways, celebrating their accessibility, including rush-hour fares for workers.[72] Tramway critics and supporters disagreed about tramways, but also about French civilization, modernity, city life, and rights to mobility. Both booming ridership and widespread talk of "crusher tramways" suggest that physical danger was a more relevant risk for most riders than corrupting "civilization," though both drew commentary and controversy.

Progressives also wielded another epithet, "phantom tramway," which meant tramways that were officially approved but not built or operating, or operating tramways whose cars were defective or scarce. Many had criticized irregular and unreliable service since the horsecar era. In 1892, the 15th district councilor Athanase Bassinet called the Auteuil–Saint-Sulpice line a "phantom tramway which passes in this neighborhood every once in a while."[73] In 1893, he sarcastically criticized the Vanves–Avenue d'Antin line as a phantom that "currently passes about every eight days."[74] The councilor for the 14th district, Georges Girou, first called the 1897 barbarians' tramway "phantom." Some progressives mocked critics of the original barbarians' tramway because the CGO line offered such sparse, phantasmic service.

The departmental councilors Pierre Cherest (Neuilly) and Louis Parisot (Courbevoie) criticized "phantom" tramways in 1899–1900, as did the 15th district municipal councilor Adolphe Cherioux in 1906.[75] A 1903 accident report titled "Phantom Tramway" reprinted in *Le rappel* and *Le XIX*^e^ *siècle* recounted an electrical fire that interrupted service.[76] In 1905, *Le Figaro* derided a Romainville–Temple line, approved a decade prior, which had never carried a tram.[77] Like other epithets, "phantom tramway" was flexible, rhetorical, shared, and strategic; even *Le Figaro*'s staunch tramway critics appropriated it. Tramway opponents were eager for rhetoric that reinscribed the tramways' famous, foreign, and fearsome reputation. In 1907, the progressive Pierre Cherest blamed critics of the barbarians' tramway for a nonexistent tram approved in 1900.[78] Overall, "phantom tramway" conveyed a demand for tramways that was tinged with impatience having waited too long for the promises of infrastructure.

This rich and varied body of tramway epithets shows that citizen attempts to invert Haussmannization were never homogeneous, but instead reflected class positions, geographies, political ideologies, and views of technology. Parisians both clamored for and criticized tramways as vehicles of Haussmannization, mobility, modernization, and motorization.

Tramways are physically woven into streets, making them socially relevant for understanding street life—for example, during public debates about tramway development, interpersonal conflicts on trams, traffic accidents, and winter storms. This chapter has therefore gathered everyday street-level stories—drivers punching passengers, feet getting soaked while waiting in the rain—which highlight the experiential and affective dimensions of living with infrastructure. Both the novelty and the hybridity of street railways sparked controversy. Symbolic oppositions like French versus foreign and civilization versus barbarism structured debates over many historical shifts: the growth of Paris's mass transit market, the energy transition of motorization, and streetcar suburbs made by new radial railways. Elites and conservatives used these symbolic oppositions to distance tramways through exoticism, calling them "monsters" and "odious," vehicles of American or Asian influence, cultural leveling, dangerous lower-class mobility, and threats to cultural practices including leisure walking and reverence for monumental architecture. Workers, progressives, and socialists hailed trams as "democratic" and modernizing tools of physical and social mobility. Tramways thus became vehicles of class conflict, city–suburb conflict, risk negotiation, and competing attempts to invert infrastructure. They inspired more, and more diverse, epithets than other infrastructures. Tramways became Jones-Imhotep's "failing" technologies, tolerated despite their faults, including physical risks and institutional fragilities. In sum, efforts to expand, motorize, and democratize Paris's tramways during secondary Haussmannization faced consistent politicization that kept Parisians waiting for infrastructure's promised future of accessible, reliable, and ubiquitous motorized mobility.

4

The Dream Life of the Métropolitain

After 1872, the metropolitan question slept for a long time.

What would it mean to live with future infrastructure? This chapter answers by examining debates about the imagined Métro between 1872's departmental railway commission and 1895, when the municipality won control and planned the actual Métro. Long before transporting people after 1900, the subway was a vehicle for dreams and desire, a device for scripting "sociotechnical imaginaries," shared hopes for interlocking social and technological futures. Railway dreams helped Parisians negotiate the "promise of infrastructure."[1] Parisians dreamed that the Métro would do much more than move people, and projected fears onto it. Their fantasies inverted Haussmannization by dreaming of a railway Haussmann scarcely imagined, under municipal control, outside his jurisdiction.[2] The Métro is the only municipal infrastructure discussed in this book, falling between private housing and the other departmental public works. This chapter shows that living with infrastructure meant not only materially being with physical infrastructure but also culturally living amid multiple contending visions of future infrastructure. As in the epigraph above, Parisians often cast this experience in terms of "sleep" and "dreams."

Paris's earliest city railway plans date to the 1840s, even before its first horse tram in 1853. Because nearby London, often imitated in France since the 1700s, opened its "Underground" in 1863, many have noted that the

Métro opened as late as 1900. By then, the engineer Adolphe Schoeller counted around one hundred Métro plans.[3] The 1872 to 1895 railway struggle, Norma Evenson's "long debate" or Alain Cottereau's "battles," reflects the work required to fit Paris with rails. For around a century after the engineer Louis Biette's 1906 account, scholars largely explained Paris's slow subway growth *politically*: locals and nationals fought for jurisdiction, reducing the railway to an object of bureaucratic battles. The leftist municipal council desired a public light-rail network for everyday mobility, while the republican state wanted mixed-use trains on standard-gauge tracks carrying passengers, freight, or troops, creating a hub for France's five national rail networks. Legally, the state sought a "general interest railway" (*chemin de fer d'intérêt general*), while the city wanted a "local interest railway" (*chemin de fer d'intérêt local).* This politically reductionist story of local–national conflict has become *the* standard account of the Métro's prolonged prehistory.[4] Histories of technology offer an alternate account, which reduces the railway's delays to waiting for the technology, namely, electric traction, needed to remove the nuisance of locomotive emissions.[5]

Cultural studies of the Métro have revised the standard and alternate accounts in four ways. First, scholars have shown that subways were controversial everywhere, though Paris's were more so than London's.[6] Many Métro problems were neither political nor technological, but sociocultural, rooted in opposing conceptions of cities and railways. Second, questions and choices about subways reached beyond technology, transforming lifeways and urban form. Cottereau's Métro "battles" not only disputed future technology but also negotiated living with infrastructure. He wrote, "Veritable choices of urbanization and of ways of life were unleashed, under cover of technical arguments."[7] Institutional questions of design, funding, operation, and regulation revealed deeper issues about living with railways. Parisian Métro dreams did what scholars call "appropriating" technology—socially reconstructing technology and the cityscape to blend them, making railways and Paris appropriate for each other. Rather than simply *adopting* railways, Parisians faced the complex sociotechnical task of *adapting* railways and streets to each other. Métro debates rewove the urban fabric, showing that existing environments, technologies, and practices must be retrofit to accommodate new infrastructures.[8]

Third, Parisians consistently linked subways with imagined futures and modernities, progress, and even utopia. Subway plans projected railways with their social and environmental consequences and contexts.[9] Fourth and finally, Métro debates show how institutional analyses limit our sense of relevant actors in infrastructure history. Those debating the Métro were not only engineers, politicians, and transit companies but also architects, contractors, journalists, intellectuals, landlords, businessowners, workers, and citizens. Carlos Lopéz Galviz revealed a "range" or "array" of publics debating subways, raising the question "whose futures?" The standard account is also incomplete, therefore, because Métro debates had more than two sides—from the departmental government's role in the local–national legal conflict to private companies, civil associations, and these multiple publics.[10]

Following these four claims, this chapter reads Paris's protracted Métro debates as evidence of "sociotechnical imaginaries." Métro dreams envisioned new technology accompanying environmental, social, and spatial changes. Railway designs also concerned commerce, traffic, street life, health, safety, environmental quality, and architectural aesthetics. Rails were not laid over the urban fabric but woven through it, because the social, cultural, and political work needed to stitch them into daily life began before any technical work. Parisians building the subway after 1898 and riding it after 1900 had already dreamed of it for decades. At the dawn of urban railways, the technologies and practices involved were not yet solidified, standardized, or socially accepted, so imagination filled the gaps. Restoring imagination to the design process helps scholars see the full range of actors—designers and users alike—who engaged in Métro debates. By weaving the Métro into visions of Paris's future, a diverse cast of dreamers gradually learned to live with railways as they vied to decide what railways would become.[11] The actual subway, under municipal jurisdiction, became one of the most successful official inversions of Haussmannization.

With a Touch of the Wand

During the Franco-Prussian War and Paris Commune of 1870–1871, Parisians who escaped to London, including the historian Hippolyte Taine and the poet Arthur Rimbaud, admired the world's first subway. Another admirer of London railways was Prefect of the Seine Léon Say, whose special railway commission (discussed in chapter 3) was looking to London's subway as an example for Paris's proposed metropolitan network (*réseau métropolitain*), which would soon be nicknamed Métro. The civil railway engineer Félix Sérafon also invoked London's leadership. London was the world's largest city and like New York motorized transport early. A French model city since Voltaire, for many London excelled in modernizing infrastructures, inspiring dreams of Paris's future. If the nineteenth century envied Paris, in 1872 Paris envied London. Parisians constructed urban modernity and the modern infrastructural ideal through such transnational comparisons of cities and their infrastructures.[12]

The liberal newspaper *Le temps* ran an anonymous letter from London written by Hippolyte Taine:

> London is the city of railways. Not only is it connected by major lines to all parts of the kingdom, as befits a capital; not only do its suburban lines open it, as in Paris, to easy communication with the belt of towns and villages which surround it, but, what is particular to London [is that] three of its major railways, the south, south-east and the west, pushed their stations into the heart of the city . . . and these new stations are connected to one another by rail.
>
> Finally, an underground railway, the *Metropolitan*, passes by these three stations and puts them in contact with all the others. The great arc it inscribes in the interior of London is complete and closed.
>
> These new railways, with their numerous intermediate stations, singularly facilitate entering and exiting London; but their principal advantage is to put the most far-flung neighborhoods of this immense city in rapid contact with one another. The interior circulation of London operates on this network of rails. Instead of going on foot, by bus, or in a cab, one transports oneself from one neighborhood to another by railway.

Unlike Paris, where horses still pulled most vehicles, London was learning to live with railways. Londoners eased the spatial and technical tensions between rails and streets by weaving rails through the city via trenches, tunnels, and viaducts. Taine admired the London railways' systemic qualities: they were centralized, connecting the whole city and the entire kingdom, "as befits a capital," forming a "complete and closed" system. He described the ubiquitous, unifying networks of the modern infrastructural ideal.[13]

As Taine shifted from London to Paris, he described his "dream" of floating over Paris, magically transforming it "with a touch of the wand," using an image often applied to Haussmann.[14] He predicted that Parisians would experience "ennui" underground. Smoke-filled tunnels would suffer poor visibility and air quality, with no daylight or view. He was perceptive about public doubts about the subway, and yet he predicted a subway. He found that London's stations enjoyed smooth, comfortable pedestrian circulation, and argued that humane underground travel demanded freely flowing light and air. A well-ordered system could smooth flows of people, vehicles, light, and air without upsetting the routines and scripts of everyday life. Similarly, the lawyer Louis Le Hir claimed that his 1872 subway plan would reduce street traffic and accidents, although Say's Commission refused to evaluate it.[15]

Yet subways posed unique obstacles by exhuming a common image of underground Paris as fearsome catacombs, famously depicted in Victor Hugo's *Les misérables* (1862) and Gaston Leroux's *Le fantôme de l'Opéra* (1909–1910). The underground stood for crime, revolution, filth, disease, and death, a nether-realm of evils banished from civilization—the historian Rob Zaretsky's "labyrinths of the unthinkable or unspeakable." One critic called the Métro a "sewer train," and the thinker Walter Benjamin called it the "underworld." While social norms of street life complicated surface railways, dark ideas of the underground discouraged subways. Conservatives in particular feared subways' effects on passengers and patrimony alike.[16]

Not all dreams envisioned a complete railway. In 1875, the Ponts et Chaussées council, France's premier engineering advisory board, argued that the Métro should integrate the Petite Ceinture belt railway, run by a syndicate of the five national railway companies. The state would gather the rail companies to make Paris a national "turning plate" (*plaque tournante*) or inter-network hub, allowing trains to transfer among France's five rail networks without leaving the rails. In November, the department offered 30,000 francs for further Métro study. Soon the Ponts et Chaussées engineer Edmond Huet and Director of Works Alphand envisioned radial lines "in a covered trench" well-vented for locomotive exhaust, tying the Ceinture to a new hub under the Palais Royal. Huet planned locomotives riding standard-gauge rails—like the national railways, whose trains "could be called to circulate on the metropolitan network." Practically, this meant a commercial and military network for freight, information, and passengers. Huet's project title, however, contained the term "local interest," hotly contested by the minister of public works Albert Christophle, writing in 1876 that the Métro "presents to an eminent degree the character of GENERAL INTEREST. Because this network does nothing but connect the stations of all the major general interest networks to a common center, its concession should emanate from no authority other than that of sovereign power." The local–national legal battle had begun, squeezing the departmental Office of Works in the middle.[17]

The municipal council, desiring a locally controlled and funded Métro, rebuffed the state and sent an 1876 study mission to London.[18] These moves and countermoves began the twenty-year local–national standoff. Despite the difficult decades ahead, Louis Biette's term "sterile," which has haunted Métro histories since 1906, is misleading. For example, take the 1877 Grande Ceinture, a suburban belt railway run by cooperating rail companies, as engineers suggested for the Métro in 1875. After the municipal council's 1876 rejection, this new 1877 belt was an alternate route to a national railway hub. The nation improved its "turning plate" while the departmental suburbs became more connected. Before Parisians reached a final Métro plan, parts and versions of the railway were built piecemeal, making the "sterile" years surprisingly productive—even in material terms, ignoring more than one hundred unbuilt designs.[19]

The architect Louis Heuzé designed an 1878 elevated Métro plan, modeling New York, to calm Paris's roaring traffic, ensuring a "double circulation" of vehicles and pedestrians. His ornate iron viaducts covered a "special street" to shelter pedestrians from weather, embellished with lamps, benches, gates, and other street furniture, or the iron and glass facades of Paris arcades and shops. Heuzé envisioned railways durably woven into social routines, traffic patterns, and architectural styles, tightly suturing viaducts into streets. For him, subways risked flooding, groundwater pollution, damaged monuments, collapsed buildings, and falling property values. Railways traversing "putrid" and "fetid" soil might threaten "epidemics," and just seemed morbid: "For the adjective *metropolitan*, Parisians will soon substitute that of *Necropolitan*, for a railway obliging the public to descend by way of long staircases into veritable catacombs!" Like Taine, Heuzé claimed that Parisians preferred riding in daylight and outdoors (*en plein air*), as on the fashionable upper decks of omnibuses and tramways. Other engineers and commentators agreed that subways might make passengers bored, anxious, or even sick.[20]

The British engineer T. W. Rammell's 1878 pneumatic subway plan rejected locomotives because of their tensions with tunnels. His compressed-air subway worked like similar examples in London and New York, or like Paris's pneumatic message service. Rather than air pressure driving engine parts, his tube-shaped trains with a rear "screen" "like . . . the sail of a ship" fit tightly into sealed tunnels and were pushed by bursts of air. This system, he boasted, avoided locomotives' environmental nuisances: "no heat, no smoke, no steam, no noise, no vibration." Ingeniously, compressed air could solve both traction and ventilation problems—or so he claimed; pneumatic or "atmospheric" railways were short-lived.[21]

Heuzé and Rammell designed contrasting solutions for the problems of subways: limited ventilation disqualified locomotives, driving the search for alternatives like compressed air. These challenges were sociotechnical. Deploying florid language and evoking public opinion, Heuzé's pamphlet was as emotive and imaginative as Taine's earlier letter. Métro dreams were appropriated for many different cultural and political projects, and many like Heuzé's and Rammell's never materialized. To advocate subways, engineers like Alphand, Huet, and Rammell mixed technical design with sustained symbolic work on the uncanny underworld. Techniques that increased tunnel safety and comfort—lighting and ventilation—also calmed public fears. In the 1870s, technical design melded with the social, political, and cultural task of appropriating railways, weaving them into the city's physical fabric and its culture and customs.

Darker dreamers like Heuzé stoked the public's railway fears with a language of disgust, disease, and death. Many expected a future Métro but feared its social consequences. Railways, a symbol for all industrial technology, often served nineteenth-century thinkers as demiurges of change. Parisians understood modernity as shaped by technology, driven by historical motors like railways and steam engines.[22] In the 1870s, the Métro's dream life made frequent dramatic scene changes from the utopian to the nightmarish. Paris architects, engineers, the lawyer Le Hir, and the historian Taine appropriated subways with ambivalence—enthusiasm and ennui—setting the scene for the 1880s, when elevated railways and organized campaigns took center stage, Taine's utopian "magic" faded, and the Métro "question" grew achingly familiar.

The Question(s) of the Métro

In the 1880s, Parisians named "the Metropolitan question" (*la question du Métropolitain*) as a recognized public debate.[23] As for the housing and water questions, this label masked the many questions involved. One common strategy for appropriating urban railways recruited them to steer urban change and solve social, environmental, or public health problems. The railway engineer Eugène Flachat dreamed that a central subway would disperse buildings and traffic, reducing congestion. Many dreamed that the Métro could answer the housing question (see chapter 2). Léon Say hoped railways could diffuse population, helping Parisians find clean, healthy, comfortable, spacious housing.[24] The engineer J. A. Théry's 1885 elevated plan described "an unrest and an inexpressible anxiety in the Parisian population, due to the expense of rents, to the hygienic conditions of the city, to the encumbering and uncleanness of the center, to the insufficiency of, and the difficulty of moving between, livable spaces located in the extremities of Paris." For him, railways were "the most urgent among many other creations being studied," as potential solutions to nagging traffic, housing, and public health woes.[25]

In the 1880s, dreamers enrolled the Métro in the Third Republic's booming debate on social reform and hygiene. One engineer proposed pacifying radicals and socialists: "Concede the workers a railway!" Many planners claimed that railways could steer urban growth by increasing mobility and improving housing access. Take Heuzé again, "The centralization of commerce and the decentralization of inhabitants depend on the good or bad disposition of rapid lines [of transport]. The more centers of manufacturing are brought together, the easier commerce is; the more the population disperses itself, the more it should find lodgings of better price, clean, sometimes with gardens and always with promenades, in the neighborhoods called 'eccentric,' and even in the greater suburbs."[26] This familiar vision of an expanded, suburbanized, globalized city integrated long-standing bourgeois fantasies of Paris cleansed of biological and revolutionary dangers by removing the working classes from sites of barricades and disease and transplanting them into suburban garden villages, facilitated by motorized mobility. This vision continued Haussmann's project of grooming the urban core for administration, commerce, finance, public buildings, and monuments, while displacing worker homes and workplaces outward (discussed in chapters 1 and 2).[27]

Two basic visions of the Métro's urban future loosely aligned with the local and national factions vying for legal control. The first (more municipal) vision sought to extend the center's dense urban fabric into the outer districts. Alain Cottereau called this vision "dominant" or widely accepted; it projected traditionally dense and mixed land use (no zoning), with people walking to work. The second (more national) vision followed London and Haussmann in pushing proletarian life to the outskirts, promoting practices of commuting, gentrification, segregation, and zoning. The first vision leaned into municipal socialism, projected renovating the city's living space, and treated "public works" as services for public use and enjoyment. The second vision dreamed of displacing residents, emphasizing national liberalism and private investment, for which "public works" meant fitting the nation-state's needs or the population's aggregate interests. As with the 1872 Opéra gaslights, leftists imagined infrastructure serving practice, while conservatives saw it steering practice (chapter 1). Unlike more statist and technocratic nationals, locals domesticated the Métro for social equality and mobility. Recalling the Commune, this debate pitted everyday Parisians and the municipality against the Haussmannizing development agenda of the department and the nation.[28]

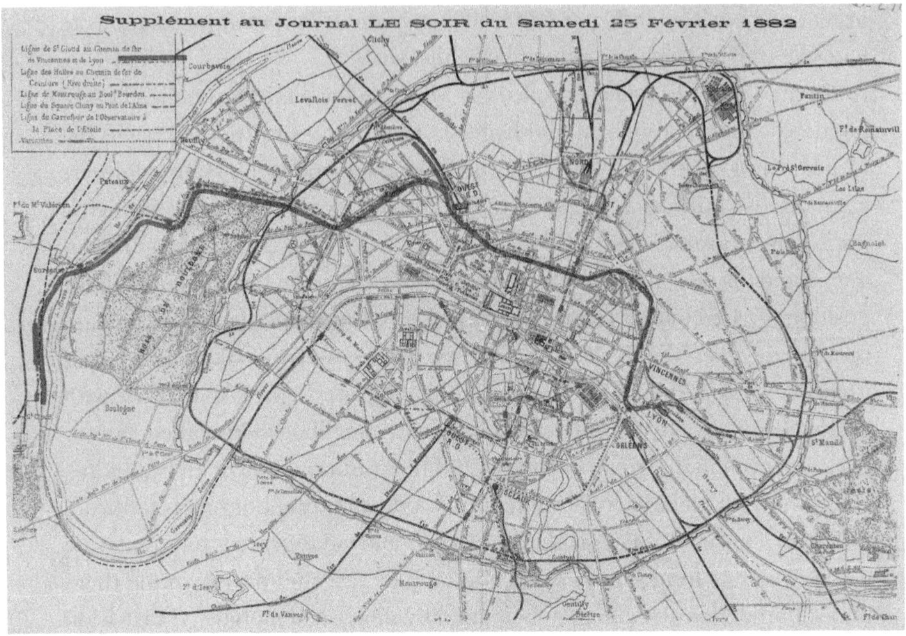

Figure 4.1 Map of subway lines proposed for the 1882 Métro public utility survey from *Le soir*, supplement, 25 February 1882. Bibliothèque Nationale de France.

The law of 11 June 1880 gave municipalities more control over local-interest railways to jumpstart development. But it fueled the legal conflict over the Métro by placing any railway entirely within a single municipality (*commune*) under local jurisdiction, while networks spanning many cities would be departmentally governed. Thus, the municipality tried to exclude its rails from the suburbs, despite constant suburban demand for more transport to and from the capital, while the nation courted departmental allies by arguing that their plan would serve suburban mobility better than any exclusively municipal plan. In 1883, the Ponts et Chaussées council argued that the Métro was neither a local- nor general-interest railway, and thus the state should exceptionally concede it to the city—predicting the 1895 outcome.[29]

In 1882, the Prefect of the Seine called a Métro public utility survey (*enquête d'utilité publique*). Registers opened from 15 February to 15 March at the mayor's office (*mairie*) of each Paris district and suburban municipality, alongside descriptions, drawings, and maps of projected works. Citizens made comments and criticisms to a commission of experts, landowners, and officials. The plans for 1882 projected a Métro with the common cross-inside-a-circle shape, with a central subway and peripheral elevated trains, powered by steam or compressed air (fig. 4.1). Public responses were favorable overall. The General Omnibus Company (CGO) president penned the only recorded criticism, stating that the Métro would violate the company's 1860–1910 "monopoly" on public transit, harming its profits.[30] It had used the same strategy to obstruct and bully the tramway companies (chapter 3). Founded on horse traction, it zealously resisted competition from motorized alternatives. Private interests like the CGO fed this Métro stalemate alongside the three levels of government.

When no investors took the 1882 project, public impatience grew as plans piled up without being realized. Governments and investors struggled to launch projects. The economist-politician Alexis Bailleux de Marisy explained that though "we dream everywhere" (*on rêve partout*) of improved transport, the Métro remained "always in project" (*toujours en projet*) because city deficits were stalling public works.[31] Haussmann's legacy made debts and expropriations contentious. Landlords and tenants alike feared eviction by a government buyout for "public utility." Yves Guyot later argued that the municipal council's Métro plans tried to protect taxpaying landowners (and their socially disgruntled tenants) from buyouts.[32]

These multiple stakeholders—the CGO, investors, landowners, tenants, politicians, and engineers—demonstrate Lopéz Galviz's diverse "array" of participants in Métro debates. They also show how far Parisians were from consensus about the railway in the 1880s. It remained a cultural artifact with unfixed meanings, and a political artifact with unsure regulation. This large technical system's design was not yet stable or closed, but showed interpretive flexibility, being many things to many Parisians.[33] The impasse spurred people to act, creating organized civil campaigns for specific Métro designs. Parisians grew impatient, Métro dreams diversified, and it became harder to imagine a railway that could meet the varied needs of nearly three million people. As officials and investors stalled, dreaming drifted toward civil society. Parisians organized and struggled to realize their dreams. Commercial interests (landlords and transit companies) were joined by associations from the republic's bustling civic culture of voluntary organizations.[34]

An early organized Métro campaign grew from the Society of Civil Engineers' journal *Le génie civil*, which printed comparative studies of Berlin, New York, and San Francisco railways. Louis Figuier and Felix Sérafon similarly compared railways in Berlin, Brussels, London, Naples, New York, Philadelphia, and Vienna. After August 1883, the society promoted an elevated Métro alongside the engineer Jules Garnier, the architect Louis Heuzé, and the contractors the Milinaire brothers. The Ponts et Chaussées engineer Paul Haag also advocated elevated trains, for example, speaking in 1883 at the French Association for the Advancement of Science, another voluntary organization in the Métro debate.[35] The civil engineer Jules Revin claimed two benefits of elevated trains. First, they were cheaper and dodged landlords and expropriations. Because viaducts could only follow existing wide streets, no new land would be needed. Second, elevated trains would "damage the aspect and the circulation of public ways as little as possible," a frequent justification for both viaducts and subways. Elevated trains' better ventilation limited traction choices less than subways, thus easing ongoing experiments with motorization. But living with elevated rails raised its own questions.[36]

Unlike 1870s subway plans, 1880s elevated plans imagined the Métro in the streets, threatening surface traffic and posing aesthetic problems as Parisians debated whether viaducts could become Haussmannized street furniture. Viaducts too large or visible to be properly concealed could be clad in concrete, brick, or stone to match Haussmann's style. The engineer Eugène Chardon wrote in 1887 that viaducts could be "veritable works of art" and dreamed of shopping arcades underneath. Planners imagined integrating elevated tracks not only into the physical streetscape but also into everyday routines like commuting, shopping, and leisure walking. The Milinaire brothers thought the spaces beneath viaducts could protect pedestrians from traffic and weather, as well as house "kiosks for the sale of newspapers, drink-halls, police posts, a place for employees to stow their tools, advertising columns, water closets, etc."[37] Engineers' drawings visualized these fantasies of social life under elevated tracks (figs. 4.2 and 4.3).

Viaducts created unique spatial problems. Slender iron viaducts suited rails routed over existing traffic arteries: avenues, boulevards, and even the Seine. By contrast, heavier structures like the tracks atop masonry arcades in Berlin and Vienna demanded cutting new paths through the city, recalling Haussmann. In 1883, departmental engineers rejected elevated plans from Heuzé, Théry, and Haag, which "caused an uproar" for requiring "major cuts through houses in the interior of Paris." As Guyot noted, officials dreamed that the Métro would upset landowners less. Viaducts or their construction might also interrupt street life or domestic life. Elevated rails would pass by the windows of houses, perhaps making noise, reducing property values, or violating domestic privacy. Engineers often designed slender viaducts above the middle of boulevards without touching the famous "planted promenade" of chestnut trees, that nexus of Paris boulevard culture. Viaducts about 3–6 meters high

Fig. 203. — Le chemin de fer transversal dans une rue spéciale

Figure 4.2. Louis Heuzé's "Transversal railway in a special street," figure 203, in Figuier, *Les nouvelles conquêtes de la science*, 2:600. Courtesy of University of Oklahoma History of Science Collections.

Fig. 204. — Le chemin de fer transversal, formant au-dessous de la voie un passage ouvert pour piétons

Figure 4.3. "The Transversal Railways, forming a covered passage for pedestrians underneath the tracks," figure 203, in Figuier, *Les nouvelles conquêtes de la science*, 2:601. Courtesy of University of Oklahoma History of Science Collections.

Figure 4.4. Albert Robida, "L'embellisement de Paris par le Métropolitain," *La caricature*, no. 338, 19 June 1886, cover. Courtesy of University of Michigan Library (Special Collections Research Center).

would allow trees to prevent passengers from peering into the houses they passed, at least during leafy seasons. The Milinaire brothers and Jules Garnier projected new rows of trees planted on both sides of elevated tracks.[38]

The eye-catching climax of 1880s elevated rail debates was an 1886 caricature by the science fiction author, editor, and illustrator Albert Robida (fig. 4.4). Combining the "urban body" and "Queen City" idioms, Robida drew Paris as a queen tangled in and penetrated by tracks and violated by viaducts. Robida's image reworked the nightmares of 1870s subway critics about elevated trains. The picture's profanity was as striking as the caption's language, which claimed to depict the 1889 exposition and mocked the idea that viaducts could ever "embellish" Paris streets. The image's biting dark humor cloaked a serious call to save this damsel in distress.[39]

In 1887 the Society of Civil Engineers organized a new association: the Parisian League for the Elevated Metropolitan (Ligue Parisienne du métropolitain aérien), the swan song of only four years of promoting an elevated Métro. Meanwhile, officials were debating systems that mixed underground and elevated tracks—as in the 1882 *enquête* dossier or Minister of Public Works Charles Baïhaut's 1886 plan. But the opponents of elevated trains had already organized. Around 1880, the Society for the Friends of Parisian Monuments formed as a historical preservation group, claiming that any underground or elevated Métro would damage Paris's architectural patrimony. Members worried about structural damage or tunnels collapsing from construction or rumbling trains, and the risk that viaducts would disrupt vistas of monuments and streetscapes. The Friends institutionalized a special wariness of the Métro among traditionalists, conservatives, and elitists, as well as fearful futurists like Robida. Wealthy, educated traditionalists, aristocrats, and academics were the only groups in 1880s Paris who not only rejected underground or elevated plans but also opposed any Métro, which the conservative Madier de Montjau called "anti-national, anti-municipal, anti-patriotic, and detrimental to the glory of Paris." Ultimately, theirs was a losing battle; investing so many dreams in the Métro created expectations for the future subway.[40] Yet many remained ambivalent about living with railways, seeing the Métro, variously, as a practical necessity, a historical destiny, or a cultural or technical roadblock.

The critics' call to "save" Paris meant saving the Haussmannized bourgeois center and west. Different norms governed east and west, center and periphery, the Left and Right Banks, and elevated and underground railways. These divided geographies made a mixed Métro system the most likely outcome. By 1889, even the Society of Civil Engineers sponsored a mixed project by the veteran railway engineer Louis Le Chatelier, illustrating Paris's uneven, divided geography. Viaducts would upset street life in less Haussmannized working-class sectors—the Right Bank's northeast and the Left Bank's outer districts—drawing tidy lines between center and periphery. Inside these lines, across the center and the west, Le Chatelier intended underground tracks to bypass the surface. Because the Petite Ceinture already served the periphery, he omitted any Métro access across the long arc of working-class Paris from Batignolles to Bercy (the 17th–20th and 12th districts). He unevenly distributed railway infrastructure's benefits and its burdens.[41]

Le Chatelier's plan bookends 1880s Métro debates, reflecting compromises among architects, engineers, and officials gradually emerging from the diverse dreams vying since 1872. Accepted and recognizable designs were distilled from decades of creativity. Mixed systems could more gently stitch large constructions like viaducts and tunnels into the urban fabric. They indicate important limits to the prescribed uniformity of the modern infrastructural ideal, which ignores the rigors of retrofitting. Many engineers suggested that the Métro be divided and built piecemeal to adapt to its shifting environment, foretelling the railway's future. *Le génie civil* argued that this "sectioning" was the only way Parisians could have the Métro "of their dreams."[42]

Le Chatelier lectured on the Métro question during the 1889 Universal Exposition:

> This diversity in the conception and the considerable differences which result in the proposed modes of realization have in no small way contributed to obscuring the ideas of the great mass of the public about the significance of the word *Métropolitain*, and in so doing, has slowed its execution. The powers that be, themselves, did not escape hesitation, because there was uncertainty, even conflict over the attribution of legal title.
>
> Whether or not the Métropolitain is of general interest is a formality without importance for the Parisian. What the population wants is that we finish with this question.[43]

Legal stalemate, slow development, and public clamor for mobility—Le Chatelier evoked these fixtures of the Métro's dream life to push ahead. Progress demanded a move from dreaming to planning, which in turn required some consensus about the future railway. Le Chatelier saw Métro debate slowing, though it remained politicized while officials could not realize their dreams. Officials knew they were sleeping on the job. The 1889 Universal Exposition was the only one from 1855 to 1900 without a burst of Paris transport development, an embarrassment for republican elites.[44] Ironically, the travel editor Karl Baedeker's 1888 Paris guide described the "Paris Metropolitan Railway, which is to be finished in time for the Exhibition of 1889(?)."[45] Baedeker's question mark reminds us that, during the 1880s, the Métro "question" became a named debate, a social reality. It also reflects lingering ambivalence as network designs converged around mixed systems, while legal and financial obstacles still loomed. Many saw that railways would upset street life somehow, during both construction and operation, for both underground and elevated rails, and no matter which neighborhoods they served.

Leaving the Domain of Legend

The urgency of realizing the Métro redoubled after the embarrassment of 1889. The chief departmental engineer Hétier wrote in 1890, "We don't have to establish the utility, one could say the necessity, of constructing a

Figure 4.5 Map of subway lines proposed for the 1890 Métro public utility survey from *Le temps*, illustrated supplement, 21 July 1890. Bibliothèque Nationale de France.

metropolitan railway." Similarly for the tramways, he continued, "For more than eight years now, we have recognized the urgent necessity of completing the network of tramways which serves Paris and its suburbs so insufficiently." The legal stalemate vexed both experts and the public. The engineer Paul Villain soon published plans labeled "the Métro we can do" (1891) and "a Métropolitan that won't cost and won't trouble anything" (1892), while Max de Nansouty noted in 1891 that "the philosophy of the Métropolitan is done." Yves Guyot later wrote, "The best metropolitan is that which will get done."[46]

A second Métro *enquête* spanned 15 July to 16 August 1890, and again *Le temps* printed the proposed project (fig. 4.5). This plan came from the North Railway Company partnered with Gustav Eiffel's Compagnie des Établissements Eiffel, fresh off the Eiffel Tower's 1889 debut. The North Railway

Company proposed reaching deeper into the city, linking the North Station, Saint-Lazare Station, East Station, and Les Halles with new subways. Eiffel added a simple Right Bank loop, tracing the Seine's quays and the Grands Boulevards and running mostly underground (Madeleine to Oberkampf), before climbing to elevated tracks for its southeast arc, where it met the Vincennes Station (Place de la Bastille), the Lyon Station, and finally kissed the Left Bank at the Orléans Station (today Austerlitz). The two networks crossed for transfers at the Opéra and Les Halles.[47]

The map joined a full-page feature by the science writer Max de Nansouty, who engaged the public about the *enquête*, though he knew their response would be as positive as in 1882, because "the utility of a metropolitan network in Paris is generally admitted." That year, over 100,000 citizens signed a petition favoring Eiffel's network, while 17,000 petitioned against the North Railway's project.[48] Nansouty was associated with Louis Figuier's science magazine *Les merveilles de la science*, the longtime editor of *Le génie civil*, and a regular writer for *La nature*, often covering railways and transportation. He was an opinion maker, science booster, and national liberal playing the public educator in grand journalistic style. He gave both Eiffel's tower and his Métro plan good reviews. For Nansouty, any Métro had four tasks: to entail no financial risk for the state and no "guarantee of interest" or "subsidy" for the contractor; to connect all train stations with rails; to show a general concern for hygiene throughout construction and operation; and finally, to mix underground and elevated rails, a move he called "sollutionist" (*sollutioniste*). The word awkwardly but astutely captured the urgent mood, a will to wake from dreaming about the Métro, which "seems called to enter into a period of realization very soon" and was "assured of public favor." Nansouty, already Eiffel's champion, thought the 1890 project met his four demands. He acknowledged the drawn-out Métro debate, repeatedly abandoned and rediscovered "due to various struggles." As demand and expectations for mobility grew, the tramways and omnibuses could not meet the public's "aspirations to progress and the need, ever more pressing, of movement and activity."[49]

Nansouty thus gave the Métro a new task: accelerating traffic. He wrote, "Movement leads to movement; the creation of a metropolitan network will not diminish circulation by omnibus, just as the tramways and the steamboats did not diminish activity in the older means of transport. Is it necessary to recall, in the same order of ideas, that circulation on the national roads and canals has not stopped developing since the creation of the railroad?" Nansouty did not imagine rails easing street traffic, as in earlier decades, but dreamed that more and faster traffic would harness the Métro to snowballing progress and pull France out of the global economic slump of the first Great Depression, 1873–1896 (not coincidentally roughly contemporary with the subway legal stalemate and a tramway lull). Even without the legal stalemate, it would have been hard to find investors given the cost, novelty, and risk of urban railways. As Louis Biette wrote, legal and financial factors and public and private actors conspired to prolong the Métro's prehistory.[50]

Yves Guyot echoed Nansouty, dreaming that the Métro would encourage economic flows: "Having placed my ideal in scientific and productive civilization, I considered that the whole economic life of a country depends on the facility and the rapidity of the circulation of people and things." Existing omnibuses and tramways were too slow for Guyot (9–12 kilometers per hour), unlike the world's first electric subway, the 1890 City and South London Railway (18 kilometers per hour). He predicted that the Métro would accelerate traffic and keep Paris's economy healthy, as railroads did nationally. In fact, the Métro would have both local and national benefits. He haughtily dismissed local light-rail plans as antipatriotic: "The first responsibility of a politician is to put the general interests of the country above the interests of his circumscription."[51]

The national liberals Nansouty and Guyot dreamed that powerful private companies could solve the legal and financial issues delaying the Métro. They also enlisted the Métro for international competition, unifying national networks and increasing fast, fluid traffic, creating healthy, vibrant mobility, and multiplying capital. The political economist Paul Vibert included the Métro in his massive study of land and sea transports as tools of "foreign competition." Around 1890, the Métro was an artifact of techno-nationalism, a lever of international, industrial, and imperial power. For national liberals, it was like the Suez and Panama canals or the transcontinental railways, another infrastructural tentacle of empire, making high-speed links to, from, and across an imperial capital.[52]

Leftists contested whose interests the Métro should serve. Always defending the little guy (*populo*), the anarchist Émile Pouget dismissed the Nord-Eiffel project as "the bankers' Métro," designed to enrich the "king of thieves" Baron Rothschild and his "lackey" (*laquais*) Yves Guyot. Pouget derided nationalist Métro dreams for their iniquity and irrelevance, asking: What do Parisians care if the train stations are connected? The charged phrase "general interest," he wrote, "always signifies the interest of fat cats [*grosses légumes*]."[53] Joseph Odelin, a socialist municipal councilor from the

Saint-Germain-l'Auxerrois neighborhood, wrote a pamphlet praising the North Railway Company's half of the *enquête* plan, but not Eiffel's. He claimed a consensus that the Métro should "bring a recognizable improvement of the material sort to the working class" and embody "a well-formed socialism, responding, in principle, to the needs of the greatest number." He sought "a work of general interest," like Pouget contesting the meaning of this legal term from contests over "public utility" and Métro jurisdiction. Odelin thought the Métro should serve the public by relieving traffic congestion, expanding affordable rapid transit, letting the poorest live outside the center, and creating construction and transportation jobs.[54]

For Odelin, recent municipal plans were too big and costly, while Eiffel's network was too small; it did not serve enough places or the right ones, and many found it "puny." It neglected the Left Bank, but for the Orléans Company station, being less a "Paris Metropolitan" than a "Right Bank of Paris Metropolitan." Others insulted it as "the rabatteur of the North Company," a parasite on its would-be business partner. "Rabatteur" also means a man with an unsavory job like a pimp, bookie, or scalper, conveying corruption. Oppositional and independent papers including *L'intransigeant* alleged a vast conspiracy of "Jewish finance" with the state. Odelin disagreed, stressing, "We don't want, therefore, to associate ourselves with the first recrimination formulated by this 'League of the Public Good' which has placed Deputy Mesureur at its helm to chase Eiffel from the Parisian temple." Odelin revealed another conservative association intervening in the Métro debate, which used the term "public good," and weaponized the antisemitism that haunted finance, real estate, and public works.[55]

Eiffel's failed Métro bid touched the antisemitism that birthed the 1893 Panama Canal scandal as prologue to 1894's Dreyfus affair. The language of chasing Eiffel from the temple recalled the bigotry of the Catholic journalist Edouard Drumont, his 1890 hate group the Ligue nationale antisémitique de France, and his 1892 newspaper *La libre parole* (Free Speech). The populist Drumont denounced big business as fiercely as he asserted that free speech protected antisemitism. Odelin condemned this mounting antisemitism, refusing to shun Eiffel for his role in Panama. Though Eiffel was not Jewish and was later exonerated, these attacks still damaged his reputation. He was caught in the French web that tightly wove antisemitic and anti-German ideas; conspiracists conflated Eiffel's German heritage with Judaism. Odelin thought it better to criticize the inequalities of Eiffel's plan: it neglected the Left Bank and the working classes, failing to connect the center with the periphery or suburbs. Odelin demanded a Métro of "central penetration" (*pénétration centrale*), to encourage radial movement, increasing worker mobility and living standards.[56] Tweaking familiar planning ideas, Odelin's Métro set moral and material improvement in a socialist key. For him, unlike Guyot and Nansouty, the Métro served local ends—steering urban growth, creating work, and improving workers' living conditions—not national or global ones.

The geologist Albert-Auguste de Lapparent cautiously responded to the 1890 *enquête* in an article series for the conservative Catholic review *Le correspondant*. He opened by noting that "this thorny and passionate Metropolitan question" returned "each year" to legislative debate. In 1887, after another failure to end the debate, he wrote, "Once more the Metropolitan sank [*sombra*]." Lapparent's verb "sombrer," as in the bilingual idiom sink or slip into unconscious (*sombrer dans l'inconscience*), was another evocation of sleeping through the long debate. Stressing the complexity of Paris's subsoils, he rejected all subways, but also carefully considered whether an above-ground Métro was the best solution to Paris's urban problems: congestion, traffic, insufficient transport capacity, suburban relations, and connecting of the train stations. He dismissed Odelin's plan to create jobs as a "resurrection" of 1848's national workshops. Finally, he demanded a "modest" and "reduced" application of elevated light rails, only where needed, which he confessed could not connect the train stations.[57]

As new routes and new motors emerged, the boundaries between existing and imagined means of transport blurred. The Métro was not cut from whole cloth, but woven through spaces already crossed by coaches, omnibuses, steamboats, tramways, national railways, and two new cable cars (funiculars) serving the heights at Belleville (1891) and Montmartre (1900). Hence Métro dreams often projected the physical integration of existing infrastructure or the financial integration of existing companies. It was not always clear whether any existing railway was part of the Métro or not. The Grande Ceinture was part of it in 1877, but not after 1895. Karl Baedeker's 1884 Paris guide had listed the Métro among Paris's other railways even though it was "still in its initial stage," while his 1891 guide admitted that the railway was "still a thing of the future." The 1904 independent Nord-Sud subway was absorbed into Métro lines 12–14 after 1930.[58]

Another debatable part of the Métro was Jean-Baptiste Berlier's "tubular underground tramway," commonly "tube," an 1887–1892 plan to connect the Bois de Vincennes and Bois de Boulogne with a central east–west subway. The shallow cast-iron tunnel under the streets would carry light-rail trains with electric lighting and traction, anticipating Métro Line 1. Berlier was a devoted promoter of underground pneumatic technology; he designed the city's subterranean mail system in the 1860s–1870s and ran trials in the 8th and 17th districts of a pneumatic sewer system in the 1880s. While Louis Figuier thought Berlier's plan solved the Métro question, and the public agreed, one critic called it a "veritable nightmare."[59] The municipal council liked the plan as a modest start to the local light-rail Métro, although his title lacked the word "metropolitan." Poubelle said it could be "connected" with "the future metropolitan."[60] Interpretive flexibility gave Berlier's tube broad appeal: municipals saw the Métro; national liberals saw another network; the departmental prefect Poubelle saw both.

Paul Vibert liked Berlier's project, but only as a welcome addition to Paris's means of transportation. Like Guyot and Lapparent, he emphasized that no light-rail system could complete the national railway inter-network. In 1896, he called Berlier's tube "a simple electric tramway, very practical, which has nothing to do with the superior interests that a Metropolitan should represent." Ironically, he cheered the same plan as the municipal council whose Métro dreams he found "antipatriotic," "criminal," "reactionary," and "hateful." The council's proposal for elevated railways on the exterior boulevards, he wrote, was "concentric," meaning that it would enclose traffic in Paris. Like Odelin, he thought an "eccentric" network should speed radial flows in and out, but he dismissed Berlier's "tramway" as irrelevant to the Métro question.[61]

By 1892 both the municipal and departmental councils approved Berlier's plan—the farthest any plan had come during the stalemate, but Berlier never found enough capital. Yet his "tube" was reborn in 1895 when the engineer Fulgence Bienvenüe modeled Métro Line 1 after it. That year, the coming 1900 exposition finally pushed the railway from a "suspended" dream into a "current" issue, as Poubelle said. The municipal and departmental councils broke the stalemate with a clever ultimatum: either the Métro would be local passenger light rail, or the municipality would withhold 20 million francs from exposition funding. The national groups folded, and the minister of public works Louis Barthou declared the Métro a "work of local interest" on 22 November 1895.[62]

The municipality quickly called a Métro planning commission under the socialist councilor André Berthelot. The year 1896 was dense with decisions about track gauge and routing, and the mix of elevated and underground works. A mission studied Budapest's new Siemens and Halske electric subway, also an international exposition showpiece. The *enquête* approving the Métro, 16 May–16 June 1896, received 689 depositions, 473 of which were coordinated responses aping a Paris Landowners Association handbill. Yves Guyot called it "shameful" that Paris still lacked a Métro, and Karl Baedeker removed any mention of it from his 1896 and 1898 guides.[63] Planners had decided most of the architectural and technological details by 1897—a mix of tunnels, trenches, and viaducts, with third-rail electric traction. The popular encyclopedist Victorien Maubry wrote, "After having necessitated laborious studies and given rise to various evaluations, which generally showed a certain skepticism, the metropolitan railway has finally left the domain of legend. The law of 30 March 1898, in consecrating its existence, has made it a reality." Construction began in October 1898, Line 1 was an exposition attraction in 1900, and the "first network" of six lines was finished just weeks before the catastrophic 1910 flood (discussed in chapter 8).[64]

From 1872 to 1895, the dream life of the Métro inverted Haussmannization by bringing motorized transport to the city outside of the departmental government's jurisdiction. The imaginary Métro was the source of bitter bureaucratic battles and rich public debates. While standard historical accounts reduce the Métro to a political object, and alternate technical accounts reduce it to a traction problem, this chapter has shown the imaginary Métro to be a sociotechnical and technopolitical artifact—an imagined technological system freighted with questions about environmental quality, urban planning, and cultural meanings to be answered with underground and elevated rails. Critics argued that the Métro would cause aesthetic, environmental, financial, legal, or social problems, while proponents maintained that it would solve such problems. But all sides agreed that new railways would transform lifestyles and the cityscape, for better or for worse, as they learned to live with railways.

Parisians debated the future railway's routing, traction, funding, regulation, operation, meanings, and purpose. Exploiting the railway's interpretive flexibility, Métro dreams asked it to articulate the underground's meanings, define safe and unsafe, guide city plans, solve housing problems, create work, and negotiate vital terms of infrastructural politics like "public utility" and "general interest." Dreamers also used the imagined Métro to articulate oppositions like national versus local, public versus private, politics versus engineering, and liberalism versus socialism. Planners included state and civil engineers, architects, politicians, journalists, scientists, and contractors. But the traveling historian Taine, the futurist illustrator Robida, the anarchist editor Pouget, and the Catholic geologist Lapparent also joined the debate. Organized interests included the Ponts et Chaussées Council, the CGO, the French Association for the Advancement of Science, the Society of Civil Engineers, the League for the Elevated Métro, the Friends of Parisian Monuments, and the Paris Landowners Association. Each individual or organization answered the Métro question in biased, interested ways, sliding effortlessly between technology, culture, and the cityscape as they foretold the future of railways, urban environments, and lifeways. Parisian dreamers appropriated railways for national greatness, social equality, urban renewal, public heath, and economic prowess. The retrofitting was difficult, as Schivelbusch suggested, and it revealed the dualities of modernity—the expectation or promise of moral and material progress, and the hopes and fears for the future that it brings. Living with imagined railways helped determine what real railways would become—technologically, socially, and environmentally.

5

Under Construction

Secondary Haussmannization, the republic's continuing urban transformations, required blanketing the city with demolition and construction sites (*chantiers*) that compromised street life. This disruption buoyed the politics of infrastructure, especially around labor issues and mismanagement or corruption in the government contract (*concession*) system, which haunted both construction and transportation companies.[1] In any city, worksites are heterotopias that lay bare the social production of space. These places of physical construction are also sites of social construction, where the cityscape's "obduracy" or "thingness" is materially dismantled and opened to scrutiny. Construction sites denaturalize built environments, revealing their contingency and malleability, showing that space is made, not given. In this way, worksites do what scholars call "inverting" infrastructures, revealing their otherwise hidden "insides." Accordingly, Paris newspapers in 1908 and 1909 described construction turning streets "upside-down." Haussmann's surgery metaphor remained apt. The urban fabric is continually unbuilt and rebuilt in David Harvey's "creative destruction," and this contingency challenges the power of people and technologies to "master" environments, unearthing the material and symbolic fragilities of urban modernity.[2]

Urban railways added layers of infrastructure to existing streets: viaducts and tram tracks aboveground and subway tunnels below. This is significant because the modern use of the term "infrastructure" first emerged in late nineteenth-century French railway engineering and the word indicates this layering.[3] The designation was an important legal artifact tailored to the French government contracting system, which often defined "infrastructure" beneath the tracks as the state's (public) domain, while contracting out "structure" above the tracks to private companies. The term retains these two distinct but overlapping registers of meaning, one layered and spatial (infra), and the other bureaucratic, whether corporate, governmental, or both (structure). French public debate thus quite organically politicized infrastructure, as material equipment of the state and as proof of the state's (potentially improper) relationships with private-sector contractors and developers.

The Métro was among the era's biggest Paris construction projects. Building the first network—six lines conceded in 1898 to the General Traction Company (Compagnie générale de traction; GTC)—lasted until 1910. The catastrophic August 1903 Métro fire (see chapter 6) inspired significant renovations of station architecture, rolling stock, and electrical equipment that lasted through 1914, adding layers to the initial construction. Establishing urban railways therefore required several decades of citywide construction for maintaining, upgrading, and expanding tramways, for primary Métro works, and finally, for Métro maintenance, upgrades, and expansions. Similar cycles of demolition and construction also characterized expanding electricity, gas, pneumatic, sewer, telegraph, telephone, and water networks. Furthermore, work on separate networks had to be carefully coordinated—but often was not—with chaotic consequences for opening and reopening the same worksites, or overlapping worksites, rending and scarring the urban fabric. Appropriating Haussmann's surgical idiom, construction's critics personified Paris and spoke of "her" pain, rhetorically exploiting the urban body to feminized and grisly effect as in Albert Robida's 1886 Métro caricature (see chapter 4). Like the urban railways discussed in previous chapters, construction sites raised familiar concerns about the beauty of the streetscape and the impact of public works on environmental quality.

A postcard from the "Everyday Paris" (*Paris Vécu*) series, postmarked 1905, depicts men building the Métro while the public watches (fig. 5.1). The image shows how construction sites invert infrastructure, unveiling artifacts and work that are often hidden, for example, the cross section of cobblestone paving spanning the middle of this picture. The card also documents dialogue between workers underground and managers above, distinguished by their position and their dress. In this scene, the spectators invert infrastructure by

Figure 5.1. L. J. & Co. Postcard, "Everyday Paris" (Paris Vécu) series, postmarked 1905, with the caption "The works of the Métropolitain." Author's collection.

looking over the fences, which here put up much less resistance than the typical tall privacy fences constructed of tightly packed boards. The photographer also took advantage of the openness of this site to peer inside, documenting the social dynamics of spectators, workers, and bosses. Both the photographer and the Parisians pictured inverted infrastructure by visualizing it. The card documents infrastructural change and how the public experienced construction in progress.

Because construction sites were often hidden behind fencing, this chapter uses fences as an organizing principle. Labor history can reveal the perspective of workers inside the fences looking out. Primary Métro construction coincided historically with a more militant labor (syndicalist) movement, illustrated by waves of *terrassier* strikes in 1898–1900 and 1905–1908. *Terrassiers* were earthmovers, pick and shovel men. Working under contractors, they built the Métro's public "infrastructure," tunnels and stations. Twelve years of construction entailed a massive amount of labor. Both transportation and construction workers were active syndicalists, who confronted French business and government with a rolling threat of general strike, pursuing demands such as the eight-hour workday. Métro contractors often failed to uphold labor regulations, which syndicalists publicized beyond the fences—to journalists, the Hôtel de Ville, even the national government. As railways threaded through the urban fabric, the subject of public works was interwoven with broader political debates. From 1872 to 1895, railways were linked with discussions about privatization and nationalization in sparring between liberals and socialists. After 1895, railways became similarly linked with syndicalist-era topics: revolution, sabotage, strikes, and unions. Infrastructures like urban railways were (and are) important political organs, powerful tools for organizing basic strands of the urban fabric like work and mobility. This is because physical infrastructures often double as social infrastructures, for better or for worse.[4]

By contrast, examining commentary on construction sites that criticized their social and environmental impacts reveals the perspective of bourgeois critics outside the fences looking in. Installing railways in a dense city posed familiar problems of retrofit. A dozen years of Métro construction effectively re-Haussmannized the avenues and boulevards, the Métro ironically clogging the very flows of street traffic it was designed to relieve. Construction upset street life significantly, slowing traffic, troubling everyday routines and itineraries, and jeopardizing the capital's reputation. The city of worksites, "*Paris-chantiers*," became strange, unfamiliar, and heterotopic to locals. It raised cultural and aesthetic issues for well-to-do Parisians who found the torn-up capital improper and ugly. Nationalists worried that important monuments, historical sites, and tourist attractions were inaccessible, compromising the capital's effective commerce and monumentality. Above all, construction posed practical issues—questions of traffic, of the urban body's healthy circulatory function. Dissent was particularly strong from 1908 to 1914, when dissenters evoked the social contract implied in the ideas of "public works" and their "public utility" to uphold a popular infrastructural politics that challenged republican elites and the social and environmental consequences of expanding infrastructures.

On both sides of the fences, construction sites were contentious sites of conflict for workers, neighbors, journalists, and politicians. Political lines around public works were redrawn: whereas Haussmann's works had pitted Parisians against the departmental government and its Office of Works, construction of the municipal Métro pitted locals against their own government. But because Paris was the capital, national and departmental interests kept close watch on this local drama. During the construction of tramways and subways, the sociotechnical and affective dimensions of living with infrastructure resonated most.

Inside the Fences, Looking Out

The two Belle Epoque decades before 1914 were foundational for the French labor (syndicalist) movement, which discovered fertile soil in the Paris transport sector. Growing transit networks created new communities of workers in railway construction and operation. These new transport infrastructures were also social infrastructures. Along with passengers, workers faced the social challenges of new electric light-rail technologies both above and below ground.

But all transport was contentious, even older horse-drawn vehicles. For example, Paris omnibus workers struck in April and May 1895, and it became a cause around which radical republicans and socialists rallied, critiquing the General Omnibus Company (CGO) monopoly and calling for broader changes in Paris transport. Although this synergy between radical liberals and socialists would birth the Radical Party in 1901, older tensions between liberals and socialists remained. For example, the liberal *Le temps* lamented that the leftist papers *La justice* and *La petite république* fanned the radical-socialist flames as they cheered on strikers in 1895.[5]

The 1895 strike also inspired more radical politics, both left and right. The protofascist social scientist Gustave le Bon saw striking omnibus workers as victims of a herd mentality, duped by powerful "rabble-rousers" (*meneurs*) in his antidemocratic theory of crowd behavior. The Marxist leader Paul Lafargue compared the CGO's working conditions to slavery and charged that they treated their horses better than their workers. The anarcho-syndicalist Fernand Pelloutier used the strikers as strawmen, attacking these reformist socialists negotiating with the local government, and envisioning a truly revolutionary general strike. Pelloutier was secretary of the Federation of Labor Councils (Fédération des Bourses du travail), which joined that same year with the National Federation of Labor Unions (Fédération nationale des syndicats) to form the General Confederation of Labor (Confédération générale du travail or CGT), which remains France's largest national confederation of labor unions. In the two decades after 1895, the CGT became the institutional hub of the dynamic French anarcho-syndicalist movement, with Pelloutier among its leading voices.[6] The histories of France's labor movement and Paris's public transit systems intersect at many points.

Workers went on strike not only for more pay, shorter hours, and better conditions. After the CGO publicly posted employee wages, the union posted notices warning workers and passengers that operating or riding CGO vehicles during the strike was considered crossing a picket line and subject to harsh reprisals. When omnibus drivers seeking work during the strike reached the Place de Clichy depot one morning, strikers blocked them, insisting that all work stop. The coachmen's union called a nationwide strike, inviting sympathetic workers from all trades. In union meetings, the conservative *Le Figaro* reported, "The atmosphere was terribly charged with electricity, even before any orator had spoken"—a glimmer of contemporary fears about electricity's dark side (chapter 6). *Le Figaro* claimed that Parisian passengers, suffering without omnibus service, were not sympathetic enough to join picket lines. But other transit workers *did* join, including cab drivers, tram drivers, and railway workers. When CGO trams continued running, violence erupted, as promised. Near the northeast suburb Pantin, strikers attacked two trams and knocked them over, passengers and all. Strikers hurled rocks at Aubervilliers–Place de la République trams, wounding two passengers. Elsewhere, strikers injured a police officer and twenty people were arrested. *Le Figaro* snidely quipped that striking ambulance drivers would abandon the wounded, and bicycle sales would soar. The police said the labor union's threatening posters had incited civil disorder and jailed the union's leaders, ending the four-day strike.[7]

Tensions around the CGO touched national conversations the next year. The political economist Paul Vibert criticized the CGO before the 1896 senatorial elections, supporting the radical-socialist alliance. Meanwhile, the Ministry of Public Works ordered the CGO to transform existing horse-drawn trams to mechanical traction, by Article 7 of its charter, which required it to use available innovations to lower fares, though the CGO had resisted for over a decade (chapters 3 and 6). In hindsight, the 1895 strike, the 1896 controversies, and the CGO's sluggish adoption of motorized vehicles signaled the company's coming death in 1921.[8] Its bloated Haussmannian baggage was particularly noxious in this era of radicals, socialists, and syndicalists, when Paris transit and construction workers unionized and protested energetically.

Métro construction was another important syndicalist theater. In 1898, the municipal council committed to building the Métro's "infrastructure"—tunnels, trenches, viaducts, and stations—while contracting out operation to the General Traction Company (GTC) and its subsidiary the Paris Métro Company (CMP), which managed rails, rolling stock, daily operation, and maintenance, and shared profits with the city. When the contract ended in 1933, the entire network would revert to municipal ownership. This arrangement maintained municipal influence over standards for wages, hours, and benefits, setting similar working conditions for municipal employees and GTC or CMP workers. It kept big finance and the national railway companies away from the major works. Like Haussmann's works, the primary Métro works required local authorities to hire numerous private contractors, who in turn hired laborers. But unlike Haussmann, the left-dominated turn-of-the-century municipal council promoted stronger social welfare, upholding relatively high employment standards for city employees and contracted construction and operating companies, and it acted from the municipal, not the departmental, level—a faint echo of the Commune's former "communal luxury."[9]

For Métro employees, work standards included: (1) biweekly paychecks, with 150 francs per month minimum, or 5 francs per day for temporary employees; (2) ten-hour days and six-day weeks; (3) ten days annually of unpaid vacation; (4) full salary during military training or service, during illness for at least one year, and up to complete hospitalization in case of accident; (5) accident insurance paid by the company; (6) full compliance with health and safety regulations; and (7) job tenure for any adult, male or female, after two years of service. This contract followed the national accident insurance law of 8 April 1898, which made employers responsible for accidents resulting from working conditions. For its time, the contract was extremely generous, particularly the gesture toward gender equality and benefits such as tenure, vacation, and paid leave. According to the American social scientist Edmund James, these working conditions "indicate the high-water mark attained by modern cities."[10] Such comparisons from foreign observers strengthened Paris's reputation as a model of urban modernity. But foreigners overlooked heated local conflicts about whether this contract could soften the harsh realities of construction work, especially given the limited effectiveness of the state's labor-inspection system and private company health systems.[11]

The Métro was more than a political trophy won by radicals and socialists in the municipal council. Echoing the Commune, the subway became a vehicle for municipal socialism, an institution governed by socialist principles, and a social infrastructure for the local welfare state. Welfare programs flourished, in part, because the municipal radical-socialist projects met approval from radicals and socialists at the national level. This welfare is significant for two reasons. First, it challenges the common argument that France "lagged behind" its neighbors Britain and Germany in developing a welfare state. The Métro work contract was a leader in its time. Second, it shifts the scale of analysis, showing that the welfare state's growth need not be measured only on a national scale, but by city, county, province, or other jurisdictions.[12]

Expanding welfare was one way that the left-leaning municipal government responded to syndicalism, but it also gave unions a foothold for demanding further benefits. A feedback loop formed between state and civil society that created both conflict over working conditions and a growing welfare state designed to calm this conflict. The city's generous and progressive Métro work contract attempted to invert primary Haussmannization's exploitative aspects, but it maintained the public-private structure that had dominated French infrastructure contracts since Haussmann as it opened new conflicts over infrastructure between republicans and syndicalists.

Even a generous modern labor contract had to be enforced, a problem that led *terrassiers* to strike for three weeks soon after Métro construction began in 1898. Tunnel contractors were not paying the full wage specified by municipal contracts. Like the CGO and tramway companies, these private contractors came under fire around 1900 as corrupt private interests betraying their mission of providing public works. Syndicalists amplified this critique. Again, labor leaders called a general strike, and this time *terrassiers* walked off worksites for the Métro, the 1900 exposition, and two new rail lines. The state mobilized soldiers to help police keep order. With the streetscape dotted by temporary guard boxes and construction sites abandoned by strikers, one author thought Paris seemed to be under a "state of siege"—a powerful image with the Prussian siege and the Commune in recent memory.[13]

Terrassiers striking on worksites for the 1900 Universal Exposition were likely among the victims of 10,527 worker injuries counted by officials during preparations for this mega event. Journalists both French and foreign blamed striking workers for slowing progress, ensuring that the fairgrounds were unfinished on opening day. Many visitors in the early weeks saw more construction sites than they had hoped to. Commentary and criticism about the exposition drew connections among construction sites, worker strikes, and delayed works. They also revealed construction flaws caused by sloppy workers, contractors, or inspectors, which seriously injured several dozen visitors and killed thirteen when two concrete walkways collapsed—but this was nowhere near the scale of over ten thousand worker injuries in building the fairgrounds.[14] The exposition's construction sites were much more dangerous places to work than they were places to visit. The Métro was similarly more dangerous to build and operate than to ride.

In October 1905, the *terrassiers* union informed Prefect of the Seine Justin de Selves that Métro contractors were again violating contracts. Unionists gave the prefect six weeks to examine their claims about wages, health, hygiene, and safety, but he rudely snubbed them in November when they visited his office; it was outside his jurisdiction. But what good were generous labor contracts if neither the city nor the department would help enforce them? The union began a strike that lasted until January 1906, with all the usual syndicalist scenery: skirmishes with police, tensions between skilled and unskilled workers, conflict between strikers and those who continued working, and a healthy helping of sabotage. Reported numbers of strikers varied widely. The leftist paper *L'humanité* counted 20,000, while the more moderate *Le petit Parisien* counted 3,373. The centrist *Le journal* reported in January that when skilled workers tried to restart work after weeks on strike, unskilled *terrassiers* used sabotage against them, flooding their worksite with water from the Seine to stop their work, which provoked a fight.[15]

The skilled workers in question were called "tubistes," men who worked inside giant, often pressurized, iron tubes called "caissons," used to dig underground tunnels in what is called "shield tunneling" in English. The technique is useful in loose, wet, or unstable ground. In Paris it was essential for Métro and railway lines crossing under the river.[16] Workers had to excavate earth, pour cement, and install sections of iron "tube" repeatedly. The workspace was often pressurized with compressed air to keep it as dry as possible, and it was lit with strands of temporary electric lights. Workers decompressed slowly after shifts to avoid the bends, also known as "caisson disease." The risks of working in caissons were among the safety issues highlighted in the 1905–1906 strikes.

Caissons were large, high-tech, and disruptive, and eye-catching enough to attract contemporary photographers and spectators. Two 1905 Neurdein brothers (ND) postcards, numbered 1703 and 1996, document one such colossal armature being constructed aboveground before it was gradually lowered underground to form the skeleton of the Saint-Michel Métro station (figs. 5.2 and 5.3). Although the public often viewed construction and caissons from outside, these two postcards spatially inverted infrastructure, peering into the worksite to reveal its interior workings and piecemeal progress. They also temporally inverted infrastructure by featuring the iron skeleton that would later be hidden by concrete, plus ladders, wooden planks, and cart tracks that workers later removed. Both postcards accentuate the caisson's size with their composition, in which the metal shell fills the frame, towering above the people toiling inside, and pushing the cityscape into the background. The Métro caissons used along the Seine were massive enough to disrupt both street and river traffic at various times, a fact which is especially visible in postcard 1703

Figure 5.2. Neurdein brothers (ND), 1905, Postcard no. 1703 (postmark illegible), with the caption "PARIS—Place Saint-Michel, Works of the Métropolitain, interior of the Caisson." Author's collection.

Figure 5.3. Neurdein brothers (ND), 1905, Postcard no. 1996, postmarked 1906 and signed by "Lucien" with the caption "PARIS—Works of the Métropolitain Railway. Line no. 4. Porte Clignancourt to Porte d'Orléans. The iron armature of the Place Saint-Michel station before sinking." Author's collection.

(fig. 5.2). The resulting subterranean iron armature is still visible today in the access stairways descending into the Saint-Michel and Cité stations on Line 4.

Caissons that worked well were marvels of modern engineering, but when they failed, accidents could be serious. In a 1906 Métro caisson accident, workers were lifting and moving a compressed air "bell" (*cloche*) weighing 4,000–5,000 kilograms when the chain snapped, dropping the bell, which broke and knocked into the crew members, injuring seven.[17] In 1907, the air rushed out of a badly sealed caisson, creating a vacuum that collapsed a nearby tunnel and fatally trapped five workers in a flooding space. Contemporaries called it a "catastrophe" or "explosion," although the chief Métro engineer Fulgence Bienvenüe pushed back against the public's association of compressed air with explosions; the term "collapse" was more apt.[18] The next year, a similar accident was narrowly avoided when a night watchman heard compressed air hissing from a caisson, found a drunk man inexplicably manipulating a valve, and turned him in to the police.[19] In 1909, after heavy rain, the flooding Seine broke a wall between two caissons, shutting down construction and train service for several days; soon after, *Gil Blas* reported that a caisson had been sabotaged, perhaps by striking workers, but also in ways that risked worker injury—either way, it was "criminal."[20] Finally, the Paris press widely covered a 1913 caisson accident during construction of Lorient's deepwater port, which killed four and wounded several others.[21] Deadly work accidents, fortunately rare, were the unhappy result of risks built into shield tunneling. *Tubistes* knew these risks, and their contracts granted hazard pay for dangerous working conditions.

The 1905–1906 strikers also publicized something else: health and safety risks resulting from adverse conditions created by negligent contractors and inspectors. Structures were not being built to specifications. The populist *Le matin* reported some cement walls 5 centimeters thick instead of the required 60 centimeters and other walls in which cement was improperly replaced with cheaper rubble (*moellons*) and mortar. Badly built tunnels risked flood or collapse, wasting money, and endangering workers and passengers. Together, unionized *terrassiers* and sympathetic journalists inverted hidden underground worksites, taking the conflict to the national level, recruiting the state into this otherwise local conflict. Minister of Public Works Armand Gauthier, a social democrat, agreed to inspect the faulty works in January 1906 with two *Le matin* reporters and a *terrassier* guide. They found many violations of materials and construction standards, and several worksites flooded by water leaking through cracks. There was enough water that Gauthier evoked

Venice, eerily echoing an allusion also used to describe the polluted river Bièvre and the 1910 Seine flood (chapter 8). They found a hole in a tunnel's roof opening into a disused sewer, a sanitary mishap. In early February 1906, the Rue de Chabrol collapsed into the ill-built tunnel below it. Entire lots of Métro construction had to be redone and, as a sewerman told *Le matin*, "It is the Parisian taxpayers who pay." Workers criticized the brittle and corrupt contracting system, leveraging the power of the press and of the departmental and national governments to fight for better working conditions.[22] Workers inverted worksites, publicizing and politicizing what happened inside the fences to those outside.

This bid for publicity was both calculated and risky: workers did not publicize it widely that shoddy tunnel work was partly a result of strategic sabotage, which syndicalists called "bad work for bad pay." The year 1905 had opened a historical season of strikes. As one author observed in 1908, "Since the year 1905, the reclamations of the building trades have been incessant."[23] On May Day 1906, *terrassiers* joined carpenters and masons in a strike that bled into that summer's famous CGT–led general strike for the eight-hour workday.[24] *Terrassiers* struck again in 1907–1908. Of France's 1,275 strikes in 1907, only 24 lasted beyond100 days, and Paris *terrassiers* mounted the three longest ones lasting 234, 273, and 320 days. At two worksites with different contractors, 252 workers on 19 June and 171 more on 13 September walked off the job until 4 May 1908. The top four industries represented in 1907 and 1908 strikes were building, textiles, transport, and metal.[25] Strikers demanded a one-franc raise for "unhealthy" (*insalubre*) working conditions, complaining of underground worksites heated by smoky coal stoves and ventilated poorly—an affront to public health, worker safety, and the culture of light and air. After this lengthy strike, a judge ordered a pay raise for all underground workers. By the summer of 1908, many Métro sections were stalled, behind schedule, or shut down by the authorities, contractors, or strikers. Laborers faced layoffs and lockouts; for example, around 18,000 of 26,000 masons had been locked out.[26] The independent paper *L'intransigeant* argued that with gas company workers already striking, Parisians should stay alert, for perhaps "tomorrow it will be electricity, then mass transit." Fear of a truly general strike, implying the risk of infrastructural slowdown, shutdown, or sabotage, was never far off.[27]

Paris transit strikes in the two decades after 1895 drew in both construction and transportation workers. Amid broader and growing worker organization and radicalism, the capital's transit struggles gained both local and national attention, helping create a climate in which a broad left-wing opposition, including radical liberals, socialists, and syndicalists, appropriated the republican language of "moral and material improvement" and turned it to new reformist or revolutionary ends. They sought to keep public works public minded, criticizing construction and transport companies for violating their contracts as well as the authorities for badly enforcing contracts and even challenging the very practice of granting state contracts to private companies by calling for public ownership. The Marxist Paul Lafargue contrarily dismissed public ownership as state capitalism. But the Second Empire pattern, still practiced by many companies including the CGO, of greasing the revolving door between the public and private sectors, continued to provoke liberals, independents, radical-socialists, and the far left into sniffing out Haussmannesque corruption under the Third Republic.[28]

A shared politics of infrastructure thus opened conflicts within the left, as workers mobilized to leverage the radical-socialists' generosity—not unlike the republicans who used Napoleon III's reforms to bring down the "liberal empire" in the 1860s.[29] Young leftist papers like *La voix du peuple* (founded in 1900) and *L'humanité* (founded in 1904) closely supported the striking *terrassiers*. *L'humanité* argued that dangerous working conditions justified extra pay, questioning the distinctions between skilled and unskilled, normal and dangerous work, contesting standards on workers' behalf.[30] Tensions also divided leftists who saw railways as progressive, and demanded more accessible motorized mobility for everyone, from other leftists, who were more Luddish, and deeply questioned the value of motorized transport. The far left sometimes critiqued public works as appendages of an already moribund body like capitalism or the state. Because the state had taken responsibility for providing public transport, transportation issues were pulled forcefully into conflicts between classes, between state and civil society, and among Paris's varied political factions: conservatives, liberals, socialists, and syndicalists. Construction sites were sites of social conflict between skilled and unskilled workers; between political groups; and among workers, contractors, and regulators. Such conflict nurtured the many-sided politics of infrastructure as Parisians learned to live with new railways gradually retrofitted to an old city. These multiple frictions inside the fences were matched by controversy from without.

Outside the Fences, Looking In

While workers contested construction sites as workplaces, bourgeois Parisians saw them as obstacles to traffic and blemishes on the cityscape. Most of the Métro's first network (Lines 1–6)—about 24 of 41 kilometers—was constructed

Figure 5.4. Unused JH Postcard 696, with the caption ""PARIS—Works of the Métropolitain. The caissons place Saint-Michel. They begin to place the metallic floor of the elliptical caisson. End of May 1907." Author's collection.

underground. To reduce land costs and building demolitions, the municipality routed tunnels under avenues and boulevards where the city already owned the land. Major streets were also the only open spaces in central Paris large enough for such construction.[31] Underground structures took two forms: tunnels, excavated and then lined with cement; and covered trenches, dug out and then reinforced with walls and vaulted roofs of metal and masonry in the "cut and cover" method. Covered trenches required removing much of the roadway, and underground works demanded large pits for worker and equipment access; therefore, both methods seriously obstructed surface traffic. The streets became shifting landscapes of detours, fences, pits, plank walkways, and scaffolds. As the novelist Jules Romains wrote, "The scaffoldings of the subway, which arose all over the place like fortresses of clay and planks . . . had ended up strangling the streets, blocking all the intersections"[32] (fig. 5.4). Construction thus significantly upset street life, which Romains in the corporeal and violent idiom of his era described as "strangling," but his word "fortresses" also reveals a vocabulary of warfare that recalls the "state of siege" claims above.

Complaints that Métro works upset traffic and commerce were common. Between 1900 and 1910, shopkeeper groups on the Rue de Rennes and the

Rue Réaumur petitioned the authorities for financial relief for decreased business, claiming that Métro works made it difficult for customers to find or reach their shops.[33] Métro scaffolds and pits often blocked sidewalks, breaking the everyday script that linked walking with window-shopping. Métro construction sites thus upset both temporal and spatial patterns, complicating routines and itineraries with obstacles and detours. Everyday practices like going to work, taking a stroll, or going shopping had to be rescripted. Many Parisians found "grave inconveniences" in the Haussmannized streets, which "force the pedestrian into the most complicated topographic exploration," making walking more difficult and dangerous, less leisurely and linear. One writer quipped that only gymnastic prodigies could survive as pedestrians in the hazardous streets.[34]

Walking—flânerie and the promenade—was therefore among the most threatened cultural practices. Key monuments (the Opéra, the Madelèine, the Louvre), streets (the Grands Boulevards, the Avenue d'Italie, the Boulevard des Capucines), plazas (Opéra, République), and network nodes like train stations were inaccessible, blocking familiar itineraries above ground as new ones expanded below it. As the leftist newspaper *L'aurore* printed, "It's a perpetual encumbrance. Detours upon detours. It's existence in zigzag," while the far-right *La libre parole* added, "For *promeneurs*, it's always an intercepted passage."[35]

Vehicle traffic faced similar problems. Paris national and municipal public works archives contain many formal requests from tramway companies and the CGO to modify their itineraries to create detours (*déviations*).[36] The left-leaning *Le radical* wrote, "Currently in Paris it is materially impossible to make a direct trip in a vehicle. One must . . . take interminable detours." The editors added: "Everywhere streets are blocked off, roadways have been gutted, sidewalks are smashed up. Everywhere—and all at once—they have turned Paris upside-down. Circulation is difficult, and the inconvenience caused for those living nearby is considerable. From all sides complaints are rising." Complaints came from three principal camps. First, merchants protested that customers could not reach their shops. Second, coach passengers found routes more circuitous, which, along with the kilometric counters recently required for all coaches, increased fares. Third, foreign tourists were "furious to encounter nothing but construction sites, and to take *fiacre* rides through a bunch of little streets which were not included in their itineraries." According to *Le radical*, discontent was "justified" and "general"; the "deplorable situation" amounted to lost money and time for locals and tourists alike.[37] Beyond traffic, Métro construction physically shook the city, causing minor damage. The Office of Works archive contains a modest number of official complaints about property damage—mostly cracks in walls, ceilings, and floors.[38] Sometimes, sections of tunnel collapsed, for example, at the Place de l'Étoile (December 1899), the Rue de Chabrol (February 1906), or the Place de l'Alma (November 1915).[39]

Besides social and spatial problems, Métro construction inspired complaints about hygiene and sanitation. First came local complaints about the unsanitary condition of specific worksites, especially Métro tunnels. For example, the newspaper *La patrie* complained in 1908 that tunnels were "stinking" and "unbreathable," and while passengers passed quickly through bad air, workers spent all day underground. In 1910, the leftist *L'humanité* reviewed a decade of Métro work and its hygiene conditions, claiming "sickness and death" had affected over 3,400 victims since 1900. Such hygiene complaints on behalf of workers and passengers tapped the cultural tradition of seeing Paris's underground as suspect, dirty, and dangerous. Health experts joined the fray, too, maintaining underground ventilation and air quality as central concerns.[40]

A second type of more general hygienic complaints bemoaned the dirt, dust, and clutter of the streets under construction. In July 1900, the Committee for the Interests of the 7th Arrondissement, boasting 150 members, wrote to the municipal council, complaining that "commerce, liberty, health, and the very life of neighbors are gravely compromised" by tramway construction serving the 1900 exposition. Braiding together moral and hygienic concerns, the letter continued, "Each day until 10 or even 11 o'clock in the evening we suffer the dust, the infernal noise, the oil, seeing even sputum [*crachats*] or cigar butts, projectiles of all sorts, not to mention the indiscrete peeping [*regards*] that violates the secrecy of our apartments or the dirty jokes [*plaisanteries grossières*] that offend the modesty of our daughters and wives."[41] Here dust, noise, and machine oil mixed with catcalls, phlegm, voyeurism, and cigar butts, forming a moral and material indictment of construction for offending the bourgeois norms, both social and environmental, of street life outside the fences.

Other hygiene complaints took a less social and more biological tack. In 1909 the *L'événement* writer Robert Darcy reported a fenced Avenue Gabriel "dump" (*dépotoir*) for municipal construction waste, whose stench caught his nose. Grabbing his handkerchief, he asked a gasman working nearby why it stank; the man replied by calling the dump a "ratarium." Darcy opened a gate with a "do not urinate" sign to find a chaos of trash, discarded tools, weeds, broken glass, and a dead cat. He sensed "the black specter of cholera and plague." These haunting evocations of pests, odors, disease, and death deployed familiar rhetorics of disgust and heterotopia. Darcy highlighted the disruptions and dangers that construction sites brought to everyday life, their consequences for health and environmental quality, even when they were hidden behind fences.[42] Fences could not contain the odor that drew him in. Filth from inside the fences always spilled out.

In 1911 the newspaper *Les nouvelles* reviewed many construction sites scheduled to open under the heading "the grooming of Paris" (*la toilette de Paris*), a phrase that again feminized Paris and made the streetscape her "face."[43] Such complaints characterized bourgeois papers like *Le temps*, which cast construction in wealthy west-side neighborhoods near the Opéra as particularly offensive. In 1909 the newspaper *L'événement* complained about construction disrupting bourgeois life on the Boulevard des Capucines, which it called a "sacred place" that should be open to visitors but was instead "an inaccessible place." The editors evoked bourgeois values—the city's "aesthetic," the ease of comfortable walking, and the neighborhood's important national landmarks, "all this a few paces from the Madelèine!" Their article obsessively detailed workers' tools strewn around worksites, as if listing offenses or disciplining children. These work objects were not acceptable scenery for public life, but shameful, inappropriate clutter to be hidden—back-staged infrastructure or a heterotopia. Another article described a group of "lumberjack women" (*bûcheronnes*) on the Avenue Niel, not far away. These women cared so deeply for their neighborhood that they snuck into construction sites early in the morning to tidy up. The term "lumberjack" came from picking up wooden tools and paving blocks. Pavement on the Boulevard des Capucines was finally relaid more than two years later, in 1911, showing how long such problems could persist.[44] While the consequences of construction spilled out beyond the fences, outsiders from journalists to lumberjack women breached fences to get inside and expose or tidy up worksites that refused to stay contained.

No site was more precious than the Place de l'Opéra, among Paris's more famous and frequented squares. A showpiece of Haussmannization known for artificial lights, it hosted one of the Métro's largest construction projects in the decade between 1903 and 1913: the massive triple-decker underground metal and masonry station where Lines 3, 7, and 8 crossed. The works were widely covered by French, British, and American science and engineering journals. The new subway station continued the plaza's high-tech role, while construction upset traffic at this busy intersection. During this decade of intensive works, images of the urban body returned to center stage as Parisians voiced their aesthetic and affective attachments to the plaza (figs. 5.5 and 5.6).[45]

Figure 5.5. Photograph of the "Opéra Hole," ca. 1903–1910. Wikimedia Commons.

The plaza's repeated unearthing became an irritant. The worksite was commonly called the "Opéra hole," using the French term "trou," which has decidedly negative meanings—a grave, prison cell, or human anus. Other bodily terms included "mouth" (*bouche*) and "orifice."[46] In 1904, the Belgian modernist architect Frantz Jourdain, the designer of Paris's Samaritaine department store, played on Louis Heuzé's earlier *nécropolitain* (chapter 4), describing the Métro station using terms "grotesque," "grave stone" (*monument funebre*), "mausoleum" (*mausolée*), and "catacombs."[47] *Le matin* lamented in 1909, "They are going to gut our dear old place de l'Opéra one more time—we don't count anymore. She must be made for that." A year later, *L'intransigeant* more sarcastically pretended to lose count of how many times the plaza had been dug up, and then feigned disbelief: "The place de l'Opéra is no longer anything but an immense hole!" Many newspapers called construction a "spectacle," *La liberté* reporting that a "crowd" gathered at the Opéra worksite "each day" simply to watch Paris be "gutted."[48]

Figure 5.6. "The Intersection of Three Underground Railways" from "The Busiest Underground Corner in Paris," *Popular Mechanics* 14, no. 1 (July 1910), 48. Courtesy of University of Oklahoma Libraries.

These spectators show that not all Parisians were so upset by the Opéra hole. Reports of spectators watching construction are confirmed by photographs of the era, postcards, and illustrations, which depict Parisians pressed up against the fences, or gathered at the fence openings for a better view from the outside in (see figs. 5.1 and 5.5). One newspaper reported that onlookers—at least on one occasion—were "excited" (*emballée*), applauding and cheering as workers loaded and unloaded trucks of rock.[49] This report suggests an affective register wholly different, and more joyous, than the graphic bodily violence evoked by other sources. While elite tastemakers like Frantz Jourdain, joining many bourgeois or conservative voices, maligned the "Opéra hole," many anonymous Parisians, and engineering journals both French and foreign, greeted Métro construction with palpable technological enthusiasm and municipal or national pride.

The long-running construction at the Opéra raised another issue, that of coordinating the multifarious works happening at various locations and supporting a variety of infrastructures. The liberal *Le temps* described "Paris under demolition," claiming the city had not looked so bad since the Commune: "And while waiting [for construction to finish], we witness the most abracadabra-like incoherence in the organization of works." Why had roads near the Opéra been repaved three times? Could layered underground works (Métro, gas, sewers, etc.) not be coordinated so that all works could finish before the road was relaid? The cause of such "incoherence," *Le temps* argued, lay solely "in the lack of coordination between administrative services." As *L'éclair* wrote, "The administration paves, unpaves, repaves. It blocks, unblocks, and reblocks streets. There is no overall program, no unity of direction." This complaint became common. The independent *L'intransigeant* called it "disorganization," the leftist *Le radical* expressed despair regarding if or when works would end, and the far right *La libre parole* added "Travaux de Paris: we'll never finish them!"[50] Would Haussmannization last forever?

Behind these critiques lay real technical problems. As a new network woven through the already dense Paris underground, the Métro upset sewer, gas, water, electric, and telecommunication lines, as well as gardens and green spaces. In 1908–1909, the municipal council president, Adolphe Cherioux, promoted a plan called "unity of worksites." He argued that with so many works in progress—some increasingly complex and intensive, others increasingly delayed—projects required better planning, so that road works would not conflict with sewer works, subway works, and so on. He recalled one of the era's key lessons: if public works are not carefully planned, they can become burdensome rather than beneficial, modern monsters instead of modern comforts. Complex issues of retrofit beset any new infrastructure layered into Paris's dense urban fabric and construction sites bared this for the public to see.[51]

The Parisian public clearly understood this, especially from 1908 to 1914, when the historical record overflows with popular critiques of those responsible for public works—the local, departmental, or national governments, the Office of Works, transit companies, investors, contractors, and syndicalists. These critiques share a palpable sense that infrastructural work was not progressing well. The tone of newspaper sources indicates journalists' zeal to stage a confrontation between citizens and the government, to steer opinion and mobilize the public. The authorities' press clippings collected in the Archives de Paris show that officials carefully monitored these powerful currents of public opinion. Secretaries at the Office of Works reviewed the press to survey opinion about living with infrastructure, creating a unique archive of yellowed clippings marked in red or orange crayon or colored pencil. This archive documents how citizens, journalists, and bureaucrats alike maintained worksites as objects of concern and criticism by inverting their contents for all to see.[52]

These various critiques of public works—hygienic, spatial, and cultural—were couched in an affective language of distress, pain, and shame. In 1909 *Le matin* wrote, "The capital is flayed, turned upside-down, gutted, broken," again evoking the urban body's distress.[53] A language of crisis emerged around a familiar trope: public works as surgery, causing pain, trauma, and disfigurement. Two catchphrases dot the press: "Paris-chantiers," suggesting a city defined by construction sites, and "Les embarras de Paris" (chapter 1), suggesting that their main impact was traffic congestion. Historically, "les embarras de Paris," the capital's storied traffic congestion, dated at least to Nicholas Boileau's 1660s eponymous satirical poem. As *Le journal* put it, "The satire of the *embarras de Paris* starts again each year in the papers and each day among Parisians; but we have searched in vain for the remedy since the seventeenth century." *Le petit Parisien* quipped, "Boileau had not predicted the Metropolitan."[54] Parisians had reasons, plural, to be embarrassed by their torn-apart capital, and perhaps none more so than traffic, a sore issue since the 1870s (or the 1660s).

The affective and aesthetic registers of "embarras," which in French connotes both congestion and embarrassment, thus hovered over practical traffic problems and colored public debate.[55] One article called "Les embarras de Paris" argued that the problem went beyond the city's celebrated beauty, touching on its proper function, fluid traffic, and sanitation. The article used medieval and military imagery to suggest that Paris was regressing instead of progressing, reverting to a barbaric state. Echoing Romains, the scaffolds and lifts of construction sites became "castles," "towers," and "forts." No longer urban, worksites resembled "villages" or "ruins." *L'intransigeant* used terms like "craters" and "mountains," conveying an estranged city, no longer architectural but geological. One paper described Paris as *gâchis* (a mess or waste).[56] The comparison of Paris under construction to Paris under attack remained central, borne by war words like "forts" and "ruins." "Les embarras de Paris" was often connected with that other traffic-related catchphrase, "intersection of the crushed" (chapter 1).[57] These critiques of worksites, using claims of barbarism and regression, ran parallel to inverted progress narratives for critiquing tramways (chapter 3), accidents (chapter 6), and water shortages (chapter 7).

These examples show how worksites became heterotopias between 1905 and 1910, plunging locals into a state of upheaval in which everyday life, which coalesces around familiar scripts and routines, was destabilized.[58] Even before 1910, Parisians were tired of this upheaval—and then, in bitter irony, came the 1910 flood, knocking out Paris's networked infrastructures for much of January and February. The flood did billions of francs of infrastructural damage, heaping more debris on the already difficult pile that had backed up at the Office of Works since 1900. Controversy continued after the flood. While some anarchists and syndicalists advocated sabotage as rebellion or resistance, Parisians more commonly used the term "sabotage" to denigrate leftists as vandals. But in 1910–1911, several newspapers reversed this usage to charge the authorities with sabotaging the capital. *L'intransigeant* moaned that there were "holes everywhere."[59] This way of likening public works officials to syndicalist saboteurs is suggestive. In the early 1900s, while syndicalist attacks on Métro construction typically targeted contractors, journalists more often targeted officials. But both were inspired by the maddening, crippling consequences of constant construction and demolition, class conflict, expanding mobility, and a government contracting system that kept corporations and the state closely connected while traffic and safety suffered. For syndicalists, journalists, and citizens, politicizing public works articulated something like a right to the city (and its beauty, comfort, or efficiency), which was threatened by Haussmannization.

This chapter has shown how construction sites tear the urban fabric, revealing the labor and social conflict that go into building and operating infrastructures. Construction unearths the often-hidden work of builders and operators who maintain infrastructures. It inverts infrastructures, turning them inside out to show their inner workings. Construction also opens a window onto the contestation of infrastructure, showing us the politics of infrastructure in action. Construction sites were heterotopias in which the urban fabric was materially and symbolically dismantled, revealing normally hidden artifacts like pipes and wires, and raising concerns about aesthetics, labor, health and safety, and traffic. Paris contests over worksites reveal the strong feelings and colorful representations of the experience of a city constantly upgrading its infrastructures. Living with infrastructures meant tolerating their construction, maintenance, retrofitting, and upgrading. Sometimes the cycle of creative destruction seemed never-ending.

This was particularly true under secondary Haussmannization, when Paris suffered a period of upgrade, during which many infrastructure systems did not yet work well but promised to work better in future. Someday the Métro would streamline traffic, but not until the 1920s, when cars began to snarl the streets above. Before 1914, an older generation had to suffer growing pains so that the next generation could enjoy better mobility. The stresses felt during infrastructural upgrades are an often-overlooked cost of attempting to realize the modern infrastructural ideal. This ideal envisions complete and finished infrastructures but ignores how infrastructures are conceived and built piecemeal. This limitation helps explain why attending closely to construction sites offers a clear view of the gaps between infrastructural visions and realities. The messy realities of construction, maintenance, technical accidents, and work stoppages trace the limits of Paris's "moral and material improvement" as well as the limits of the "modern infrastructural ideal." Construction sites offer a particularly striking example of urban networking as an unstable process, rather than the networked city as a stable product.

In the personified and feminized "Queen City," bourgeois assessments of construction took the form of bodily and surgical language. Like surgery on living bodies, construction opened the urban body, revealing its inner workings, and then scarred it afterward. This scarring registered in Paris texts about construction as repeated references to wars, sieges, and the Commune. Revealing what was otherwise hidden could be fascinating, frightening, or frustrating. Demolishing and rebuilding the urban fabric evoked wartime damage, ruins, and memories of primary Haussmannization, the Prussian siege, and 1871's civil war. Parisians outside the fences also likened construction to other disruptive events, including strikes, sabotage, and storms, both rain and flood. Meanwhile, workers inside the fences felt the impact of construction on their own bodies, in the form of work accidents, lockouts, and the sweat of daily toil. On both sides of the fences, living with construction sites was a conflictual and turbulent experience, saturated with the politics of infrastructure. Construction sites, hidden behind the fences, were ripe for inversion no matter which side of the fences one stood on.

6

Living with Risk

In July 1904, the Boulogne–Champ-de-Mars electric tram no. 1905, powered by overhead wires (trolley), stalled after its motor failed on an iron viaduct crossing the Petite Ceinture trench in west Paris. As the driver tried to clear the bridge, the overhead conductor jumped the wire, contacted the viaduct, and sent a surge through the Ceinture's electrical system, instantly melting telephone and safety signal wires. Electricity also raced down a metal staircase in the West Ceinture Station (Gare de l'Ouest-Ceinture) to the rail line from Montparnasse, again damaging communication and signal wires, disabling all trains to Versailles and Cherbourg. Twenty kilometers east at the Rapée-Bercy station an unlucky signalman was shocked but not seriously injured. A large western arc of the Ceinture closed for two hours. In this cascading electrical accident, metal structures conducted current from one railway to another, far beyond its intended bounds across the city. Causing more technical damage than human injury, this incident reveals risk as a feature of living with expanding electric infrastructure.[1]

Fittingly, tramways and trolleys inspired an idiom that cast risk as a pending danger. The idiom appeared in 1879 amid tramway financial troubles, when one author described the threat of a government tramway buyout as a "sword of Damocles."[2] In 1896–1897, during heated debate about the General Omnibus Company (CGO) monopoly and motorization, the tramway engineer Hillariet called the company a "sword of Damocles" over Paris, while other critics cast overhead wires themselves as a hanging threat.[3] In 1902 tramway operators lamented legal limits on the trolley as a "sword of Damocles."[4] In 1906, the women's columnist Floriane used the image for the disease risks of shared vehicles including trams.[5] As Parisians negotiated the health, safety, financial, and regulatory risks of trams, this classical image represented risk as potential or future danger.

Risk's potentials are actualized in accidents, which, like construction sites, invert infrastructures, revealing the built environment's contingency. Major or mortal enough accidents become "catastrophes" or "disasters." Defined by both contingency and injury, scholars traditionally understand accidents as "unmotivated" and "unpredictable," and yet we anticipate them.[6] The cultural theorist Maurice Blanchot noted that disaster language is future-oriented, while risk historians analyze the predictive "culture of temporality" surrounding risk.[7] This chapter examines the experience of accidents, mostly electrical, that occurred between two landmark fires: the 1897 Charity Bazaar (Bazar de la Charité) blaze, which killed 125, and the 1903 Métro tunnel inferno, which killed 84. Like the 1904 tram accident, these were cascading failures, chain reactions in which electric technology sparked outward-spiraling problems. But in the extraordinary 1897 and 1903 accidents, electricity sparked fire in crowded, enclosed spaces, causing mass casualties. Unlike in 1904, Parisians cast these deadly accidents as catastrophes or disasters to challenge the idea of electrification as progress. These extraordinary accidents tend to overshadow the fact that minor accidents were frequent, even "everyday," and yet never quite normalized. This chapter therefore focuses on minor accidents, including several Métro and tramway accidents in 1900–1901, and their politicization and debate.

Normal Accidents?

Since 1885, the Bazar de la Charité, an annual high-society charity gala, occupied various sites in Paris. A temporary wooden structure housed the 1897 event, which included entertainment from an exciting innovation, moving pictures. This shelter with few exits was crowded with aristocratic, bourgeois, and Catholic women when the projector caught fire, ignited the building, and caused a "panic" in which many people perished. Research on this painful moment in French modernity shows how class, gender, and religion shaped public responses. Catholics, though divided on whether the fire was punishment for modernity's sins, agreed that its wealthy women victims were honored martyrs to Christian charity. Republicans, by contrast, celebrated working-class emergency responders as macho models of secular solidarity and patriotic public service. All sides had heroes.[8] This fire became a template for future disasters, remembered textually and visually during later 1900 and 1903 accidents.

Though such major disasters were rare, smaller accidents were constant, revealing diffuse and pervasive risk, an operating cost of living with technologies such as electricity, motor vehicles, and film. This cost recalls the sociologist Charles Perrow's famous term "normal accidents" to describe the built-in dangers of societies dependent on complex technologies. Since the Second Empire, French industrialization had sparked an awareness of what the sociologist Ulrich Beck would later characterize as modernity's "risk society," which accepts, manages, and tolerates social, environmental, and technological risk.[9] Importantly, not all Parisians accepted risk. Cultural histories of technology and modernity around 1900 highlight accidents and electricity as central to articulating ideas of the risk society and normal accidents. Actual electrocutions gave concrete meaning to the metaphorical shock of modernity.[10]

A 1903 municipal study counted 1,339 injury accidents among Métro passengers and workers since 1900, while tramways killed 112 and wounded 937 across the Seine department. It bears repeating that construction accidents preceding the 1900 exposition injured 10,527 workers. Paris accidents ranged from theater fires and exploding lampposts to unwittingly electrified pavements and two metal plates (1892 and 1905) launched by small underground explosions caused by electricity igniting gas. Fittingly, water often provoked such cascades, causing short-circuits by corroding equipment and conducting electricity. Accidents thus revealed the hybridity of urban environments. Their causes combined physical forces such as water, weather, and electricity with human sociotechnical choices regarding design, use, and risk. Their consequences were also hybrid, harming humans and horses alike in the streets.[11]

Amid motorization and electrification, a counter-discourse to that of infrastructural progress emerged to tally progress's price. As accidents inverted infrastructures, Parisians inverted progress narratives. One writer noted, "All is not well with electricity, and quite often the modern fairy brings with her singular drawbacks [*désagréments*]."[12] Among the era's enduring popular epithets for technology, current from 1889 to 1930, was "the ravages of electricity" (*les méfaits de l'électricité*).[13] Like "crusher tramways" (chapter 3), this catchphrase became a common journalistic "rubric" or headline. Electric accidents joined construction and traffic accidents, embodying the dangers of evolving streets, railways, and electrical networks. For the public learning to live with electricity and motor vehicles, accidents revealed infrastructure's dangers as well as faults in the republican ideology of moral and material improvement and the modern infrastructural ideal. The counter-discourse on accidents juxtaposed technological optimism with the dangers and failures of modern machines. Paris's complex public-private infrastructure administration led journalists and other critics to blame both the operating companies for violating contracts and the state for weak contract enforcement. Charges of corruption and neglect flew far and wide. Parisians discussed "responsibilities" after accidents, and explicitly addressed the role of electricity, horses, water, and weather. Urban hybridity raised questions about whether technology, human agency, or natural forces caused accidents. Thus, moral and political questions of blame intersected with causal questions about infrastructures as social-technical-environmental assemblages characterized by complexity, risk, and uncertainty. Ultimately, were accidents "normal," and was risk acceptable? What were the costs of living with infrastructure?

Electrifying the Tramways

In 1890, Clermont-Ferrand opened France's first electric tramway, while the French Electric Accumulator Company (Société française d'accumulateurs électriques) tested four accumulator (battery) trams on Paris's Madeleine–Levallois line. Motor tests conducted in the 1880s never delivered the desired savings, and Paris's struggling tramway companies believed electricity would be a likely fix. The engineer Léon Francq, known for creating compressed steam motors, penned several articles from 1886 to 1892 supporting an electric Métro. He called electricity a "dream," arguing that it would be cheaper than horses, cover start-up costs, drive down prices, expand access, and quash environmental quality concerns about tunnel ventilation and horse manure. Francq, like many engineers and intellectuals, saw electrification as an urgent upgrade and potential panacea.[14]

In the 1870s, electricity became an important cultural object, a prismatic signifier of cleanliness, speed, and decreased distances, but also of modernity, progress, and the future. During the 1878 Universal Exposition, for example, the Place de l'Opéra, the Place de la Bastille, and Les Halles displayed Paris's first electric streetlights, prompting a sell-off of gas stocks. Paris's 1881 Exposition of Electricity and 1889 Universal Exposition increased electricity's popularity and glamour, plus the awe attending this modern "marvel" as a magical or sublime force. The historians Alain Beltran and Patrice Carré have shown that electricity was a "fairy" (*fée*), French for inspiration or muse, and a "servant," an all-purpose, limitless prime mover that would revolutionize society as steam had before it. Electricity became a metaphor for energy, drive, excitement, inspiration, and life. It sparked fear, fascination, and fantasy as in Albert Robida's science fiction.[15] Thus, the conservative *Le Figaro* fearfully described the 1895 meeting of striking omnibus workers as "charged with electricity," while in 1896 the liberal *Le temps* gleefully greeted the Métro-to-be with "hail the fairy electricity!"[16]

Electrification seemed somewhat magical, as in the discourse of modern technology's "wonders" or "marvels." The travel magazine *À travers le monde* (Around the World) noted how far electric traction had traveled since Siemens's trolley demo at Paris's 1881 Exposition of Electricity: "Who could have foreseen, before these modest beginnings, seeming more curious than really practical, the extraordinary development that we are witnessing today? Electric locomotion emerged from Germany and was studied and rendered practical in the United States, and now, thanks to the powerful impulsion it received in America, it is establishing itself definitively in Europe and propagates itself in all parts of the world. . . . Thus, electric locomotion propagates itself with a prodigious rapidity, which seems without equal in industry." Electric trams, like horsecars, belonged to imported "American" modernity and evoked visions of the future. Electrification was a "universal development" and "the order of the day."[17] Jules Verne called electricity "this agent which will one day be the soul of the industrial world." Dreaming about the future Métro, the military engineer Narcisse-Nicolas Petitjean wrote, "From the point of view of choosing a motor, it is evident that electricity, with its innumerable resources, and the numerous forces it can put into action, will finish by triumphing over steam, and that one day, maybe soon, all forces of locomotion will be produced by electricity." Paul Vibert added in 1896 that "a great national movement is taking shape today, which will impose itself imperiously tomorrow, thanks to the truly miraculous progress of electric traction." Georges d'Avenel claimed in 1905 that "the future and even the present already belong to electric traction." Electricity's associations with triumph, miracle, the future, and progress resounded from periodic expositions that showcased electric technologies, especially lighting and motors.[18]

Behind this vibrant sociotechnical imaginary scripting the promise of electrification, electrical technology was still young, raising practical efficiency and safety issues for both designers and users. During the infancy of electric motors, compressed air and steam were more common, and horses remained dominant. Paris's electrical network was limited until preparations for 1900—equipping the exposition jump-started Paris's electrification. Many exhibits showcased electricity. The tramways and Métro were at the forefront of electrification, among the earliest electric technologies that Parisians experienced frequently and firsthand.[19]

There were three available technical ways to power trams.[20] Dominant globally and enabling the Anglo-German tramway boom was the "trolley" system of Werner Siemens and Frank Sprague. Here, electricity followed wires hung above the rails, which touched the "trolley," a conductor on a slender arm on the car's roof. Anglophones named the system "trolley" or "trolley car" after the trolling conductor, and Francophones borrowed the term. A second system set cables in the pavement between the rails and hung conductors under the trams. Paris engineers and entrepreneurs tried three in-ground systems: a trolley-like device dragging in a charged furrow (*canniveau*), a dragging "cursor," "foot," or "ski" (*frotteur*) contacting charged plates between the rails (*contacts superficiels*), and a third rail, as in the Métro. All surface and overhead conductors were wired to a power plant. Paris workers usually buried electric cables under pavement, hiding them from the streetscape, but outside the sewers for safety.

A third and final traction system employed rechargeable accumulators, or batteries. This was the first electric system tested in Paris, and engineers quickly found its faults. The engineer Émile Vignes and the writer George d'Avenel saw it as "the worst of all." As Paul Vibert explained, a battery as big as a top hat produced about one horsepower. At least two horses pulled most horse trams, but two batteries were not enough. The railway engineer Louis Mékarski had observed that tramways used variable power for frequent stops and starts or climbing hills. More batteries could provide such headroom, but heavy batteries ironically increased the power needed to drive trams and caused more wear to the rails. Worn rails could provoke derailments and accidents. Finally, batteries were fragile and frequently replaced. This clumsiest form of electric traction was bulky, heavy, fragile, limited, and risky.[21]

In 1892–1893, the Tramway Company of Paris and the Department of the Seine (Compagnie des tramways de Paris et du département de la Seine; TPDS) adopted accumulators for three lines to Saint-Denis, and within two years hit financial trouble. TPDS leaders complained to shareholders about "an era of transformations, of uncertainties, of studies" yielding "mediocre results." Competition from railways and omnibuses, rising expectations for equipment, and the growing cost of keeping horses had tightened belts. Expanding horsecars to compete with motorized ones was as costly as testing new technology—either way, meeting expectations for both electricity and mobility was tough. Though the ailing tram companies reorganized in 1887, the TPDS's problems mirrored those of its predecessor the North Tramway Company. While electricity ultimately proved cheaper than other power sources, results before 1900 were mixed. TPDS trials in 1893–1894 found that steam was five times cheaper than accumulators; the company thereby falsely generalized that steam could be cheaper than all electric systems. Like earlier motor trials, accumulator tests were costly, solving neither financial nor technical concerns. Experts had known since the 1870s that motors were theoretically cheaper than horses, but twenty years of batteries, cables, compressed air, gas, and steam had not succeeded.[22]

Then there was the trolley. The Siemens trolley at the Paris 1881 Exposition of Electricity helped ensure the system's global popularity, and yet, in Paris, trolleys remained controversial and exotic like "American" and "barbarians'" tramways. Tramway companies knew that trolleys were cheap to install and operate and wanted them where possible across the department—often in mixed systems, with trams switching from central city accumulators to peripheral trolleys. Some municipal councilors accepted trolleys on the exterior boulevards, but most agreed that the center city should remain inviolate. Others rejected overhead wires everywhere. The council did not relax this stance until 1902, after tramway surface contacts had electrocuted a series of humans and horses. When councilors allowed limited trolleys, "cries of pain and anti-trolley outrage surged up." One member shouted in session: "If they ever install trolley wires in my quarter of the city, I will cut them down with my own hands."[23] Trolleys also posed their own safety risks—one critic maligned the "mortal trolley" as a "sword of Damocles."[24]

The photographer E. Mouchelet used perhaps the most rancorous anti-trolley rhetoric. He called trolleys a "scourge" (*fléau*) and overhead wires their "sinister accessories." "Spider webs" of wires compromised photography by interrupting sight lines. Plazas and intersections were "covered with a sort of trellis," and became "vast chicken cages." Mouchelet mobilized the stock anti-railway terms "brutal and barbaric" and "vandalism" alongside animalizing imagery. He extended conservative attacks on iron construction, maligning "modern tin-work" (*ferblanterie*) and "diabolical iron-work" (*quincaillerie*, which could also be translated as "hardware" or "junk"). He bemoaned cities "flooded by the horrible scrap iron [*ferraille*] of tramways," and "contaminated by contemporary industrialism," because "Americanism is flooding us." For him, tramways and all modern infrastructures were like a disaster (epidemic or flood) menacing Paris. Mouchelet fixated on metal, aesthetics, and "low" Americanism, conveying a keen sense of shame about modernizing infrastructure.[25]

The engineer Paul Dupuy stated flatly, "Electric tramways supplied by overhead wires [are], so to speak, impossible in major cities, because of various reasons, the principle [one] of which is the ugly aspect of the wires, which destroy the aesthetic of streets." His use of the word "impossible" is hyperbolic given the trolley's contemporary success in the US, Britain, Germany, and even provincial France, but for the era's Parisians, it could feel like fact. The cultural, aesthetic, and patriotic priority of keeping Paris beautiful often challenged infrastructure developers. Conservatives especially struggled to keep transportation infrastructure (subways, viaducts, trolleys, and locomotives) out of the center. Affection for Paris's cityscape (its "physiognomy," in the corporeal idiom) inspired opposition to overhead wires, making Paris exceptional in France, where many smaller cities used trolleys. But Paris fitted an international pattern. Despite similar trolley resistance in New York, Washington, DC, Toronto, Brussels, Berlin, Dresden, and Vienna, the north Atlantic led tramway growth in this era because trolleys spread rapidly in smaller cities. By 1889, there were 1,032 kilometers of US electric trams. Europe's boom came later. France went from 37 kilometers (1892) to 1,995 kilometers (1902), and Germany went from 102 kilometers (1893) to 3,692 kilometers (1903). Across the north Atlantic, regional and national capitals resisted trolleys; they had special standards for architectural aesthetics and environmental quality.[26]

By excluding accumulators and trolleys, only one way was left to electrify Paris trams: in-ground conductors. Hence, preparing for the 1900 exposition, operators upgraded many tramways after 1895 from horse traction to "surface" or "superficial" contacts. The East Tramway Company employed the Italian engineer Alfredo Diatto's system patented in 1894. In his system, a magnetic conductor (*frotteur*) dragged beneath each tram, collecting current from charged "plots" in cement boxes between the rails (fig. 6.1). In each box, an electrically charged metal pin (*clou*) with a magnetized head floated in a mercury bath to form a closed circuit. As a tram passed over a plot, its conductor magnetically lifted the pin up to the lid, breaking the circuit formed by the mercury, and delivering the charge. Diatto's complex and delicate design was vulnerable to rain, salt, and short circuits. The reservoir holding

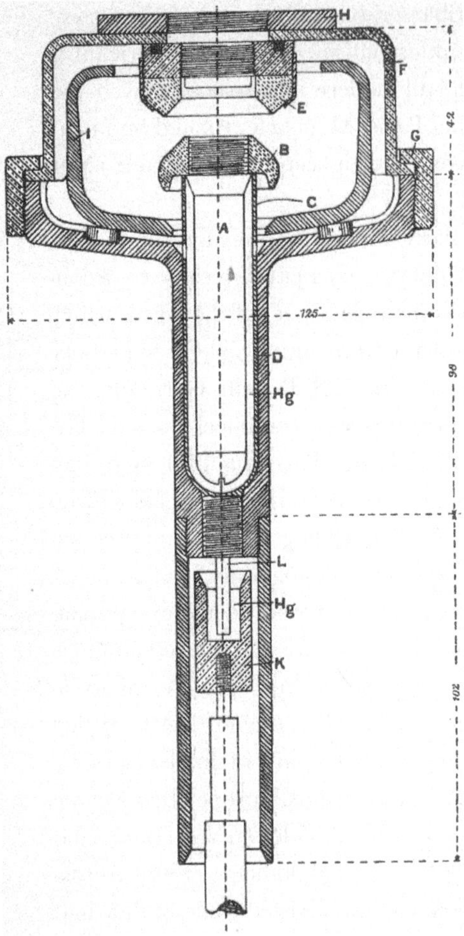

Figure 6.1. Cutaway illustration, Diatto plot mechanism, from Ernest Wilson and Francis Lydall, *Electrical Traction*, vol. 1 (London: Edward Arnold, 1907), 173. Courtesy of University of Oklahoma Libraries.

the mercury and the pin could fill with water, letting mercury escape, carrying charge with it, or buoying the pin up to contact and activate the plot's lid without a tram.[27]

While savants like Francq, Gauthier, Vibert, and d'Avenel followed international views of electrification as a decisive, even destined, victor over previous energy systems, not everyone agreed. Local technical experience since the 1870s suggested that all motors—compressed air, steam, cables, and electricity—were promising, but difficult. Yet the capital's reputation rested on its beautiful streetscape. Preparing for the 1900 exposition therefore unleashed contradictory impulses: both restless, futuristic zeal for new infrastructure, which would transform the existing city, and pride and vanity in Paris's beauty, which suggested its preservation. To compete with the perceived modernity of Germany, Britain, and the US before 1900, elites applied various systems of traction after 1896, mostly surface contacts, but also compressed air, steam, batteries, cables, and mixed systems.

Engineers, entrepreneurs, and politicians made quick technical decisions while preparing the exposition. Electric light railways were new and "high tech," still unstable and unscripted, as were the public and private institutions that managed them. The TPDS administrators clearly recognized their own era of experiment, futurism, and risk-taking. But different social groups faced different risks. Investors risked losing money on expensive experimental equipment. Government contracts risked breaching the public-private boundary, signaling corruption and damaging legitimacy. Finally, for the increasingly diverse Parisians who lived with the new means of transport—construction workers, conductors, drivers, and passengers—the physical risks of accidents sparked controversy. Mass transit aboard light rail confronted the public with a choice: were the physical risks of speed, fire, collision, and electrocution infrastructural costs that Parisians could live with?

"The Ravages of Electricity"

Since 1890, periodicals across the editorial spectrum raised technical concerns about electricity. Popular science magazine *Cosmos* and industry magazine *L'éclairage électrique*, alongside newspapers *Le matin* and *Le XIXe siècle*, noted poorly insulated wires.[28] Early electric networks also lacked many emergency circuit breakers. Dailies *Le matin* and *Le petit Parisien* later joined more rebellious *La lanterne* in highlighting water's conductivity, under the recurring headline "the ravages of electricity."[29] The 1900 exposition shined a bright light on these issues.

Métro Line 1 opened on 19 July 1900, midway through the exposition. One observer described ticket booths being "assaulted" by visitors and Parisians, and trains running at double capacity for several days. Newspapers claimed that everyone enjoyed the fast, affordable ride. The cool tunnels, a refuge from the summer heat, were clean and well lit—not the dark, dank *nécropolitain* haunting public debate since 1878 (chapter 4). *Le rappel* evoked the civilizing mission, saluting this "admirable agent of moral and material progress." There was cause for enthusiasm. The subway was a symbolic, if unofficial, exposition showpiece. Born from imagined futures, its debut was a happy ending to its dream life, a wish come true. Visitors enjoyed a fantastic, futuristic ride, the world's fourth electric subway after London (1890), Budapest (1896), and Boston (1897), while Parisians previewed the railway that would transform mobility in the twentieth century.[30] The 1900 exposition, like its precursors, celebrated industrialization in a great open-air museum for teaching that science and technology were motors of history, keys to social progress, and pillars of civilization. This farewell fete for the "century of steam" also inaugurated the "century of electricity." The *Le Figaro* writer Petitjean personified the coming era: "My name is the century of electricity, and I come to the world to continue the work of peace and progress started by steam. . . . It is thus in plain social and industrial revolution that the great 1900 Exposition will open its liberal doors to all nations."[31]

Such technological enthusiasm partially hid the Métro's technical errors and minor accidents—short circuits, small fires, lighting failures, delayed trains, and what journalists called passenger "panic." *Le petit Parisien* tried to calm public fears, describing "little accidents without any seriousness, whose importance should not be exaggerated." But minor accidents show that both users and operators found electric technology novel and unfamiliar. One critic belittled the subway as a mere exposition ride, arguing that to serve local interests, "the toy should become a tool."[32] The historian Jean Tricoire portrayed the Métro's early years (1900–1903) as experimental, groping gradually "from improvisation to method."[33]

Tramway accidents continued apace. June of 1900 brought 125 injuries and 3 deaths. In July, *Le petit Parisien* observed, "It's decidedly a series," reviewing more than a dozen accidents in August, mostly collisions with motorized trams, under the recurring headline "the murderous tramways." *Le petit Parisien*, again wrote: "They make us talk about them too much—and in such an upsetting manner—these tramways with mechanical traction, whose creation was welcomed with such favor by the Parisian population."[34] That summer, trams hit children, cyclists, and other trams, a pneumatic canister exploded, and some passengers "panicked" after their tram derailed. Finally, a malfunctioning surface contact shocked a dozen coachmen and killed a horse—a grisly preview of the next winter.[35] Accidents were common enough that the left-wing *L'aurore* called it "astonishing" to pass several days without printing an accident report, while the humorist Pierre Wolff joked, "There is at least one tramway accident per day," a ridiculous understatement—the public expected more than one.[36]

Accidents continued that fall, when *Le petit Parisien* wrote, "It is undeniable that the new mode of mass transit with which we are afflicted constitutes a public danger." The problem, the editors argued, was that mechanical and horse-drawn vehicles shared streets and rails, writing: "Tramways and carriages could never mix well. The one excludes the other." Another observer used the stronger verbiage of a French cliché: "I know well that mechanical traction is destined to replace animal traction: 'the one will kill the other.'" Uneven development caused friction. Horse traction was becoming obsolete, unfit for intense modern traffic, while faster motorized trams were unsafe in crowded spaces—unsuited, the editors argued, for all cities. Recent accidents resulted "from the very existence of tramways with mechanical traction in cities where circulation is too intense."[37]

On 11 October, when a Métro train reached the Tuileries station, the conductor opened the door to find a fire from another short circuit. *Le temps* tried to normalize such electric accidents: "This type of traction is recent; so, we must give it credit for the inseparable difficulties of the debut." The Métro's first collision was on 19 October near the Concorde station, causing twenty-nine minor passenger injuries and seriously hurting the driver. Parisians fanned this news into a roaring blaze of rumors, claiming that hundreds were hurt. Prefect of Police Lépine acknowledged public safety concerns, quickly revised Métro signaling rules to prevent further collisions, and opened an investigation of tram accidents. Sensibly, given frequent accidents, many wanted to discuss responsibility. But investigators merely repeated the press's claims, facilely accusing drivers of causing accidents by speeding on busy routes.[38]

Hindsight reveals other causes—incomplete scripts for using novel electric technology, which drove engineers and operators to improvise. Electric traction chafed against Paris's dense spaces and traffic, creating risks for drivers, passengers, pedestrians, and horses. Driving trams was hard work—navigating crowded streets while operating cutting-edge equipment of uncertain design and use. Progress ideology also mattered. By blaming drivers, progressives who linked technology with civilization and improvement were able to deemphasize its risks, to avoid friction with enthusiasm for electricity. By implicating human error or improper use technology and its designers were protected. Indeed, squaring the dangers of electrification and motorization with reigning technological optimism posed a central, difficult cultural challenge in the era of the universal expositions and the Second Industrial Revolution. Anglophone guests in 1900 also tackled this challenge, hence the odd, thrilled-and-chilled tone of the American historian Henry Adams in his reaction to electric generator exhibits in "The Virgin and the Dynamo," or the Scottish social theorist Patrick Geddes in his confident, yet ambivalent reflections on humankind's ability to control nature through science and art. While engineers and intellectuals defended new infrastructures with some ambivalence, drivers and riders slowly learned about them through experience and debate with more ambivalence.[39] Periodically, popular anti-tram violence erupted. In October 1900, fifteen drunk young men attacked a tram and beat the driver and conductor. *Le temps* reported that some peripheral tramlines were attacked so often that they shut down at night, provoking passenger ire and prompting Lépine to post police on trams.[40] The risks of evolving infrastructure inspired uncertainty and ambivalence.

During these early Métro accidents, the Diatto system's risks reached public attention. Charged plots were electrocuting horses and pedestrians, and East Tramway Company stocks plummeted.[41] Newspapers across the political spectrum used the epithet "murderous plots" for the next decade, one

Figure 6.2. Lucien Metivet, "Joujoux dangereux et joujoux cassés," *L'assiette au beurre* no. 35 (30 November 1901), 553. Courtesy of University of Michigan Library (Special Collections Research Center).

paper specifying: "murderous, homicidal, canicidal and hippocidal" (*meurtriers, homicides, canicides et cavallicides*), deadly for people, dogs, and horses. More often right-wing papers like *Le Figaro* and *La croix* used the phrase, but so did left-wing papers like *L'aurore* and *La justice*.[42] The liberal *Le temps* reported that accidents "strongly interest the public . . . and with reason!" and contrasted engineers' idealized drawings with the real dangers of streets. The leftist paper *L'aurore* called police "powerless" against accidents and the nationalist *Le Gaulois* described eastern neighborhoods as "terrorized" by electrocution risk.[43]

In November, the radical satirical *L'assiette au beurre* called trams and early automobiles "dangerous toys" (fig. 6.2). In Lucien Métivet's illustration, vehicles wreak havoc above in a playful mockery of modern traffic, contrasted with the macabre yet cartoonish dismembered "broken toys" lying in the morgue below. Shocked by a plot, the horse's legs twist surreally, reworking a popular image: months earlier, the populist *Le matin* had noted that motorists who hit plots danced "in their seats like veritable marionettes." Failing tramway plots turned humans and horses into uncanny puppets dancing in the street. This idiom of shocked horses "dancing" dated to 1889, the dawn of electric trams. As with the Métro in 1900, Métivet and *Le matin* belittled a railway as a "toy."[44]

Plots killed eight more horses in December 1900, and fifty-seven horses and one cow in January 1901. Like *Le temps*, the councilor Louis Duval-Arnould blamed the Diatto system, as both journalists and politicians inverted electric technology. Duval-Arnould showed limited knowledge of electrocutions but provided a sophisticated technical discussion to justify his claims. He marveled that walking on plots killed horses but not humans, not recognizing that iron horseshoes conduct charge. His report marks a moment in electrical history when Europeans began closing the historian Bernhard Rieger's "knowledge gap" between lay and expert understanding of technology through experience and experiment. Duval-Arnould also complicates Rieger's concept, showing shades of meaning between lay and expert. This politician, who saw himself and his constituents as lay tramway users, nonetheless deployed technical knowledge. He apologized for "considerations of a technical order," promising to be simple and brief so councilors would not be "drowned," hinting at the role of water. He knew that the council's job was officially politics, not engineering. This was not just a matter of rhetoric, custom, or law. Unlike recent tram collisions, electrocutions could not be blamed on drivers but were caused by engineers or technology. But because the government had ordered the CGO and tramway companies to adopt motors (especially electric) in 1896, it took responsibility for regulating them. Electrical engineering entangled with governance as electric safety became an urgent task for municipal and departmental officials, who increasingly inverted technology, scrutinizing its inner workings to protect workers and passengers.[45]

Duval-Arnould specified three technical issues. First, the conducting "pin" (*clou*), designed to leave its mercury bath to meet the tram's conductor (*frotteur*) often remained engaged after trams passed, leaving plots improperly charged. Second, tramway companies had fitted trams with a second dragging device to deactivate plots after use. Ironically, these safety devices caused new short circuits, starting small electrical fires that damaged plots, increasing the risk of electrocutions. Finally, Duval-Arnould noted the problem of water. The cement box holding each plot could fill with rain, corroding hardware or carrying mercury and electricity outside their intended bounds and inviting further short circuits. Future accidents loomed after rainy months. Politicians, engineers, journalists, and citizens realized that these accidents showed the entanglement of humans, machines, animals, and the weather. The socialist councilor Armand Grébauval warned that, "no one knows how the surface contact system will behave with snow and rain," while a report months later concluded that, "these accidents coincided with rainy periods." The police study commission found two major causes: (1) Diatto's fragile design; and (2) rain-soaked soil and pavement. The Diatto system was cheap and quickly installed, and its modest profile preserved Paris's streetscape—crucial for the 1900 exposition. But it chafed against Paris's humid climate, intense traffic, and continued reliance on horses. Water and electricity connected trams, horses, and people in complex, unpredictable ways. The writer Gustave Chauveau described "plots, whose functioning can be rendered defective by the humidity of the soil, the violent shock of a car, by salt strewn to melt snow, etc." Similarly, *Le matin* later observed of the Métro, "We know that short circuits can be produced both in one of the motive dynamos, or at the point of underground conduits [*canalisation*] and can be triggered by the interposition of a metallic object, hot and humid atmosphere, [and] the presence in the air of conducting dust, metallic or otherwise, in suspension." Electric accidents underscored that light railways operated outdoors amid rain, dust, horses, and salt.[46]

Despite strong language in the press, the 1900–1901 Métro and tramway accidents did not inspire the terms "disaster" or "catastrophe." Dead horses apparently mattered less than dead humans. Mostly normalized as technological or human failures, only the staunchest critics saw these accidents as indictments of infrastructure's inherent dangers.[47] But the Diatto accidents became an institutional breaking point, throwing the companies operating plots—the East and Left Bank Tramway Companies—into a crisis like that of the late 1870s. The department embarked on systematic tramway "reorganization" from 1902 to 1914, slowly opening Paris to the trolley, which proved safer than surface contacts, though never risk free.

The 1903 Métro Fire

While the Diatto accidents were a tramway turning point, the Métro had its fulcrum, and entered a full decade of architectural and technological overhaul before 1914, in the railway's worst accident yet, on 10 August 1903. On Line 2's elevated tracks over the exterior boulevards, water and deicing salt had corroded electrical equipment, exposing conductors to the elements and inviting short circuits. Riding east, Train 43's engine short-circuited and ignited near the Barbès station. After passengers had safely evacuated, Métro employees with limited electrical knowledge repeatedly tried to snuff out this electrical fire with water. Next, they tried to push the burning train to the terminus, Place de la Nation, with Train 52. In the commotion, they forgot to lift Train 43's conductor off the third rail—a second human error that fed more current into the damaged motor. Fire quickly engulfed Trains 43 and 52, whose bodies made of heavily varnished and painted wood created accidentally efficient incendiary devices. The burning convoy stopped soon after the place that Line 2 descends underground, where the cascading failure unfolded. The crew fled the train as fire quickly filled the tunnel with smoke and melted the wiring for tunnel lights, setting a terrible trap for Train 48, loaded with extra passengers from Trains 43 and 52, which entered the darkened, smoky Couronnes station. Popular illustrations captured the scene's chaos and carnage in a recurring image of piled-up bodies. In the ensuing commotion at this single-exit station, dozens became asphyxiated—some were waiting at the ticket window to demand a refund and refused to evacuate when urged by employees; others tried to escape at the platform's dead end opposite the exit and became trapped. Police and firefighters removed eighty-four bodies the next day, as a curious and concerned crowd filled the street above (figs. 6.3 and 6.4).[48]

The accident taught regulators, engineers, and operators many things. The first lessons were about architecture: railings should be light, easy to get around and under; entrance and exit gates should swing both ways to avoid restraining pushing crowds; furniture in stations should be minimal, and important pieces like benches should be bolted down outside major pedestrian paths; exits should be indicated with lighted signs; lighting cables should not

Figure 6.3. C. M. Postcard, postmarked 26 August 1903, with the caption "Catastrophe of the Métropolitain. The crowd awaiting the removal of the bodies of the victims from the Couronnes station." Author's collection.

Figure 6.4. Undated, unbranded, unused postcard, with the handwritten caption "Catastrophe of the Couronnes Métro, 11/08/03 [11 August 1903], the rest of the [train] cars." Author's collection.

follow the roofs of tunnels; tunnels needed generous ventilation; and stations needed multiple entrances and exits. The second lessons concerned electricity: circuit-breakers should be numerous and accessible; the third rail should be covered; an emergency lighting system should have its own circuit; lastly, all current to a section should be cut during emergencies. The third lesson involved material: stations, trains, and other equipment should be made from fireproof materials, especially metal. Finally, several communication lessons were addressed: frightened passengers needed guidance, hence the new posted emergency instructions, alarms, and visible exit signs; employees managing a crisis needed reliable real-time communication between stations, and hence stations were more generously wired for telephones. Overhauled operating rules set new standards for train timing and spacing, and revised employee emergency procedures.[49]

Popular responses were more visceral. Ticket sales dipped by half the day after the accident, and when the news reached London, a similar "panic" broke out on its subway. The *London Times* found the accident interesting because Parisians were already so loyal to the subway: "The Metropolitan now forms part of the daily life of many thousands of Parisians." Its lines "completely changed the mode of living of the working population." The public "got accustomed to this new means of locomotion in an amazingly short time." Thus, Parisians' pride and fondness for the Métro made the accident traumatic. The accident challenged the otherwise hegemonic notion that the subway was a technological advance and a practical convenience, causing friction between attitudes about the network and its realities. Mohandas Gandhi also watched Paris from afar; for him, such accidents "show a grim tragedy behind all the tinsel splendour of modern civilization. . . . Nothing that modern civilization can offer in the way of stability can ever make any more certain that which is inherently uncertain. . . . The boast about the wonderful discoveries and marvelous inventions of science, good as they undoubtedly are in themselves, is, after all, an empty boast." Thus, the accident recalled the limits of modern Western science, technology, and progress.[50]

Like the Charity Bazaar fire, the 1903 tragedy combined electricity, fire, insufficient exits, public "panic," and mass casualty. The liberal *Le temps* made this connection, before suggesting that Métro operators reduce risk by removing third rail traction, making trains fire-resistant, and increasing tunnel ventilation. The Catholic paper *La croix* also evoked 1897 but towed a traditional line about accidents as divine punishment for the hubris of modern secularism and technoscience, criticizing the Métro company as "notoriously beneath its task." The populist *Le matin* also recalled the Charity Bazaar and saw hellfire in this underground blaze: "And there is something Dantesque in the spectacle of a city equipped with all the refinements of civilization, of all the perfections of science, and which, for seven hours . . . was obliged to let fire devoir the entrails of one of its neighborhoods, and [let] death take human beings, [who were] unable to try resisting, unable to put up a fight, unable to risk a rescue." All three papers called the fire a "catastrophe," but *Le matin* also revealed the term's contingency, writing: "It was not a simple accident, as we thought, as we hoped; it was a horrifying catastrophe," and later, "the accident was a cataclysm."[51]

The radical *L'assiette au beurre* highlighted the ironies of progress, depicting a uniformed emergency responder exiting the darkened subway crying for candles, in a parable about the failures of modern electric technology and the triumph of ancient candles. In another cartoon, a working-class everyman at a café lamented, "The Métro below, the autos up here, and the blimps up there. Damn it! I'm not going out anymore." Like *La croix*, which it would have chided as clerical claptrap, *L'assiette au beurre* criticized the General Traction Company for cutting costs and amplifying risk.[52] While *La croix* blamed technology for the dangers of living with infrastructure, *L'assiette au beurre* blamed capitalism. These critiques staked science and modernity upon technological control of nature. But water's corrosion of outdoor wires likely caused the initial short circuit, and water's conductivity surprised the workers who threw water on the fire. When environmental and technological forces conspired to go haywire, their hybridity called the human technological mastery of nature into question. The accident disrupted the dominant technophilic idea of the Métro, raising questions about the risks of living with electrification and motorization.

In both 1901 and 1903, water infiltrated electric systems and caused short circuits, disrupting railway infrastructure. As a dense and conducting fluid, water was a risky player in hybrid systems. Together, water, electricity, and fire endangered flammable and conducting bodies, whether train cars, horses, or humans. Water could enter the system through natural forces like rain or through human intervention that caused technical failures, design flaws, or the throwing of a bucket—with deadly consequences. Accidents happened when systems meant to control electricity and exclude water rudely proved their hybridity. Accidents were traps Parisians had unknowingly set, triggered when the slow, corrosive action of water and salt met the sudden spark of short circuits in potentially lethal ways.

As a common subject for news stories, illustrations, and picture postcards, accidents were an important inspiration for inverting and politicizing infrastructure. Accidents inspired public debate about "responsibilities."[53] Seriously, and without its common sensationalism, the press posed the question of "responsibilities" in the plural, sometimes without answering. Beyond blaming people or institutions, journalists evenhandedly explored the agency of humans, animals, technologies, and physical forces like electricity, water, and weather. Although human choices or technical failures explained some accidents, others occurred when artifice failed to discipline natural forces as intended. Around 1900, Parisians recognized risk as pervasive but had not yet fully accepted it as normal. Their affective reactions suggest that both hybridity and risk were profoundly troubling. For contemporaries, the everydayness of accidents and the hybridity of infrastructure brought public controversy and contentions over modernity, technology, and civilization. Safety concerns fed broader worries about the notion of modernity as technology's triumph over nature, which was spectacularly displayed in Paris in 1900. Those with Luddite or anti-modern leanings—the far left, populists, Catholics, and Gandhi—saw accidents as symptoms of modern problems: motorization, electrification, capitalism, or science.

Paris accident language reveals three important patterns. First, disaster terms regularly preceded stock adjectives: *épouvantable* (appalling, dreadful, horrific), *effroyant* (frightful), *meurtrier* (murderous), *terrible*, and *tragique*. Stock visual imagery also recurred in dramatic illustrations of heaps of victims. By contrast, minor accidents inspired terms of estrangement and misunderstanding, including "bizarre," "curious," "extraordinary" (*peu ordinaire* or *extraordinaire*), "uncommon" (*peu commun*), and "singular," which indicate the limits of normalization.[54] Second, serious accidents provoked three forms of public behavior: (1) public "panic," especially of crowds during fires, linking the 1897 and 1903 tragedies; (2) public spectatorship, as in the crowd assembled outside the 1903 Métro fire; and (3) public debate about the disaster's causes, consequences, "responsibilities," and remedies. The word "responsibilities" helps explain why disaster language characterized some accidents but not others. The rhetorical use of disaster terms supported indictments of specific historical forces—Catholics blamed secularism, leftists blamed capitalism—or the indictment of specific social or interest groups, such as engineers, republicans, construction workers, or coachmen. Using disaster language was a tool of inversion and politicization. I have elsewhere described "the politics of disaster," the political implications of determining which events count as disasters.[55] While "accidents" could often be normalized as routine, "disasters" were more rawly affective—more likely to highlight social issues, compel reform, or delegitimize a regime.

Risk shaped a third pattern in disaster language. Smaller-scale events, otherwise mere "accidents" (*accident, sinistre*), could unleash catastrophic cascades of consequences. For example, an angry 1910 editorial about unfinished construction projects explained, "This is becoming disastrous," diagnosing impending disaster. Earlier that year, the oppositional journalist Ernest Judet argued that these same unfinished works caused an "artificial flood" that brought the "catastrophes" of the 1910 inundation. In 1913, the progressive paper *Le rappel* described a Cardiff coalmine explosion that killed more than four hundred as "[a] horrifying accident [*sinistre*] that could happen tomorrow in France." When the violent storm of 15 June 1914 (see figs. I.1 and 9.1) opened several sinkholes—one of which swallowed an entire auto-taxi—*Le petit Parisian* blared, "Yesterday's storm, by its consequences, took on the proportions of a catastrophe."[56] Although eighty-four died in 1903, several hundred in 1913, and only twelve in 1914, the term "catastrophe" was equally applicable. Even deadly enough routine short circuits, storms, or traffic accidents could be considered catastrophic. Some newspaper accident reports also mentioned fortunate moments when accidents did not cascade into catastrophe—for example, when electrical failures near gas lines did not ignite gas.[57] Other reports candidly predicted disaster, for example, a 1900 tramway accident story that ended, "We are waiting, on this line, for new and numerous accidents. Coachmen, watch out for yourselves!"[58] Commentary calling accidents "everyday" or "frequent" also reflected risk awareness.[59] Anticipating coming threats and rhetorically separating "mere" accidents from "true" disasters were strategies for living with risk, tolerating infrastructure's costs.

The 1900–1901 tramway accidents show that some accidents—even some criticized in the press—were not represented as full-fledged disasters. By contrast, a 1900 walkway collapse with only six casualties was immediately cast as a disaster like the Charity Bazaar fire for rhetorical reasons. The landmark 1897 and 1903 disasters shared the confluence of new technology, fire, public "panic," and mass casualty. Catholics saw both events as proof of God's wrath for the failings of modernity and the republic. Republicans, by contrast, saw the heroism of ordinary patriotic citizens as proof of the republic's democratic legitimacy and solidarity. But the politics of disaster went beyond the era's central political struggles over the republic, reflecting debate about broader and deeper transformations such as secularization, electrification, and motorization. These processes are commonly identified with urban modernity, and hence, disaster politics abutted familiar conflicts over the meanings of modernity. As Parisians negotiated which social and technological changes defined their era, they wrote scripts that would later represent larger-scale disasters like 1910's Great Flood and World War I. Natural disaster metaphors had represented Paris's urban crisis since the 1820s; Balzac and Tocqueville famously used the term "volcano." Such imagery would later become common in describing the Great War across Europe.[60]

Debates over whether social, technological, or environmental factors caused accidents show how electrification and motorization inverted infrastructure, revealing its hybridity, and inspiring Parisians to rethink relationships among people, technologies, and the environment. Claiming that public works were disastrous served to contest republican visions of moral and material improvement, for example, when conservatives and decentralists tried to cast the entire 1900 exposition as a disaster. But that example also shows how debatable and rhetorical "disaster" was.[61] To wit, in 1897 France's minister of education and fine arts, Alfred Rambaud, called Haussmannization the "Haussmannian cataclysm."[62] Accidents were visceral reminders of the fragility of urban modernity, both the material fragility of safety and risk, and the symbolic fragility of science, modernity, and civilization. Accidents taught Parisians that living with infrastructure meant managing, suffering, or tolerating both pervasive risk and periodic disaster.

7

Making a Stink

In 1858, the London press coined the phrase "Great Stink" to describe both the cloud of environmental odor from river pollution and the public outcry against the pollution that pushed officials to transform water infrastructures. Historians recognize five Paris stinks: 1880, 1895, 1900, 1911, and 1926.[1] Unlike London's single Great Stink event, Paris's water crisis persisted for half a century. This chapter presents Paris's long-running water troubles not as five discrete "Great Stinks," but as a several-decade "Slow Stink," comparable to the ecocritic Rob Nixon's "slow violence" and the historian Scott Knowles's "slow disaster." These scholarly terms emphasize the long lead-ups and lingering legacies that surround apparently discrete disaster events and the social and technological aspects of apparently "natural" disasters.[2]

Paris's summer stinks recurred when heat and drought reduced water sources enough to cause whole or partial cascading failures of water infrastructures, including failing taps, toilets, and sewers; low pressure; water shortages and shutoffs; river pollution; halted navigation; disease outbreaks; and environmental odor.[3] Parisians made a stink about these failures, inverting and politicizing infrastructure and environmental quality, using catchphrases like "the odors of Paris." Although often in opposition to each other, here journalists and engineers agreed that the water "crisis" returned "each summer" or "each year."[4] Parisians described the experience of water crises in *affective* terms, which they called "emotion": heat, thirst, stink, anger, disgust, disease, fear, promises, responsibilities, trust, blame, and even black humor. Unlike the experiences of housing, railway, and electric infrastructures, which diverged by class and place, as previous chapters have shown, living with failing water infrastructure inspired similar (though not uniform) responses across social lines, perhaps because water is a biological need. Water supplies clarify the biological role of infrastructures as "life supports." Depending on infrastructures for vital resources creates risk because their failures bring critical shortages.

Paris water histories show a striking tension. Several authors cast water precarity as a perennial problem until Haussmannization; for example, according to the historian Paul Léon, "By a strange paradox, Paris, born on an island and built on the banks of a river, lacked water for twenty centuries."[5] Other sources, following the water engineer Georges Bechmann, claim that Paris first "lacked water" in the 1880s, once it depended enough on piped water to suffer its absence (about two-thirds of houses then subscribed).[6] Thus, did modernized water infrastructures and practices *solve* Paris's water scarcity after the 1850s or *cause* it in the 1880s? Was scarcity ancient or modern? While droughts caused by environmental conditions were perennial, humans and technology created modern network shortages and outages. Artificial lights create blackouts and water distribution networks create shortages. But water supplies have natural sources, embodying Sara Pritchard's "envirotechnical systems," which makes drought and shortage hard to disentangle as facets of what she calls "envirotechnical disasters."[7] This chapter therefore examines an ensemble of human and nonhuman causes of water shortage: water sources lakes, rivers, rain, and springs; hot, dry weather; and increasing population, pollution, enclosure, water infrastructure, and standards for water quantity and quality.

Parisians made a stink by replaying keywords and scripts whenever water ran short: "lack" (*manque*), "scarcity" (*pénurie*), and "shortage" (*disette*). After 1880, Parisians used these words to describe water system failures resulting from weather, population, pollution, and infrastructure. In 1882, recalling recent dry summers, Paris's chief water engineer Edouard Couche noted "what we have come to call 'the lack of water' [*le manque d'eau*] of this past July," thus marking this catchphrase's birth.[8] Engineers sometimes used the term "lack of water" coolly or clinically to naturalize and depoliticize shortages, dodging their associations with disease, disgust, and disaster, and conflating them with drought. But the public politicized shortages using hotter, smellier language. Parisians condemned officials for washing streets, watering gardens, and flushing toilets with precious potable water instead of available non-potable water, while distributing cholera and typhoid in filtered river water, which the water advocate politician Paul Strauss compared to "negligent homicide."[9] Engineers responded that Parisians were hacking water meters to cheat their landlords or their bills in a wasteful water "orgy."[10]

These debates about who or what caused shortages weighed deep questions about the risks of depending on fallible, state-regulated, corporate-operated envirotechnical systems for securing basic biological needs. Gradually, the centuries-old water commons of rivers, canals, fountains, and wells ceded to water's centralized distribution as an enclosed, commodified, quantified, and technologically mediated "industrial product." This transformation lasted until around 1930, gradually reshaping water access and politics, and finally ending the Slow Stink after one last scorching, stinking summer in 1926.[11] This chapter examines the peak of making a stink about water shortages, the thirty-year blame game in which the public and departmental engineers traded accusations of "wasting" water, hosted by the press from the early 1880s through the early 1910s.

The Water Question

Paris's "water question" was a named social debate in the 1860s, like the Métro question, but much longer lasting. Until Haussmannization, Paris had no water distribution system but a patchwork of water commons: rivers, canals, wells, and fountains. Haussmann and the engineer Eugène Belgrand founded a Water Service for engineering and a Water Company to sell service, with an 1860–1910 monopoly matching the General Omnibus Company's. The smaller 1865 Dhuis Aqueduct added 20,000 cubic meters of water daily, while the larger Vanne Aqueduct took another decade to build. Despite great efforts, Paris remained thirsty. In 1870, 30,000 groundwater wells were in operation.[12] Belgrand's expanding "dual piping" network, which separated "public" (non-potable) from "private" (potable) pipes, reached 1 million meters in 1871. Yet the British philanthropist Richard Wallace found Paris so parched that he donated around 70 public fountains from 1872 to 1879, extending the water commons while the distribution network was built. Local officials continued adding fountains, thereafter, reaching 110 by 1904. Over 100 Wallace fountains still stand today (figs. 7.1 and 7.2).[13]

From the birth of the republic, therefore, water access was riven by two social divides: between public water gatherers and those with private taps, and within the second group was a divide between subscribers to potable and non-potable water. Shortages affected all three groups. In 1873, half of Paris's

Figure 7.1. Unbranded, undated, unused copy of L. J. & Co. Postcard, "Everyday Paris" (Paris Vécu) series, with the caption "At the fountain," depicting a family at a small-model Wallace fountain. Author's collection.

Figure 7.2. Unbranded, undated, unused copy of L. J. & Co. Postcard, "Everyday Paris" (Paris Vécu) series, with the caption "At the Wallace," depicting two people at a large-model Wallace fountain. Author's collection.

70,000 residential buildings enjoyed piped water, and a third used the costlier potable service. Maxime du Camp described housewives with casserole dishes catching non-potable water from street-cleaning standpipes.[14] Water quality varied widely; that year, cholera killed 855 Parisians and typhoid 869. Hygienists celebrated these declining mortality numbers as public sanitation achievements and blamed disease on unsanitary private conditions, which the hygienist Apollinaire Bouchardat called "choleraic places."[15] While cholera steadily declined, typhoid still infected hundreds annually into the 1900s. Émile Zola's 1877 novel *L'assommoir* (slang for "bar," perhaps "the watering hole"), dripping with water imagery, dramatized working-class water struggles. The tale follows laundresses working in sodden, polluted shops, living under leaky roofs, and gathering water in public or drinking alcohol, which was cleaner and more accessible than water, but carried the (often fatal) risk of alcoholism.[16]

The Vanne Aqueduct opened in 1874, twenty years after its planning, its daily 60,000 cubic meters of water inadequate for public needs. Two new technical measures followed in 1876: water quality testing at the Montsouris Observatory and water meters, which were not yet mandatory.[17] Without a meter, the Water Company offered two subscription types: *au robinet libre* ("by free tap"), unlimited water for a higher fee, and *à la jauge* ("by gauge"), a fixed amount of water daily based on the amount paid. Property owners subscribed by address, giving landlords power over tenants' water access, quality, and cost. But landlords, concierges, and tenants alike could cheat the company through a hack to "prop" or "stall" (*caler*) the tap. Users placed small pieces of wood or string between the pipe wall and the membrane (*diaphragme*, *lentille*) inside that limited flow, holding open the conduit for extra water. In 1878, Alphand reported that free tap subscribers consumed 125,000 cubic meters of water but paid for 90,000 cubic meters. Engineers and politicians steadily denounced such "abuse" and "waste."[18]

In 1880, officials reformed water regulations to universalize water meters and limit waste, but problems persisted. In 1884, Couche reported that subscribers by gauge sometimes received 50 percent more water than they paid for.[19] Growing subscriptions reached nearly 50,000, over two-thirds of houses, by 1881.[20] Meanwhile, nonsubscribers collected rain from gutters, balconies, and windowsills or bought water from illicit sellers.[21] Subscribers steadily innovated hacks, widening their houses' intake point or boring holes in pipes *before* the meter. As meters spread, others learned to carefully prop taps to a trickle, securing seeping access without triggering the meter. Meters also inspired new conflicts between tenants and landlords because the latter could not reckon tenants' usage without control and surveillance. So, landlords pressured tenants with extra fees, limited water hours, shutoffs, or, as Émile Pouget alleged, locks on taps (chapter 2). Engineers and technical writers often hyped self-closing taps (called "intermittent" or "deterrent") as anti-waste technology, but they had limited impact on water use.[22]

These conflicts around water distribution paralleled the 1880 Great Stink, 1881's major shortage, 1882's typhoid (3,352 deaths), and 1884's cholera (986 deaths).[23] In 1881, residents of the Temple neighborhood "in fury" (*en colère*) tore down Water Service notices urging water conservation to prevent a shortage. One writer described Parisians as "suffocating" (*étouffons*) and "very irritated" (*fort irritées*), demanding, "Is Mr. Alphand mocking us?"[24] Alphand called the "extraordinary persistence" of shortages "alarming."[25] Two authors later suggested that subscribers withhold water payments in protest.[26] Tensions simmered during the 1880s as disease, odor, pollution, and shortage melded in the morass of the Slow Stink.

In 1884, Prefect Poubelle proposed "unity of piping," distributing only potable water to dwellings, then a distant aspiration.[27] Around that time, the Water Service began substituting filtered river water for potable water during shortages, which doctors and scientists quickly correlated with typhoid.[28] Thus, Louis Figuier and others advocated maintaining Belgrand's "dual piping," rigorously separating river water from potable water.[29] In 1889, the *Bulletin médical* noted that districts 13–16 received river water before hot weather. Municipal councilors debated whether that year's Universal Exposition caused the shortage, which public works officials disputed.[30] Debates in the 1880s over water access and quality tapped decades of controversy about drinking water from the Seine.[31]

In 1892, cholera killed almost 1,800 across the department, half in Paris.[32] That year, although officials declared distribution mains "complete," and new regulations required landlords to provide potable water, at least half of buildings lacked it. In 1893, the municipal councilor Paul Strauss promoted "this grand project of the future: free and obligatory water for all."[33] Access to clean water was also critical. By 1893, routine testing measured water's hardness, plus levels of bacteria, oxygen, chlorine, nitric and sulfuric acids, magnesium, iron, aluminum, potassium, sodium, chalk, organic matter, volatile matter, and fuzzier "variable matter." Scientists analyzed data by arrondissement, by aqueduct, by water testing station, and even by street addresses where disease, especially cholera, appeared.[34] But spot checks could never monitor the entire system, which had grown to 11 water plants (*usines*) and 17 reservoirs. In 1894, typhoid killed around 700 Parisians.[35] *Le matin* noted the public's "legitimate emotion" and later linked the outbreak with 28,919 dead animals that police removed from the Seine and the Marne in 1893, though filtered river water and a corrupted Vanne Aqueduct emerged as likelier culprits.[36] The pairing of river water and typhoid became central to the Slow Stink's continuing controversies.

Crisis Years, 1895–1898

September 1895 brought a five-day shortage after the seasonal July–August heat (*canicule*), which followed a record thirty-nine rainless days.[37] River levels dropped 70 centimeters as odors spread. Paris's sewers reached the Seine in the suburbs of Clichy and Saint-Denis where the enfeebled river was choked with waste. *Le petit journal* used lurid language: water was now "muck" and "sludge" "thickened" and "blackened" with "multiple islands" of "fecal matter." The river was "nothing more than an immense cesspool . . . because the current doesn't exist anymore to speak of," and yet, the disgusted reporter noted, people still swam in the river to cool off.[38]

At 37°C, 99°F, 10 September was Paris's hottest recorded day for that month. On 11 September, the Water Service announced that six central districts (1–4, 9, and 10) would receive river water for twenty days. Press protest erupted, calling the water fetid, nauseating, slimy, thick, unclean, warm, and yellow. Though river water was drawn upstream of sewer outlets, journalists insisted sensationally that customers now drank "this blackened, unclean liquid that they draw . . . from the mouth of the sewers."[39] *Le petit journal* demanded "Do You Want Some Microbes?" because taps dripped "purée of microbes," not water. The populist *Le matin* described the public as "panicked" by fears "of microbes, of fevers, of epidemics of all kinds," expressing "horror" and "a certain emotion." The paper later noted that if shortages happen every time heat and thirst are the greatest "it concerns not only hygienists but also logicians." Writers indicted officials for mismanaging infrastructure, neglecting human needs, and spreading disease.[40] *Le petit Parisien*'s persistent water critic Jean Frollo called it a "crisis."[41]

Amid the crisis, commentators likened Paris to colonial Africa, the "land of thirst" (*pays de la soif*).[42] *Le petit journal* blamed the shortage on the "Senegalese heat" (*chaleur sénégalienne*), a catchphrase used until the 1940s.[43] Appropriating the language of bacteriology, writers described microbes "colonizing" water or as host organisms. The anarchist and social Darwinist Émile Gautier found the shortage ironic, asking: How could "even civilized" France produce electric lights and trains, but not supply clean water, something the Roman "master colonizers" could? For Gautier, Paris's "purée of microbes" inverted narratives of civilization and infrastructural progress.[44] The leftist Maurice Charnay wrote, "The suppression of water during heat waves has become commonplace, a subject for summer stories, quite useful for journalists short on political news." Accordingly, he denounced the municipal council: "Hundreds of millions are wasted to open new streets, generally useless, except for landlords who receive the indemnities from expropriation . . . to construct sewers that plague the city and the suburbs. And Paris has neither means of transport, nor even water to drink!" He joined Gautier in inverting progress narratives and accused the Water Service of wasting potable water on posh west-side gardens, while also wasting money and time.[45]

Politicians and engineers defended themselves by blaming the weather and the public. The chief water engineer Edmond Humblot denounced "a veritable aquatic orgy." Beyond bathing and drinking, consumers let water run to cool and then held a bottle or carafe under the stream to chill drinks. Fulgence Bienvenüe, the head Métro engineer, suggested that his neighbors let water run all night to cool the air—precisely when water was scarce.[46] Such remarks only inflamed already heated public debate. While some sources confirmed Bienvenüe's claims, others doubted them: "Well!" another journalist wrote, "we don't know what kind of building Mr. Bienvenüe lives in, but it must have a very special disposition." Most faucets, he explained, are in kitchens or bathrooms. Bedroom taps and tenants running kitchen taps to cool bedrooms were "quite exceptional."[47] Such sarcasm was common that summer. He cast Bienvenüe as wealthy, privileged, and out of touch, while delicately dodging the topic of propping taps.

In "Water and Politics," the firebrand independent writer Ernest Judet argued that the Water Service failed to control nature: "The powerlessness of our perfected civilization shines before the phenomenon which overwhelms us: the habitual changing of seasons." With Charnay and Gautier, he inverted progress narratives, contrasting civilization with Paris's "barbaric and absurd" water rationing. For him, fresh water "is a condition of well-being and health which can only be denied in case of siege . . . [and] any contrary proposition is simply monstrous." Judet also suggested that France's tradition of revolutionary violence might end the water crisis. Such polemics clarified water politics: because the authorities had taken responsibility for providing water, citizens could claim water—even clean water—as a right. Frollo explained, "The hygiene of water thus enters into the domain of the state's obligations." Water commodified, enclosed, regulated, and piped by the state was thereby politicized.[48]

Judet also evoked empire, writing, "There is plenty of connection between the faults committed in preparing for Madagascar and the disposition of the Paris Water Service." That year, malaria killed nearly 6,000 of 15,000 French troops invading Madagascar, among the deadliest European colonial campaigns of its time. Juxtaposing colonial setbacks and nagging local problems, Judet implied that French officials worldwide failed to master disease and water in life-threatening ways.[49]

Despite heated controversy, 1895's shortage lasted five days, not twenty. As weather cooled and reservoirs rose, newspapers reported that potable water would return, but users should boil tap water for several more days.[50] Amid infrastructural crisis, environmental change, and political uproar, critics evoked Africa, bacteriology, and colonization to signal heat, drought, disease, and danger.

The Water Service was still weathering the 1895 crisis when the 1896 shortage struck, prompting Strauss's charge of negligent homicide.[51] Just as potable water ran out, a leaky river water intake pipe threatened to stop all running water.[52] Chief engineer Humblot studied population growth and growing water use, planning to gradually meet public needs by 1930, while defending filtered river water. Increasing supply, he argued, was "a pressing necessity considering the approach of the Universal Exposition of 1900," so that "numerous foreigners" would not witness "muddy streets" and "dried-up fountains." Like transport and energy projects, water projects were spurred by upcoming expositions. Engineers were not only failing to satisfy a growing populace; they were *bound* to fail, because water-use standards were also rising. In 1861, Paris's roughly 1,700,000 people used an average of 115,000 cubic meters of water daily (68 liters/person/day). By 1895, 2,500,000 people used 550,000 cubic meters daily (220 liters/person/day). Population grew by a factor of 1.5 while water consumption increased by more than a factor of 3. Expert estimates of individual daily water needs were rising, too; Poubelle estimated 150 liters in 1884; Humblot claimed 220 liters in 1896.[53] In 1896, plans to tap the Loing and Lunain Rivers ended talks among the engineers Huet, Humblot, and Bienvenüe that had started in 1892. The goal, Bienvenüe wrote, was to overcome "the precarious equilibrium that the arrival of the waters of the Avre has permitted." Each new source since the 1860s, including the Vanne and Avre Aqueducts, was inadequate upon opening, thus, "if we want this result to be attenuated in the summer of 1898, we must act without delay."[54] Following rainy months, the summer of 1897 offered a "lucky" break from shortages, which *Le matin* called the luckiest since 1886, and yet its editors still addressed this "major question always under study" (*grosse question toujours à l'étude*).[55]

Confirming Bienvenüe's warning, an 1898 shortage brought renewed debate in the press. *Le matin*'s somber headline "Shut your Taps" preceded Frollo's familiar lament: "The question of potable water comes back periodically to the order of the day for the public authorities, without ever being resolved." *L'intrasigeant* added that "odors" menaced Paris "every year."[56] Meanwhile, the search for new water sources raised tensions with the suburbs and the provinces. In 1889, the legal scholar Léon Aucoc wrote, "The combination which best serves the interests of city dwellers and brings them precious advantages, the derivation of *eau de source*, entails serious drawbacks for country dwellers who until then had enjoyed the derived waters." Avoiding conflict between the city and the hinterland, Aucoc argued, required carefully studied and planned water supplies. The following year, the towns of Verneuil and Nonancourt began resisting Paris's search for sources. The Slow Stink provoked regional water conflict.[57]

After 1898, the Water Service projected three plans for new sources: first, to connect existing aqueducts with nearby sources; second, to tap Lake Léman on the Swiss border; and third, to draw water from the Loire and Loiret Rivers, then sources for the city of Orléans. The second plan pushed Paris's thirst as far as it had ever gone, into the Alps, while the third plan collided with the centuries-old rivalry between the Bourbon and Orléans dynasties.[58] From 1898 to 1903, the city of Orléans furthered the family feud by disputing water with Paris. One critic predicted that Paris would drain the Loiret, leaving Orléans dirty and diseased. Dead, decaying aquatic plants would create a bog and "dangerous miasmas." Farming and navigation would suffer along with the "freshness of the soil or even the means of irrigation." He ended by noting recent incidences of typhoid. His fear of Paris's environmental impact echoed suburban concerns (more below).[59] Though the Water Service hoped the 1893 Avre Aqueduct would stop the distribution of filtered river water, the 1895–1898 shortages dashed this hope, and the continuing practice likely increased typhoid rates until 1900.[60] Despite no shortage in 1899, as in 1897, the press continued making a stink about shortages, typhoid, and water "administration."[61] Bienvenüe's "precarious equilibrium" was clear; engineers and journalists correctly predicted coming risks.

A Stink at an Exposition

After 1895, the Water Service hurried to build the Loing and Lunain Aqueduct by 1900. But the envirotechnical water supply responded to weather, and the summer of 1900 brought drought.[62] Water outages began in late July on high ground, the Boulevard Rochechouart at Montmartre's southern foot.[63] This elevated location matters because Paris often suffered low water pressure. During droughts, there was not enough liquid mass to raise water to upper floors and higher elevations. Similarly, pressure varied in the cheaper non-potable network. These physical facts, compounding Paris's social geography, meant water precarity for the poorest, who lived up high.[64] Water quality also suffered. In 1900, yellow and odorous river water emerged from taps in the 11th district. The stink raised in the press indicates widespread public mistrust of the Water Service and journalists' desire to amplify this mistrust. Newspapers warned that a shortage could deliver disease or hinder firefighting—both fears came true.[65]

At 38°C, 100°F, Friday, 20 July was Paris's hottest day since 1874, and the fairgrounds were deserted. Though Friday was the "chic" day to attend, only about 150,000 visitors came (versus 700,000 the previous Sunday). Ticket prices fell from 45 cents to 20 cents. The Water Service distributed filtered river water but also used another strategy: each night from 21 July to 5 August, engineers shut off the potable service from 11 p.m. until 6 a.m. to prevent its use, almost tripling 1895's five-day shortage.[66] Shutoffs reflected officials' ongoing struggle to control water use.

The press replayed familiar scripts: politicians and engineers accused users of wasting water and journalists accused officials of waste. Echoing 1889, *La presse* asked why officials ignored that the exhibits and millions of guests increased demand. The editors blamed neither heat (nature) nor user waste (society) for the shortage, but water system mismanagement (engineering) and misgauging demand (politics). The *Revue générale de chimie* estimated that the exposition's machines alone required 8 million cubic meters, which could be non-potable. Exhibitors also used non-potable water for washing sidewalks and watering gardens and could purchase potable water from the Water Service at 15 cents per cubic meter (Paris then used about 700,000 cubic meters daily, of which about 300,000 were potable).[67] The nationalist *Le Figaro* blamed foreign exhibitors, namely, the popular Swiss Village model alpine waterfall, but the Swiss had official permission to dig a well. Ironically, many, including the French Society of Civil Engineers, wanted to tap Lake Léman for Paris.[68] The exposition certainly helped drain reservoirs and motivate shutoffs.

La presse received an "avalanche" of reader letters. This disaster image naturalized citizen critiques, bolstering the practice of making a stink. One reader suggested taps for potable and non-potable water with separate counters on each floor of apartment houses. Another asked engineers to cool the reservoirs with enormous ice blocks. A third advised that consumers boil their water or use purification tablets, which advertisers busily promoted alongside bottled water.[69] The municipal council debated overhauling Paris's water distribution system, but Poubelle's "unity of piping" was controversial because some of Paris's scarce potable water would serve to flush toilets. The council discussed toilets flushed with non-potable water, which entailed repiping thousands of buildings, but these ideas dried up because landlords resisted.[70]

By July 23, journalists were spewing bitter sarcasm, angered by official accusations of waste. The Bohemian poet Saturnin snidely joked that his tap ran all day to simulate a vacation by a babbling brook.[71] *La presse*'s cartoon "Victim of Duty" depicted Justin de Selves (prefect of the Seine 1896–1911) in the bath like David's Marat, complaining, "Parisians waste water and they're taking it from me!"[72] But the *Le Figaro* writer Cornély supported charges of public wasting because his neighbors let taps run to chill drinks or the air. He argued that water metering by apartment and paying by volume would discourage waste.[73]

The water shortage cascaded through interdependent infrastructures and practices, compromising food service and firefighting. The baker Lebel explained that Paris's "briny" (*saumâtre*) tap water was too warm to control dough fermentation, thus threatening bread quality. *Le Figaro* scorned this "suspect mixture of *eau de source* and filtered river water."[74] Scarce ice dogged butchers and beverage vendors. Prefect of Police Lépine demanded vigilance from all firefighters and the use of steam pumps in case water pressure dropped during a fire, and soon the fire department warned that pressure was too weak (1 kilogram rather than 3 kilograms) to assure fire protection.[75] Lebel and others complained that the water problems returned each year, despite decades of politicians' promises.[76] *La croix* argued that the shortage showed politicians' lack of "foresight" (*prévoyance*) and "sincerity"—not only concerning water, but "on all occasions and in all things," forming a pattern of broken "promises."[77] Cartoons mocked bathing in beer, drinking one's footbath, shaving with ink, firefighters using bottled water, and resorting to other unsavory water substitutes.[78] The press raised a stink about the disastrous conjuncture of heat, water shortage, and their cascading consequences with the exposition's intensifying effects.

Drought noticeably changed the Seine. Newspapers decried the river's "poisoning" and "infection," with dead cats, dogs, and fish riding a slow, oozing "blackened sludge" (*boue noirâtre*). Receding water and slowing current created a vast stinking "bog" (*marécage*).[79] These vivid descriptions, echoing 1895, reflected public disgust and outrage. Overnight shutoffs kept toilets from flushing; less water used meant less water evacuated, threatening the water-to-waste ratio in toilets and sewers as in the river. The Academy of Medicine warned that compromised sewers could spread disease, and soon newspapers reported toilets that would not flush.[80]

Given increasing pollution plus the exposition, engineers closed all sewer outlets into the Seine inside the city limits in 1899 (chapter 8).[81] Parisian sewage was piped to the suburbs before being dumped in the river, steered into sewage farms, or spread over waste filtration fields (*champs d'épandage*). In 1900, waste from the riverbed was concentrated, fermenting, and stinking, while in the suburbs extra dumping joined low water.[82] In Saint-Denis, "the sewer has formed, by its dejections, a body of viscous and stinking matter which, at a length of thirty meters, advances in the river whose current is so feeble that it cannot break it up." Other suburbs, Clichy, Suresnes, Epinay, Saint-Ouen, Argenteuil, Bezons, Marly, and Pecq offered equally graphic reports: "From Asnières to Saint-Germain, the river is black like ink, lacking limpidity. Around the [wastewater pumping] machine at Marly, the air is unbreathable, dead fish float along the banks."[83] In response, suburban politicians organized a Union of Shoreline Municipalities (Syndicat des communes riveraines) to protect their natural resources from Paris. Achères sanitation workers testified that in July surplus wastewater was dumped into the Seine, not spread on the ground. Technically speaking, the Achères waste fields could handle 40,000 cubic meters of wastewater per hectare per year but were flooded with up to 100,000 cubic meters in 1900.[84] The symbolism of Paris defecating on its neighbors stoked the "decentralist" opposition, a provincial autonomy movement of the 1890s, which made Maurice Barrès famous, resisted the 1900 exposition, and guided Orléans's defense of the Loire and Loiret from Paris.[85]

Then typhoid spiked. In August, two Montsouris analysts found abnormal typhoid levels in water samples. While engineers defended filtered river water, these municipal scientists saw risk: "What do they think now about their abundant distribution 'of purified Marne and Seine water?' How many more victims do they need to convince them that their frightful mixes are nothing but a violent poison?"[86] The hygienist Dr. Thoinot warned medical professionals in 1899, but the public stink broke out in the shadow of the 1900 exposition. Thoinot found that typhoid fell sharply from 1894 to 1898, affirming the Water Service's hopes for the Avre Aqueduct. But typhoid rose after 1898 from 556 cases and 105 deaths to 3,148 cases and 568 deaths (912 including the suburbs) in 1900, which Thoinot attributed to filtered river water.[87] Laypeople and experts alike recognized increasing river pollution. Headlines about the Seine's "sanitation" (*assainissement*) and "poisoning" (*empoisonnement*) resonated with expert warnings about sewerage and typhoid. The heat threatened health directly. Hospitals and ambulances were crowded with heatstroke victims, 75–200 of whom died that year.[88] In late July, hospitals exercised aggressive triage, often denying treatment. But new dangers awaited patients inside hospitals, which also used river water. One journalist blamed this fact for rising patient mortality; suggestively, most of Thoinot's data came from Paris hospitals.[89]

The water system's cascading failure disturbed the exposition's practical and symbolic goals, in an embarrassing international showcase of environmental pollution alongside technological delays, errors, and inadequacy. Heat and disease interacted with water technologies to undermine the exposition's message of progress via technoscientific control of nature. In 1902, Humblot's successor, the engineer Georges Bechmann, wrote, "We have not forgotten the crisis that the service of potable water distribution in Paris . . . passed through in 1900."[90] Amplified by millions of exposition guests, the now twenty-year water crisis inspired critics to make a stink about persistent issues—environmental quality, public health, water infrastructure and use, and political failings—as they inverted the exposition's narratives of progress.

A Cooler Decade, 1901–1911

After the hot summers of 1895–1900, the early 1900s were cooler, with highs rarely over 30°C, 86°F. Thirst faded and reservoirs rose, confirming the pattern that the worst shortages accompanied hot, dry weather. Even as a piped commodity, water was a natural resource bound by atmospheric changes. The historian Sara Pritchard has shown how engineers strategically naturalize and technologize envirotechnical systems. In Paris, because the water supply required nature's cooperation, engineers naturalized shortages to deflect criticism and blame. But by publicly claiming that ecology, not engineering, caused shortages, engineers undermined their own mastery of nature. The more they naturalized infrastructure, the less they could claim to control it technologically, but both were useful (if risky) strategies.[91]

Cooler weather did not calm heated politics. In 1901, journalists and municipal councilors attacked Bechmann, the director of water and sanitation, for mismanaging infrastructure, misreporting wastewater quantity and quality, and increasing typhoid while inhabiting a tax-funded 34 Rue Copernic mansion overlooking the Passy Reservoir with a "famous golden bathtub" (recalling the 1900 cartoon of De Selves above).[92] Bechmann, like Eiffel, was hounded for years by accusations from socialists, nationalists, and antisemites. In 1902, water engineers scrambled to expand service, confirming one author's claim: "The provisioning of the gigantic agglomerations that constitute the great modern metropolises becomes more difficult each year."[93] In 1903, flooding at the source briefly corrupted the Avre Aqueduct.[94]

Journalists predicted 1904's summer shortage. In the spring, *Le matin* urged readers to boil water because of typhoid, courtesy of "castle-dweller" (*châtelain*) Bechmann.[95] In July, *Le petit Parisien* asked whether rising heat befitting "the tropics" would cause a shortage. This anticipation shows that Parisians saw the pattern: hot weather consistently brought water shortage, river water, or shutoffs, and potentially disease and odor, too. Days later, *Le petit Parisien* printed: "So many works of adduction and capturing [of sources] have not yet allowed [us] to predict heat waves and each summer a new disturbance comes to hurl itself at the Water Service!" That summer combined hot weather, river water, overnight shutoffs, and a boil notice from Dr. Thierry, director of the departmental hygiene service, because of recent typhoid cases.[96] In 1905, the Water Service cut water every night from 31 July to 10 August without any heatwave, shortage, or public uproar. Reservoirs never ran low; engineers just padded the supply to reduce risk and prevent shortage. Even more than 1904, 1905 illustrates the tedium and political weight of recurring shortages and typhoid, as Bechmann left office for private-sector work with the Nord-Sud subway.[97] The routine repeated in 1906 when one popular science author wrote, "Each year, at the same time, the question of potable water poses itself. This year, the situation is already aggravated due to an exceptionally torrid summer."[98]

Such summer shortages only affected Water Company subscribers. People facing poverty, landlords, or shoddy housing struggled daily to find water. Despite claims that piping was "complete," modern infrastructure's tendrils still neglected marginal spaces. In 1906, *Le petit Parisien* likened a neighborhood in the Zone at Pantin without running water to the "Sahara."[99] The Zone was a 250-meter-wide ring of land between the city limits and the aging

Figure 7.3. G. F. Postcard no. 1, postmarked 1923, with the caption "PANTIN—General view." Wikimedia Commons. The foreground of the photograph gives a sense of the Zone's undeveloped land.

fortifications. Technically under army jurisdiction as a "non-construction zone" (*zone non aedificandi*) to preserve long-range visibility, it was abandoned after 1870, and became a marginal realm of barren spaces, squatters, shantytowns (*bidonvilles*), and struggling immigrants, many from Southern and Eastern Europe or North Africa (fig. 7.3).[100]

Le petit Parisien interviewed the local barkeep Mme. Pornine, who explained that fifteen households inhabiting this space repeatedly approached the authorities with a petition for water, but "each one has been useless." They gathered water by ruse, fraud, barter, or bribe and shared among neighbors, "and what water we find now!" Every few days, a street cleaner opened a standpipe on the Rue du Chemin de Fer for an hour. Pressure was weak and this non-potable water required boiling. She said, "You've got to hurry to get what you need. Housewives go running with their urns, and they line up like at the theater." Another tap across the street belonged to the suburb of Aubervilliers, whose agents threatened "onerous fines." Some suburbanites shared water, but if they were too overt, agents might also fine them for misusing water. Thus, she explained, "You must take care to go at night, like criminals, keeping yourself well hidden."

This neighborhood's desperate tenants, property managers, and landlords collaborated to petition the state—in significant contrast to more typical landlord–tenant conflict (chapter 2). Mr. Clermont, a lease-holding property manager, detailed a plan to distribute water in the Zone: workers would install taps at the homes of two leaseholders, where all subletting tenants would gather water. The Water Service submitted plans for the army's review, but zoniers still awaited response when *Le petit Parisien* was printed, and infrastructure was still lacking. As the local inspector (*agent-voyer*) Courtois coldly noted, "The zoniers would have water when they want it! For that, all they need to do is to obtain the piping. It's one or the other: either they establish it at their own cost, or they guarantee the Water Company a sufficient minimum of subscriptions." So much for universal access!

This story from the Zone illustrates how enclosure excluded nonsubscribers. Despite recurring shortages *within* the network, times were harder *without* a subscription. Zoniers struggled to find water now that it was enclosed, commodified, and regulated. Gathering water from common sources was effectively criminalized, creating a longer, slower shortage than any temporary summer outage for subscribers. Engineers often exaggerated claims of infrastructural progress, conveniently neglecting people and places that lacked modern infrastructure. This exaggeration explains journalists' zeal for exposing parched places like the Zone. Poor and marginal Parisians experienced the everyday practice of gathering water, which had been familiar since the early modern period. Unlike summer shortages ("every year") among subscribers, which ended like discrete disaster events, enclosure and pollution continually reduced water commons (every *day*) across the Slow Stink.[101]

In the coming years, controversy slowly simmered. In 1907, *L'intransigeant* revived the river water issue; in 1908, *Le Figaro* bemoaned the shortage (*disette*) that had been raging since 1896; and in 1909, *L'autorité* called more typhoid an "administrative scandal."[102] These three summer fixtures—shortage, river water, and disease—then burst forth off-season in the winter 1910 Great Flood, which brought water crisis because floodwater filled the sewers, mixing fresh water, wastewater, and storm water; floodwater also shut down water pumping and treatment, in the water system's biggest cascading failure yet (chapter 8). Citizens without a water subscription as well as the 1910 winter water shortage underscored the reach of the water crisis beyond recurring (predictable) summer outages to form a slow, continuous process.

No Progress Has Been Realized

Heat and water shortage occasioned another summer stink from July to September 1911, including 13 consecutive days of temperatures over 30°C, 86°F, and a new September record of 36°C, 97°F. Typhoid had flatlined since 1900, killing 567 in Paris in 1911—a disappointing but not alarming number. But the heat itself proved unusually deadly, killing 40,000 across France, three-quarters of them children under the age of two.[103] In Paris, the Water Service distributed river water by day and shutoffs from midnight to 6 a.m. from early June to late July, the longest rationing period yet. Then, as the Water Service planned to end shutoffs, the drought caused low pressure and routine shortages.[104] In late July, the press painted a familiar scene: water waste warnings, followed by advice to ignore officials and boil tap water, flanked by references to "Senegalese" heat and the "unbelievable, abominable, criminal" fact that firefighters were failing without enough water.[105] The departmental inspector O. Boudouard claimed that 1911's "odors" were "particularly bothersome [*incommodes*]," while *La nature* called odor "a chronic ill."[106]

Again, the press spouted accusations and sarcasm. A cartoon depicted a shaggy, bohemian—perhaps *hydropathe*—café customer: "Waiter, an absinthe . . . pure! . . . so as not to waste water!"[107] One journalist feigned surprise that a dead dog did not slip from his tap, adding, "Cholera, here? Yeah right! It would die of thirst." He joked that his neighbors were in turmoil, reading the book *A Glass of Water* or bathing in mineral water. One neighbor took a sponge bath with Eau de Mélisse, a therapeutic lemon balm tincture, while the concierge at number 30 accidentally killed her canaries with Eau de Vichy, a famous therapeutic mineral water. Ironically, as in 1900, other things called "eau" were available, but no pure water.[108]

Politics rode humor's heels. Press criticisms of the 1911 shortage drew connections with the January 1910 flood (chapter 8). The painter Louis Latapie wrote: "Thus are Parisians destined to perish under the water or to die of thirst. The administration is incapable of sparing us the floods and the shortage. . . . For, is there a city in the world worse governed, with a population at once worse served and more docile? Paris is dirty, Paris stinks, Paris [traffic] is encumbered, Paris has too much water, Paris has a water shortage . . . and Parisians, smiling, are going to give millions to the taxman and throw ballots into the electoral urns as if to say: Thank you!"[109] Later he wrote, "The Republic is incapable of grand conceptions and major works. Since the

Empire, in effect, it has done nothing great or useful in Paris," deliberately undercutting republican moral and material improvement.

The seasoned critic Ernest Judet also connected the 1910 flooding with the 1911 shortage. For him, the parched city presented "the lamentable spectacle of crowds suffering in the heat, with no comfort, no refuge anywhere, deprived of water to drink, water with which to wash and bathe." This situation "is the shame of a capital like Paris."[110] Journalists also depicted poor people seeking water: "The Wallace Fountains—what few are left—don't have enough . . . for the eager clientele jostling around them." Montmartre residents roved for handouts with bowls, buckets, pans, pitchers, and pots. Others collected "yellowed" gutter water to boil.[111] *Le radical* noted that the shortage defied expert advice: "You are thirsty; doctors tell you to drink often to supply your sweat glands and avoid kidney obstructions; doctors can spew their advice, [but] the city can't spew its water, and to the great detriment of hygiene and the desires of its inhabitants, it is particularly greedy at the same instant when they have the greatest need."[112]

In August, a major technical problem surfaced: the Vanne Aqueduct had cracked in mid-July. During three days of repairs, Paris lost about a day's worth of water (340,000 cubic meters). But as the repaired aqueduct reopened, the drought overtook it, and low flow stopped water 130 kilometers from Paris.[113] Other papers noted that the river water filters had been renovated during the shortage. The administration "a bit hastily" demolished existing basins in need of being "modernized," but unfortunately, "when one undertakes to modernize them all at once, this takes us back to a hygiene worthy of the Middle Ages." In this familiar counternarrative of progress, a temporary replacement basin did not impress: "We have today but a wee little basin, a child basin, constructed in twelve hours." The editors called it a "toy basin," filtering only 35,000 cubic meters of water daily, and they challenged the officials, writing: "We're keeping track of the rapidity with which this unclean liquid flows!" The author added that current chlorine (*eau de Javel*) disinfection was crude and corrupted the water's taste, demanding 5 million francs for ozone purification, still a cutting-edge technology.[114] Other sources recommended sodium hypochlorite. Continuing debate about disinfection joined another resurfacing sociotechnical issue: water metering by apartment or by tenant to curb waste.[115] By mid-August, newspapers joined engineers in demanding use of Lake Geneva or the Rhône for Paris.[116]

Compounded causes—deliberate shutoffs, technical problems, and summer drought—created a longer shortage in 1911 than in previous summers.

On 11 September, two months after shutoffs started, newspapers still wondered if shutoffs or river water would return. Later that month, the Seine was low enough to halt navigation for a week.[117] Compared to previous years, 1911's critique of the Water Service was more pointed, because deliberate shutoffs, weather extremes, and technical failures conspired to intensify water precarity. This time, the engineers could not blame nature or the public for the shortage.

The water system remained strained on the eve of World War I. At their 29 April 1912 lunch, the Catholic political economy society, the Union of Commerce and Industry, heard the municipal councilor Emmanuel Evain's defense of municipal ("socialist") infrastructure ownership, as justified by recurring water shortages since 1885. For the private service, a shortage in 1885; water failures in the 7th and 15th districts in 1886; upper floors lacking water overnight in 1887; distribution of river water began in 1888; "two superb epidemics of typhoid fever" in 1888–1889; reduced street cleaning, in 1893; several more days of river water in 1894 and 1898; and finally, overnight water shutoffs in 1900, 1904, 1905, and 1911. For the public service, Evain counted the annual number of times reservoirs had run dry since 1900: 13 (1900), 10 (1901), 20 (1902), 9 (1903, 1904), 10 (1905), 47 (1906), 7 (1907), 4 (1908), 11 (1909), 73 (1910), and 19 (1911). Although the sources and accuracy of his data are murky, the overall pattern is crystal clear: three decades of water precarity in both the public and private networks, which indicted the Water Company's regime.[118] Journalists and engineers agreed that since 1900, "no progress has been realized."[119] A daily average of 290,000 cubic meters of potable and 350,000 cubic meters of non-potable water was available, totaling 640,000 cubic meters, but well short of official estimates that Paris now needed 1,000,000 cubic meters daily.[120]

From 1880 to 1911, the Water Service aggressively expanded infrastructure for collecting, purifying, storing, and distributing water, but it never outpaced demand for clean water, in a rolling water crisis that resurged every summer. Meanwhile, republican politicians made water and sewer service standard for residential spaces, resisting regressive landlords with regulations, as they tried to invert the inequalities of primary Haussmannization. Still the crisis continued. During this Slow Stink, Parisians used affective language to script the type of infrastructure failure we call "water shortage," due to drought, low pressure, or deliberate shutoffs. The water crisis scaled up as population, subscriptions, and per capita demand grew. This was a slow disaster, indeed: not five summers of acute scarcity, but thirty years of persistent precarity, with no final answer to the water question.

The ways that Parisians conveyed and contested experiences of water shortages show that living without water went beyond heat and thirst. The public valued environmental quality ("odors") and reliable infrastructure and services ("promises"), and it objected to both rationing (water quantity) and distributing disease (water quality). Using lurid language of disgust, sarcasm, and political sniping, journalists—alongside scientists, geographers, poets, bakers, firefighters, and a barkeep from the Zone—decried citizen suffering, administrative hypocrisy, technological dependence, and infrastructural inequalities. They detailed embodied experiences of living with scarce water, hot weather, and waterborne disease.

In making a stink, Parisians inverted infrastructure by scrutinizing water fixtures of all kinds—aqueducts, filters, meters, pipes, pumps, reservoirs, and taps. Making a stink was also a practice of citizenship, which contested state biopolitics by scrutinizing the management of fresh water, wastewater, and disease. Critics especially used French political languages of rights and revolution, alongside the broad Western idiom of modernity and progress versus tradition and regression, to argue that Paris was historically backward, "thirsty" like Africa, and that the Third Republic betrayed the nation's international leadership, inverting republican narratives of "moral and material improvement." In politicizing water shortages, or making a stink, Parisians stressed the unintended consequences of expanding water infrastructures, and mobilized their embodied, affective experiences as evidence of failure or even disaster. As they had regarding the tramways, Parisians learned to live with expanded water supply infrastructures despite their lasting failings: shortage and typhoid.

8

Resources and Risks of Rivers

Scholars have long recognized what the historian Cornelis Disco has called cities' "affinity" for the multifarious resources and services of rivers.[1] Like cities and technologies, rivers produce energy and work, but unlike them, rivers are not ultimately human-made. The historian Richard White famously called the Columbia River an "organic machine . . . an energy system which, although modified by human interventions, maintains its natural, its 'unmade' qualities."[2] Following White, studies of technology and environments have cited engineered waterways as exemplary nature-culture hybrids.[3] Although "infrastructure" often means human-made sociotechnical systems, scholars from anthropologists to ecologists have recently interpreted nature, landscapes, and ecosystems *as infrastructure*, stressing the resources and support that humans draw from their surroundings.[4] Similarly, although urban studies have traditionally claimed that cities embody artifice, they now view cities as natural-artificial hybrids, which humans build on, into, and out of the environment. River cities like Paris clearly illustrate these insights.[5]

Long before Haussmannization made Parisians dependent on rails and pipes for basic needs like food, mobility, and water, they relied on the Seine. In 1875, Maxime du Camp included the river among Paris's infrastructural "organs," calling it "one of the major avenues by which the capital supplies itself," adding, "it completes the ensemble of our organs of communication, and in addition it has a special existence, represented by the industries which live on it and by it." Seine shipping supplied foodstuffs including coffee, fish, grain, oils, sugar, vinegar, and wine; materials consisting of coal, cotton, metal, paper, pottery, soap, and wood; and finished goods from animal fodder to furniture. The river also supported boating, fishing, and swimming. Du Camp noted that it provided water and removed waste, "at once the watering trough and the general sewer." For centuries, the river had enabled what scholars call "urban metabolism," providing resources and services that illustrate the biological dimension of living with infrastructure.[6]

Beyond resources and services, rivers also produced risks such as flooding and pollution. Environmental histories show that nineteenth-century France saw growing public and official awareness of both "natural" and "technological" disasters, but also increased flood mitigation by engineering canals, dredging, levees, and reservoirs.[7] The historian Margaret Cohen examined primary Haussmannization's physical erasure of wild, polluted, or "working" waterways from Paris—chiefly covering the Canal Saint-Martin and the Bièvre—embodying what the historian of Paris parks Richard Hopkins has called "engineering nature."[8] Such works did little to prevent pollution and flooding—in the capital, these problems were delicate, difficult, large-scale, and long-term throughout the era of the "water question" (chapter 7).

This chapter contributes to research on rivers, cities, and their relationships by examining how Parisians lived with rivers as infrastructure that brought resources and risks. As sewerage expanded, river water quality was threatened, driving a search for new methods of waste disposal and treatment. Simultaneously, flooding showed how disasters invert infrastructures during major 1876 and 1883 inundations and 1910's catastrophic "Great Flood," which illustrate the river's physical power. Finally, Parisians debated the "death" of the Bièvre, a Seine tributary severely polluted by tanneries, dye shops, and laundries in an un-Haussmannized industrial "slum," which the Office of Works gradually buried, ending in 1912. These topics show that rivers, formerly open commons, were increasingly enclosed, regulated, and controlled by the state. As with accidents (chapter 6) and water shortages (chapter 7), Parisians debated who or what caused the city's infrastructural problems as they inverted Haussmannization. Sanitizing the Seine, burying the Bièvre, and weathering floods inverted riverine infrastructure, revealing infrastructural hybridity, dependance on infrastructure, environmental pollution, increasing enclosure and regulation, and cascading failures—all of which inspired politicization.

Sanitizing the Seine

Fitting the engineer Georges Bechmann's description of "artificial life," nineteenth-century cities outgrew the waste-absorbing capacity of local land and water. The Seine could no longer be Paris's main drain. As waste scaled up, its meaning and value reversed. In 1891, the agronomy professor Achille Muntz explained that feces are valuable fertilizer in rural, agricultural, and subsistence societies. All human and animal food came from the soil, and all feces composted back into the soil recycled nitrogen and other nutrients. Industrializing cities broke this cycle, importing resources and exporting waste, which, according to Muntz, "far from being a source of wealth, as in the fields, becomes a cause of difficulty." As waste's economic value changed, its meaning shifted from fertility to filth, which the geographer Sabine Barles insightfully links with "the invention of garbage." Unused waste was discarded, as nineteenth-century cities found new, technologically mediated ways to remove and process it. Like Bechmann, Muntz understood urban modernity as the human interruption of natural cycles through infrastructure.[9]

Paris's earliest waste disposal practice was *tout à la rue* (everything to the street). All animal, human, and commercial waste entered this resource commons, a theater of recycling, where an underclass of ragpickers (*chiffonniers*) collected specific materials—bone, cloth, glass, manure, metal, paper, tobacco, and wood—sorting and processing them for redistribution. Remaining residues were driven by rain or broom into sewers or rivers, eaten by animals, or swallowed by the soil. An early modern system for disposing of human feces grew around "fixed cesspits" (*fosses fixes*) under houses. From the mid-1700s to the early 1900s, workers called *vidangeurs* emptied them and carted waste to suburban fertilizer plants, where farmers gathered manure to grow the capital's food. In 1880, 80,000 cesspits were still in operation. The capital's growing hunger, thirst, and waste made urban metabolism dependent on infrastructures for resource distribution and processing as their ecological impacts spread in widening circles.[10]

Georges-Eugène Haussmann and Eugène Belgrand designed Paris's first unified sewer network and expanded piping from 143 kilometers to 773 kilometers. Their "combined" system handled toilet outflow, street sweepings, and storm water, thus threatening the livelihoods of *chiffonniers* and *vidangeurs*. The technological mediation of waste disposal devalued centuries-old waste-management practices and cut economic circuits linking the capital to its hinterland, possibly dumping masses of human and horse feces into the Seine for the first time. Haussmann's ambivalence about urban technology and his documented concern for *vidangeurs* and the Seine colored an unease about human waste in the sewers. Studies of filth note a tension between repression (hiding or evacuating waste) and sublimation (transforming waste into fertilizer).[11]

The dynamic and difficult 1860s began the Prefecture's experimental sewage-farming program under the engineer Alfred Durand-Claye, which extended into the 1900s. He shared Belgrand and Eugène Poubelle's vision of every dwelling's toilet draining into the sewers without polluting the Seine. But the Clichy and Saint-Denis sewer outlets merely moved waste downstream, fouling suburban water with little regard for suburban dumping.[12] Already by 1866, soon after Paris sewage first soiled the Seine, engineers and politicians, who managed waterways as both water sources and waste sinks, recognized pollution, calling it "alteration" or "corruption."[13] Commentators tracked pollution, too; the humor magazine *Le Charivari* noted that for fish, now that the river was a sewer, "Haussmannization is a synonym of cholera."[14] From the 1850s to the 1910s, sewer debates paralleled the question of the Seine's sanitation, querying the proper relationships between people, sewers, and the river. Water infrastructures evolved while population and waste grew, raising pressing environmental quality issues.

A related problem was public river usage. Emerging in expert debates on river sanitation were persistent questions about who could safely use the river and how it could be done. Although the Seine was long a commons for fishing and swimming, an 18 May 1872 police ordinance (no. 4433) outlawed swimming nude, to maintain moral decency, and swimming outside of designated areas (in "open water"), to ensure safety. Parallel to water metering (chapter 7), increasing enclosure shaped persistent enforcement problems. An 1899 law further tightened swimming regulations, but the police "river brigade" did not emerge until 1900 (to be discussed below). From 1867 through 1923, swimming within the city limits remained both illegal and common practice.[15] Limits on fishing also expanded beyond existing rules for fishing in parks, with 1889 prohibitions on fishing at night, fishing from boats or bridges, or fishing with any tools other than a "floating handheld line."[16] Conflicts over swimming and fishing paralleled earlier concerns about floating washhouses where laundresses worked, as voiced by Haussmann and by Émile Zola in *L'assomoir.*[17]

The minister of public works called the first Commission for Sanitation of the Seine (Assainissement de la Seine) in 1874, estimating that Paris was dumping 300,000 cubic meters of solid waste in the river daily. Commissioners discovered citizen concerns about drowned dogs, horse bathing, factory emissions, and odors sullying the river.[18] In 1882, the departmental government formed the Paris Sanitation Commission (Commission de l'Assainissement de Paris) to improve housing hygiene and ensure the evacuation of "impure products." Commissioners estimated that Paris now made 400,000 cubic meters of waste daily, noting that key water sources, the Seine and Marne Rivers, were soiled by suburban sewers and industry upstream.[19] That year, Charles Dickens recommended the Seine for safe boating, fishing, and swimming, but not drinking.[20] The impact of the upstream suburbs on Paris mattered as much as Paris's impact downstream. By 1899, Paris was creating nearly 500,000 cubic meters of waste daily. Waterways across the department needed sanitation.[21]

The sewers that the republic inherited from Haussmann were both physically and conceptually unfinished. Engineers and hygienists disagreed about whether waste was better recycled wet, as in sewage farming, or dried, like powdered manure (*poudrette*); whether it should ultimately enter the soil, the Seine, the suburbs, or the sea; and how comprehensive waste disposal could succeed, given opposition from commercial interests landlords and *vidangeurs*. This manifold uncertainty shaped what French historians call the "battle" of the *tout-a-l'égout*, the struggle among the Water Service, the municipal council, *vidangeurs*, landlords, and hygienists to promote or prevent direct-to-sewer drainage. The sewers, like the Métro (chapter 4) and the water supply (chapter 7), endured a long dream phase between the Second Empire and World War I, during which the slow development and unstable design of infrastructure shaped and were shaped by broad and contentious public debate.[22]

Experts promoting the *tout à l'égout* stressed its modernity and necessity, contrasted with barbarous, stinking cesspits, without agreeing about where sewage should go. In 1884, the inventor-chemist Hippolyte Marié-Davy, director of the Montsouris Observatory 1873–1887, called for "everything to the sewer, nothing to the Seine" (*tout à l'égout, rien à la Seine*) and "everything to the sewer, everything to the sea" (*tout à l'égout, tout à la mer*), via the never-completed dream of dredging the Seine to make Paris a seaport. All agreed that water should cover and carry waste, but where should it go?[23] Many worried that Paris's strained water supply could not reliably flush the sewers. Poubelle's decree of 14 November 1883 required each apartment of two or more rooms to have its own toilet, draining directly to the sewers, with a hydraulic flush of 10 liters per person per day and an exit pipe siphon to stop waste and odor from backing up. Coupled with his proposed 1884 "unity of piping" decree, this would have flushed toilets with potable water, sending up to 10 percent of this scarce resource down the drain. Senator Paul Strauss asked in 1895, "Are we going to lose, indefinitely and constantly, through the *tout à l'égout* . . . all the *eau de source* that we supply at great cost and so inadequately?"[24] Modern sewerage, like changing bathing and street-cleaning practices, helped drain the reservoirs. The consistent coupling of water shortages with recurring stinks (chapter 7) indicates the problem's depth: when taps failed, drainage often failed. Summer drought could provoke a cascading failure, as could its opposite, flood. "Everything to the sewer" included storm- and floodwater, which menaced the hidden potable and non-potable water distribution pipes hugging the sewer walls. Critics, from Catholics to science writers, urged cautious sewer development until these varied technical issues were solved.[25]

As for other infrastructures, cultural issues compounded technical issues. The sewers carried contradictory meanings: long seen as places of abjection, crime, disease, filth, and revolution, Haussmannizing public works officials sought to recode them as tools of biopolitics, sanitation, and progress. One effort to recode sewers was public sewer tours, which began during the 1867 exposition in the later years of primary Haussmannization and continue today. Sewer tours were officially sanctioned exercises in inverting infrastructure, welcoming the public into a normally hidden (and fearsome) infrastructural space, where laypeople riding boats through wastewater received firsthand experience and official messaging. Tours domesticated, sanitized, and defanged these otherwise threatening spaces.[26] A Neurdein brothers postcard, no. 2181 from the "Underground Paris" series and postmarked 1909, depicts the tour guides of the Service de la Visite posing in their boat in the Collector of the Quai de la Mégisserie with clearly visible pipes and wires on the right side (fig. 8.1). This posed image, created with flash photography, reflects the official public relations goal of tours: to lend legitimacy and loyalty to subterranean infrastructure.

Such attempts to recode sewers met resistance on the fringes of Paris political culture, where the sewer's symbolism and negative meanings overshadowed technical issues. Slang dictionaries document the word "sewer" as meaning "human mouth," used when silencing someone or insulting their breath.[27] The scatological anarchist-humorists Jules Jouy and Émile Pouget suggested sending the rich and powerful "all to the sewer," while the antisemitic *La libre parole* argued that flushing out Jews was the "true" *tout à l'égout*.[28] Perhaps even more forceful was organized resistance from suburbs affected by expanding Paris sewerage. Sewage farming grew steadily from 5 hectares of irrigated land in 1866 to 6,300 hectares in 1899.[29] Plans by Durand-Claye, Adolphe-Auguste Mille, and Belgrand for an 1875 *enquête* estimated that

Figure 8.1. Neurdein brothers (ND) Postcard no. 2181, postmarked 1909, with the caption "UNDERGROUND PARIS—The Sewers, Sanitation Service; Collector of the Quai de la Megisserie. Boat serving the Visit Service." Author's collection.

handling all Paris wastewater in sewage farms would take 6,659 hectares, and suggested land near the suburbs of Gennevilliers, Nanterre, Rueil, Argenteuil, Sartrouville, Le Pecq, and Achères. Suburbanites greeted these plans with what the historian Gérard Jacquemet called a "general uproar." In the Seine-et-Oise department 27 of 32 municipalities rejected the plan, having collected 8,500 signatures in opposition. One journalist questioned the capital's "right to sacrifice Gennevilliers."[30]

So, engineers carefully designed sewage farms to moderate suburban impact. Large solids from wastewater went to fertilizer plants, while water charged with smaller particles followed pipelines out to farms raising artichokes, cattle, flowers, grass, peas, trees, and tomatoes. To reduce foodborne illness, crops including strawberries and lettuce that grow near the ground and are consumed raw, were excluded. By the 1890s, experts thought that Paris needed 7,820 hectares of sewage farms, exceeding the city's own land area. Adolphe Carnot, the high commissioner for Paris sanitation, studied 40,000 hectares of suitable space as far as 60 kilometers away.[31] The city's growing waste output, mirroring its search for fresh water, sent out widening ecological and political ripples that stoked regional conflict.

Amid controversy, sewers gradually replaced cesspits. *Vidangeurs* treated cesspit waste with sulfates of iron and aluminum, dehydrated it as fertilizer, or incinerated it, but changing waste disposal practices transformed waste materials.[32] Two 1881 sources estimated cesspit waste at 1 million cubic meters annually.[33] Thereafter, cesspits declined steadily, from 80,000 (1880) to 25,821 (1913), as did waste that reached suburban fertilizer plants. Meanwhile, direct-to-sewer drainage was phased in, from 213 hookups (1885) to 52,053 (1913), while wastewater in sewage farms grew from 7,212,928 cubic meters (1873) to 226,544,409 cubic meters (1902). From 1873 to 1913, therefore, wastewater and sewers gradually replaced solid waste and cesspits while waste processing migrated from northeastern suburban fertilizer plants to northwestern suburban sewage fields. Wastewater treatment shifted from Durand-Claye's 1860s plans for sewage farming and filtration fields to testing chemical and bacterial disinfection after 1900.[34]

While sewerage expanded, waste continued to reach the Seine, which still exuded its reputed odors. In 1892, Belgrand's plan to bar untreated sewage from the river inside the city limits was meant to take effect soon.[35] An 1894 law closed Clichy's main sewer, which finally happened on 8 July 1899. Even while the *tout à l'égout* remained incomplete, awaiting both infrastructure and subscriptions, achieving *rien à la Seine* was important sanitary work. In 1892, the progressive economist Ernest Brelay called the river's pollution "nauseating" and "revolting," noting reduced swimming, and neighbors complaining amid an outbreak of "sub-cholera" (*sous-choléra*).[36] In 1895, the water expert and writer Ernest Judet railed that the sewers "plague the city and the suburbs."[37] Friedrich Engels wrote in 1899, "The Seine seems to exhale the air of Paris all the way to its mouth," while other authors deemed Paris "the emptying point of all sewers" and "the *cloaca maxima* of the world."[38] Closing Clichy's main sewer was a republican public works celebration like the road and rail inaugurations (discussed in previous chapters), with neighborhoods decorated by citizens, plenty of police for crowd control, and prominent politicians, doctors, and engineers in attendance. As for the Avenue de le République, Bastille Day was near (chapter 1). Municipal Council President Lucipia called the closure "a veritable triumph of hygiene over routine." Ironically, Prefect of the Seine Justin De Selves suggested promoting it at the 1900 exposition, without predicting the water system's collapse the following summer (chapter 7).[39]

With Paris's main drain closed, the question of how departmental sewers altered the Seine remained. Two weeks after the closing ceremony, Clichy's city council noted that suburbs were dumping into the Seine and demanded the Prefecture's intervention. In 1900, the prefect of the Seine and the minister of public works created a project for the suburbs, which were then evacuating an estimated 357,000 cubic meters daily.[40] The police's "river brigade" (*brigade fluviale*), launched during the exposition, made parallel efforts to cleanse rivers. This team of men and dogs, specially trained for swimming and diving, monitored boating and swimming, rescued drowning or accident victims, and removed trash, debris, and dead animals.[41] The enforcement of swimming regulations remained problematic. In a 1906 municipal council session, the physician Dr. Georges Rosenthal called for children to swim in the Seine only upstream from the city.[42]

To manage increasing waste processed by the soil, the Office of Works experimented with waste treatment methods, from British septic tanks to chlorine disinfection and the Pasteurian Dr. Albert Calmette's bacterial system between 1904 and 1907. By then, the Seine sanitation begun in the 1870s had grown into projects for improving the Seine, the Marne, and the famously fouled Bièvre (more below).[43] The debate Belgrand started in 1855 queried Paris's impact on the Seine; decades later, Parisians discussed the entire department's effect on several waterways. Just as Paris's search for fresh water expanded its environmental impact, and urbanism emerged to manage cities at metropolitan scale, so the lasting question of sewerage and river sanitation scaled up. The plans devised in 1907 saw much administrative work, but no realization, before 1914.[44] Parallel to the Métro and sewer "battles," the Seine's sanitation moved slowly in this era. Paris's three governments—local, departmental, and national—struggled to manage growing water pollution. With both fresh- and wastewater enclosed in pipes, experts sought to purify both types of water, intending that science and technology would control nature. But engineers and politicians faced the limits of water, soil, and human control, as well as resistance from *chiffonniers*, *vidangeurs*, landlords, and suburbanites. Journalists, and commentators from Catholics to chemists and communists, inverted official sanitation efforts by chronicling citizen complaints and sensory experiences of pollution.

The Flooding Seine

Rivers are not human-made but human-modified parts of cities. Floods defy notions of human control, showing that even the most humanized and technologized waterways have explosive physical powers with unpredictable, sometimes catastrophic, impacts. Flooding inverted the modern narrative of humans controlling nature, and this inspired ambivalence. Floods inspired

Paris engineers, painters, novelists, and journalists to reflect on how humans relate to environments. Engineers tried to repair humanism, by materially modifying environments and symbolically shoring up mastery narratives. Floods defy us and they drive us to defy them—with dams and levees, but also with words and images.[45]

Paris floods are shaped by the Seine Basin, a large, teardrop-shaped region in northeast France whose rain and waterways drain into the Seine. Rising in central France near Dijon, the river enters the English Channel at Le Havre. At Paris, the snaking river cut a valley ringed by hills, a bowl with a floodplain at the center. From 1700 to 2000 it flooded 75 times, averaging once every four years; 48 of these never topped 6 meters, which today's Prefecture of Police considers the threshold of significant impact. Within Paris, the river is humanized and technologized in its stone-clad channel; its quays stand far above typical water levels around 2 meters. Hence, most floods are not catastrophic. The massive 1658, 1740, and 1910 events were catastrophes, however, topping 8 meters. Floods are typically winter events occurring between November and April. From 1891 to 2001, 73 percent of floods over 6 meters came in January, and the other 27 percent in December or February. But these loose rhythms do not make flooding more predictable, because patterns shift over time. March floods higher than 6 meters were more common before 1800 but resumed in 1876. During the early republic, the Seine topped 5 meters in 1872, 1877, 1879, 1880, 1886, 1889, 1896, and 1897. Floods punctuate the city's normal rhythms, and the Seine's fortified riverbed and the soil's absorption usually keep water channelized. But even smaller floods that spared street life often halted navigation and left dockworkers and mariners without work. Predicting Paris floods requires rain and flood reports from across the Seine Basin. As hot weather predictably brought drought, so heavy rain meant possible flooding.[46]

Extraordinary floods left deep archival traces. Between 1870 and 1910, the Seine topped 6 meters in 1876 and 1883. Floodwaters reached 6.69 meters on 18 March 1876. In some places the river swelled to 5–6 kilometers wide, leaving people without work, food, or shelter. *Le temps* visited the hardest-hit suburbs, especially working-class Alfortville, and reported that "the spectacle is much more terrible" and "the number of flooded houses is incalculable." Reporters also described the disaster with the words *effroyable* (frightful), *terrible*, *spectacle*, *monotonie*, *immense*, and *lugubre*. In Paris, omnibus service slowed on the flooded Quai de Passy, the Palais Bourbon's flooded basement print shop closed, and flooding reached the Esplanade des Invalides and the warehouses at Bercy, which stored raw materials and foodstuffs, especially wine. *Le temps* demanded "a radical reform in the regime of the Seine with the aid of constructions worthy of a people truly master of its destiny." Despite the "gigantic" works required, some preventative engineering would cost less than rebuilding after another flood. Meanwhile, the popular encyclopedia *Le magasin pittoresque* offered tips for drying and repairing humid walls in flooded homes.[47] Artists and writers also tracked flooding. Émile Zola's fast-paced tragic short story "The Flood" (*L'inondation*) captured the drama and violence of the 1875 Garonne flooding, which killed more than five hundred people. The story's narrator-protagonist, an aged farmer from the riverbank near Toulouse, a father of three and great-grandfather of two, is left alone when floodwaters swallow his family, house, and farm.[48] Seine floods including those in 1875–1876 and 1882–1883 were fond subjects for the Impressionist painters Claude Monet, Camille Pissarro, and Alfred Sisley, who were fascinated by light on water, fleeting sensory "impressions," and suburban landscapes outside Paris.[49]

In the winter of 1882–1883, the Seine flooded twice, in December 1882 and January 1883, cresting at 6.24 meters. These floods damaged little life or property, but suspended navigation for thirty-seven days.[50] Ponts et Chaussées engineers found that these floods followed respective heavy rains, which offered a chance to test Belgrand's 1854 flood warning system and his influential 1876 flood study. Ideally, they hoped rain and flood reports could provide four to five days' warning. They concluded that flood prediction was difficult, but promising, and they dreamed that rigorous empiricism and copious data could crack the river's code.[51]

No flood of the era outdid the Great Flood (*Grand Crue*) of January 1910, Paris's worst inundation since 1658. Swelled by heavy autumn rains across the Seine Basin, the river surpassed 8 meters deep. Despite newspapers reporting human and animal corpses floating on floodwater, people trapped in buildings, food shortages, and familiar typhoid fears, human casualties were luckily light. The flood's biggest impact was damage to infrastructures worth hundreds of millions of francs, which brought the bustling capital to a near standstill for weeks and compromised it for months. The government flood commission of influential Parisian scientists, engineers, and officials identified five main types of affected infrastructure: (1) transportation networks: roads, railways, tramways, and the Métro; (2) sanitation networks: sewers, trash collection, and street cleaning; (3) communication networks: telegraph and telephone; (4) energy networks: electricity, gas, and compressed air; and (5) the built environment: roads, bridges, quays, and houses. A chain of concerns follows in the commission's report: flood prevention, Paris's environmental impact on suburbs downstream, resource shortages, public health and sanitation, unemployment, homelessness, and politicized infrastructure controversies. Faced with these many concerns, commissioners still stressed their optimism and the city's resilience, resolutely defending the modern infrastructural ideal.[52]

Floodwaters reached Paris on 22 January. East Paris Tramway employees closed the network preemptively before their Ivry power plant flooded. Trams required working streets and electric systems, but water quickly overtook both infrastructures. *Le petit Parisien* tracked a bourgeois commuter forced to transfer from the closed Gare d'Orsay to the Gare d'Austerlitz across the Left Bank. Normally, he would transfer by tram, which was closed due to flooding, forcing him into a coach: "Coach! To the Gare d'Austerlitz, and at a gallop! What a story! I'm gonna miss my train, I am! Damn flood! Damn weather! Hurry! Coach!" Over the next week, tramway companies maintained service by rolling older horse and steam trams from storage onto the rails.[53]

Failing electric tramways signaled problems deeper underground. After Haussmann, engineers used sewers to bundle and hide infrastructures. Sewer mains doubled as service corridors where the city's technical components were back-staged from the monumental streetscape. The science popularizer Camille Flammarion called Paris a "vast molehill" and "anthill network." He argued that "all of Paris rests today on a worm-eaten base." One angry handbill that circulated after the flood stated, "The city of Paris brought the flood to you with its sewers."[54] As the Franco-American writer Helen Davenport Gibbons noted, "The very skill that added to the material comfort and well-being of the city made Paris vulnerable when the unexpected and unprecedented happened."[55] Water not only infiltrated the city laterally, via the riverbed, but also climbed deep, layered structures of basements, cesspits, catacombs, groundwater, quarries, sewers, and subways. Flooding threatened buried infrastructures for light, heat, water, power, and sanitation. Even infrastructures buried outside the sewers for safety—gas, electricity, telegraph, and telephone—flooded. Compressed air, for example, moved messages in Jean-Baptiste Berlier's pneumatic network and powered machinery. On 22 January, synchronized clocks across the city stopped between 10:48 p.m. and 10:53 p.m. when the Victor Popp Company's Quai de la Gare pneumatic plant flooded. Standard, human time stopped. Floodwaters inverted the humanized environment, revealing the sociotechnical, envirotechnical, and

Figure 8.2. Neurdein brothers (ND) Postcard no. 15, postmarked 1910, with the caption "PARIS—The Great Flood of the Seine (January 1910). Collapse of a sewer on the Boulevard Haussmann." Author's collection.

infrastructure-dependent character of everyday life and the human decisions that increased flood impacts.[56]

Underground Paris hosted heavy construction during the rapid transportation growth of 1898–1914 (chapter 5). Two subway tunnels, the Nord-Sud and Métro Line 4, were underway when flooding hit. On January 22, floodwaters startled workers who were building the access stairway for the Nord-Sud station at the Place du Palais Bourbon. A major Rue de l'Université sewer line feeding the Bièvre main sewer (*collecteur*) crossed the space planned for the stairway. Water rose high enough to reverse the pipe's flow, pushing water away from the river beneath the Left Bank. Soon flooding overtook the worksite, wrecking scaffolds, carrying away materials, and collapsing the roadway above. Workers stood agape at the subterranean torrents roaring between the quays of the unfinished station, carrying away wood, brick, stone, and tools. The next day, the Boulevard Saint-Germain above also collapsed (figs. 8.2 and 8.3).[57]

Underground flooding found the Métro that same day. Lines 1 and 6 were shut down when the Quai de la Rapée power plant flooded. Power plants were often built on riverbanks or low ground, so, like underground networks, they

Figure 8.3. Unbranded, unused postcard, series "Flooding—Paris, January 1910," no. 445 with the caption "Corner of Boulevard St-Germain and the Rue de Lille." Author's collection.

flooded quickly. The same fate awaited all major train stations for the national networks. The Gares d'Austerlitz, de Lyon, des Invalides, and d'Orsay all stood on or near the Seine's banks, and those that did not—the Gares de l'Est, du Nord, and Saint-Lazare (de l'Ouest)—had multiple connections to Paris's flooding underground. For example, the Nord-Sud subway flooded the Gare Saint-Lazare. Beneath that station, the Nord-Sud tunnel passed under Métro Line 3. By January 24, the Nord-Sud tunnel was full, and water had begun infiltrating the Métro. The engineers Bienvenüe and Hétier reported that workers rushed to erect barriers to protect the Opéra, via the recently completed triple-decker Métro station there. The Opéra station sat above the underground "lake," an ancient arm of the Seine fictionalized in Gaston Leroux's *Phantom of the Opera*, which was quick to flood. Other flooded Métro tunnels included Line 4's Rennes station, near the Rue de l'Université breach. In all, Hétier and Bienvenüe counted flooding in 19.4 kilometers, almost one-third of the Métro's 63 kilometers. These engineers concluded that interrupted Métro service "has been without a doubt one of the most obvious facts for the immense majority of the Parisian public, regarding the trouble, as profound as it was unexpected, that it brought to the relations of everyday life." Even

Figure 8.4. Paris as Venice. Unbranded, unused postcard with the caption "Souvenir of the Flooding of January 1910. Paris. The Rue Jacob." Author's collection.

for engineers, the degree to which modern everyday life now depended on such networked infrastructures was "profound" and "unexpected." The flood inverted infrastructures for engineers and the public alike.[58]

Of the city's four main garbage-processing plants, only the Romainville plant was elevated enough to escape flooding. Waste backed up quickly and by January 23 the prefects of the Seine and of the police agreed that all excess waste should go into the Seine. Contemporary postcards and illustrations show men lining the Auteuil Viaduct shoveling waste from carts into the torrent. This only aggravated the continuing conflict between Paris and its suburbs over river pollution. In Paris, by contrast, freshwater lines ran in the sewers. As they flooded, the freshwater supply slowly mixed with sewage, river water, and storm water. *Le petit Parisien* ran the Water Service's familiar boil notice the same day in this rare winter water outage.[59]

Infrastructural breakdown disrupted everyday life. Flooding put many people out of work and home and scrambled the scripts that Parisians used to navigate time and space. Urban metabolism, absorbing resources and evacuating waste, slowed. As infrastructures failed, normal routines and rhythms dissolved, the cityscape transformed surreally, and Parisians were forced to develop hacks and workarounds, new ways to inhabit familiar urban spaces, now estranged. Many remarked that Paris had become Venice (fig. 8.4). Streets became canals, second-story windows became doors, and out-of-work men and women became volunteers, roaming the sodden streets to help others. Sandbags, rowboats, and lumber were suddenly in high demand. Without modernity's comforts, conveniences, and confidence, Parisians turned to older technologies: rowboats replaced railways, candles replaced natural gas, and animal traction replaced electricity. As they improvised new ways of living, Parisians watched the edifice of urban modernity—normally so attractive and imposing, even hegemonic, and offering benefits like convenience, efficiency, and productivity—dissolve, leaving behind an uncanny other of everyday life, and turning the intimate spaces of home into an obstacle course. Signs of upheaval abounded. Driven from underground, thousands of rats roamed the streets. Infrastructural inversion was so fascinating that Parisians actively sought new ways to experience and remember the flood. Crowds of Parisians, awed by the roaring river, gathered on bridges to watch. The Eiffel Tower reported unusual sales on 22 January, selling over one thousand tickets to people excited to survey the flooded city from above. Similarly, Helen Gibbons climbed the tower of Nôtre-Dame. Finally, ingenious photographers walked the city taking pictures, building an enormous market for mementos like their postcards and photo albums, now among historians' best sources for reconstructing this historical moment. The disaster-stricken city, like accidents and construction sites, was a curious spectacle, a temporary heterotopia at metropolitan scale.[60]

Extraordinary circumstances made engineers of Parisians and challenged the city's engineers. The disaster demanded ingenuity, embodied by the *passerelles*, ad hoc wooden walkways perched on sawhorses and scaffolds over flooded streets (figs. 8.5 and 8.6). Several contemporary postcards document improvised rafts (e.g., fig. 8.7). As Parisians tried to keep dry, engineers, soldiers, workers, and volunteers scrambled to erect barriers, pump water from buildings, and make quick decisions. The press questioned the authorities' flood management. *Le petit Parisien* decried "the powerlessness of the engineers" (*l'impuissance des ingénieurs*) to control nature.[61] Writing for *Le soleil*, the journalist Henry De Larègle argued, "the functionaries of the State and of the city of Paris are in part responsible, not for the flooding itself, but for its disastrous consequences." He offered several critiques, for example, including that Left Bank train stations hugged the riverbank, or that the Nord-Sud company should cover damages caused by its tunnels. Rumors circulated that the sewers

Figure 8.5. Unused Neurdein brothers (ND) postcard no. 117, with the caption "PARIS—THE GREAT FLOOD OF THE SEINE (January 1910). Ways on sawhorses constructed in the low-lying Rue du Bac to assure the circulation and the supply of inhabitants." Author's collection. The man in the top hat second from the left is a photographer with his camera hanging around his neck.

Figure 8.6. Unused Roussel postcard, with the caption "The Flood of January 1910—Paris—Rue de l'Hôtel Colbert," which depicts elaborate passerelles. Author's collection.

Figure 8.7. Unused J. H. postcard, with the caption "The Flood of the Seine (January 1910). An improvised raft in the garden of the Rue Van Loo." Author's collection.

had "burst." Although the engineer Colmet-Daâge from the flood commission argued that they had not burst, but merely reached maximum capacity, De Larègle thought that officials had shirked responsibility by claiming that the city's pipes had not failed; instead, it was the privately owned pipes connecting houses to the sewers.[62] Questions of "responsibility" resounded.

The conservative firebrand Ernest Judet wrote: "Everyone is beginning to understand that the current catastrophes are not accidents without cause but are the screaming proof that the innumerable *travaux* in which Paris is covered have been engaged at random, thanks to the engineers and entrepreneurs, whose ideas and plans have never been subjected to the appropriate critique of the collective interests of the city. Both groups shot blind, running after solutions and progress whose charm their imaginations have increased tenfold, and which the public ignores, but enchanted by certain seductive innovations, it has unfortunately become love." Judet's virulent criticism of contractors (*entrepreneurs*), engineers, and politicians cut through the modern infrastructural ideal and accused them of being seduced and enchanted by technology, more magic than science. He built on press dialogue about delinquent contractors and the cityscape disfigured by construction. The city that flooded in 1910 was also the city under works (chapter 5) and the risky city whose normal accidents could scale up into catastrophes (chapter 6).[63] Judet expanded this critique during 1911's September stink (chapter 7).

The *Le matin* writer Henri de Jouvenal generalized the criticism: "One says to have faith in science: one learns that it contains goodness, morality, and peace. We believed what was said. But today everyone is asking the same question: How has this science, so sure of itself, been defeated by the primitive waters? How was it not capable of protecting the most beautiful city against the capricious river?" For him, the flood put local authorities and science itself on trial. He contrasted science and modernity with the "primitive waters," calling the flood "the 1870 of the engineers," a humiliating defeat like Prussia's victory over France.[64] Activating progress counternarratives circulating since the 1870s, the press called out Office of Works engineers, defending citizens against the Haussmannesque failures of his republican successors.

By upending central Paris, the flood undermined the modern infrastructural ideal promoted by engineers since the 1850s. It showed that Paris's communication, energy, timekeeping, transportation, and waste disposal infrastructures were fragile. Everyday life had come to depend on these modern icons, and on the Seine's cooperation and control. The flood revealed their city's awesome complexity and hybridity, as it inverted infrastructures, bringing their technical components, commonly hidden for aesthetics and safety, back into view. Social routines depended on tightly coupled systems that bound humans, their tools, and their environments in modern, mediated, hybrid ways, bringing both resources and risks.

Disaster research consistently shows that many "natural" disasters are truly hybrid, caused by both human and nonhuman agents.[65] The 1910 flood's impacts resulted from both the physical behavior of rains and rivers and the previous half-century of Haussmannization. The urban fabric's hybridity and fragility inspired Parisians to use the authorities' flood management difficulties as levers for critiquing modernity, science, technology, and government. The flood taught that what engineers and the public recognized as Haussmannized urban modernity was fragile, complex, and contingent. As society depended on technology, so technology depended on nature. Stress anywhere in this hybrid triad could produce failure in other parts. Flooding inverted the simple, optimistic, and teleological narratives of technological progress familiar from forums like the universal expositions. Since the summer of 1900, these were repeatedly called into question, by water and housing shortages, typhoid outbreaks, tramway and Métro accidents, and delayed public works—hard realities of a city living with fallible infrastructures, lasting inequalities, and intensive works. The Great Flood, a colossal cascading failure in which the river inverted the city, was a lesson in how "profound" and "unexpected" the entanglement of society, infrastructure, and the environment had become.

Burying the Bièvre

The Bièvre River begins near Versailles. Historically, it entered south Paris near today's Porte de Bicêtre. Crossing the 13th district, passing Place d'Italie and then the 5th district, meeting the Seine at the Pont d'Austerlitz, it split into what one observer called "two or more" branches—two major branches, plus several smaller diversion canals (*biefs*).[66] The east branch, a medieval diversion, was the "living" *Bièvre vive* or *bras vif,* while the west branch or original riverbed was the "dead" *Bièvre morte* or *bras mort.*[67] For the Bièvre, this common idiom reflected deeper questions about the place of life and death in modern cities. Environmental histories and geographies often answer these questions by examining "urban nature," or the complex role of biological life in the human-built environment.[68] After the 1870s, the once living Bièvre slowly disappeared, as the Office of Works finally covered the river, achieving its "death" in 1912.

Historians portray the Bièvre as a long-standing industrial lifeline, an example of natural infrastructure (figs. 8.8 and 8.9).[69] From medieval mills to the early modern proto-industry of the Gobelins textile works and the Mobilier National, the river powered machines and helped produce beer, dye, flour, glue, leather, paint, paper, starch, and textiles. Its valley was home to proud artisans, grinding poverty, and heavy pollution, which often stoked bourgeois fears of ragpickers, rats, and disease. Officials long monitored the river's public health dangers. The famed hygienist Alexandre Parent-Duchâtelet began his career with an 1821 report on ways to "ameliorate" the river, which was updated by departmental officials in 1847.[70] These reports treated the river as a living thing worth cultivating. Its industries revealed the political ecology, and what the ecocritic Nicole Shukin calls "rendering life," behind Paris's cafés, hotels, and boutiques. Along its banks, workers turned animals and plants into the white linens and soft hides that upheld Paris's status as a capital of dining, hospitality, and fashion. These reports also show that the river was already struggling to support these industries; its "amelioration" was economically and hygienically urgent. Ponts et Chaussées engineers planned to channelize the river and build an exurban reservoir to steady its flow. The Prefecture bought and demolished four historic mills, de facto barrages that slowed the river. Instead, new rationalized barrages would control flow and ease cleaning of the riverbed.[71]

Figure 8.8. Unused J. H. postcard no. 550, "Old Paris" (Vieux Paris) series, with the caption "A picturesque corner of the Bièvre," ca. early 1900s. Author's collection.

Figure 8.9. BC Postcard no. 113, postmarked 1905, "Old Paris" (Vieux Paris) series, with the caption "Gobelins—View of the Bièvre in the passage Moret." Author's collection.

Haussmann's Office of Works began the river's deeper envirotechnical transformation. In a major 1868 milestone, workers diverted the Bièvre's output into the Left Bank main sewer (*collecteur*), removing it as a tributary and polluter of the Seine. Maxime du Camp wrote that the "muddy stream" with its "multicolored and nauseating liquid," realized its destiny as a sewer. Sewerizing the Bièvre advanced three Haussmannian projects: controlling and domesticating urban nature, removing industry from the city center, and clearing slums. Seeking a functional division of space, today's "zoning," the Office of Works often cleared similar early industrial, mixed-use neighborhoods where slums, factories, and workshops huddled tightly. The Bièvre survived until it fell victim to its own success.[72]

By the time of the early Republic, the river's industries had concentrated. Serving on an 1875 police commission, the doctor and councilman Antoine-Baudoin Poggiale assessed its banks, from the fortifications to the exterior boulevards, observing: a glue factory; "a great number" of tanneries; several leather finishing shops; laundries specializing in upholstery, linens, and bleaching; and plants making parchment, paper, and cloth. Inside the boulevards were more makers of hides, dyes, paints, and starch or flour, along with brewers, butchers, slaughterhouses, and the nearby horse market. Finally, many small, local gutters and sewers emptied into the river. The resulting mass of animal and chemical pollution corrupted water quality. Poggiale reported,

> The bed of the Bièvre, normally sandy upstream, becomes filled with a thick, black sludge [*vase*] at Antony, Cachan, Arcueil, Gentilly, and Paris. The bottom and sides of the river are permeable; its course is slow, the volume of water is insufficient; the dejections and detritus carried here are not carried away. The water is charged with a considerable quantity of organic matter in dissolution or suspension; it gives off bubbles of fetid gas, composed of hydrogen sulfide, ammonia, ammonium sulfide, carbonated hydrogen, and nitrogen. At certain points during summer heat, this off-gassing is so abundant that the water seems to be boiling. White scum/foam [*écumes*] and black grime [*crasses*] float on the surface. From Cachan to the Seine, the emanations are intolerable; iron, lead, copper, silver, and lead-based paint take on a leaded color and build up a coating of metallic sulfur. Bleaching linens becomes almost impossible, and neighbors complain of often being sick. . . . The sludge contains a considerable quantity of animal debris; it dries out during summer heat, cracks [*se fendille*], and

Figure 8.10. Unused J. H. Postcard no. 542, "Old Paris" (Vieux Paris) series, with the caption "A picturesque corner of the Bièvre," ca. early 1900s. Author's collection.

> lets off unhealthy fumes. One can thus say without exaggeration that the Bièvre is no longer anything more than a dangerous and foul sewer.[73]

Poggiale's recommendation, as Haussmann and Du Camp had suggested, was to cover the river, realizing its destiny. A 1907 municipal report called it *fossé*: cesspit, ditch, or grave. Literary sources are equally lurid. In 1854, Alfred Delvau wrote, "a water gooey, black, red, impossible, I know . . . it's an open sewer." In 1898, Georges Cain called it "hideous to see. Tinted in all tones: yellow, green, red, it carries unclean detritus and rolls at our feet, gluey, as if immobile, without reflections under the grey sky, nauseating, the color of congealed blood." Despite Poggiale's "without exaggeration," such accounts *were* exaggerated, because their rhetoric was meant to provoke disgust, shock, and outrage. Yet even black-and-white photographs confirm that such texts fairly accurately describe water quality (figs. 8.10 and 8.11). Authors struggled for language to describe it, even faced with rather concrete concerns. In 1901, a Dr. Farabeuf warned the National Academy of Medicine that mosquitoes from the Bièvre had spread unfamiliar skin lesions that lasted weeks, though no other source confirms his claim.[74]

Compounding pollution was the problem of flooding. Major storms often caused floods and accidents. The Left Bank foot of the Pont de l'Archevêché

Figure 8.11. Unused J. H. Postcard no. 544, "Old Paris" (Vieux Paris) series, with the caption "A picturesque corner of the Bièvre," ca. early 1900s. Author's collection.

buckled under floodwater in 1885, killing two. An 1897 humor piece, drenched in irony, claimed that the Bièvre's thick, gluey water precluded flooding, asking, "Can a jar of jelly transform into a cascade?" But the same publication soberly admitted there was flooding when water disrupted construction of the Gare d'Orléans for the 1900 exposition. A May 1901 storm partially collapsed buildings and roadways. Soaking in polluted water nearby industry was damaged, forcing a few hundred people from work. Locals often blamed the Bièvre's channelization—and thus Haussmannizing bureaucrats and engineers—for worsening flooding, as the historian Jeffrey Jackson documented during the 1910 flood, but the Office of Works archive holds dozens more cases.[75]

Together, pollution, flooding, and health concerns drove the Office of Works to cover the river, a controversial choice. Engineers began studying which neighbors supported covering in 1874, and resident groups signed petitions against covering in 1882 and in favor of it in 1883.[76] Property owners who feared filth and flooding supported covering, but many local workers and entrepreneurs resisted because the river still supported industry, work, and life. Petitions in 1894 and 1895 complained of pollution that stank and made the water less useful for industry.[77] If the river's pollution was a tragedy of the commons, even the altered waters were shared infrastructure worth defending. By 1899, of 8 kilometers within Paris, only 1.36 kilometers were open-air, spread among 13 *biefs* (sections of canal) whose residents resisted its covering. Government sources reveal a heated battle for the Bièvre. Resistance, thickest near the Gobelins, inspired the Prefecture of the Seine's 1901 Special Commission to obtain settlements (*traités amiables*) from locals. By 1907, two-thirds had settled, leaving 28 residents who refused to sign over 300 meters of open water.[78] National government intervention in 1909 and the Great Flood in 1910 pushed covering toward completion in 1912.[79] The Office of Works at last buried the river, physically transforming it from open to enclosed, from alive to dead.

During this battle for the Bièvre, discourse blossomed around the river's meaning. John Culbert and Leonard Koos have each charted the river's role in Paris's urban imaginary.[80] Cast as the other or antithesis of modern urban industrial civilization, the Bièvre was identified with the wild, savage, natural, subaltern, female, premodern, and provincial. It was a classic heterotopia, a pastoral *rus in urbe* or garden. Some called it a paradise lost to modernization. It was also mourned as an environmental concern, often the "pauvre Bièvre," a figure of pity and loss reflecting the harm done by modernization to environments, to the poor, and to women. Both wilderness and wasteland, it was rife with haunting rot and fantastic projection.[81]

Joris-Karl Huysmans's famous 1886 essay personified the Bièvre as a familiar nineteenth-century Parisian stereotype: the poor, young provincial woman seeking work in Paris (like Zola's Denise), who is corrupted and exploited by city men, and often ends up with an unwanted pregnancy or in prostitution. He called the river a classical water nymph, evoking "all by itself an entire antique province." The river also chronologically inverted Haussmannization; here, the old city peeked out between "modern avenues." The old city was home to declining artisans, whose poverty, he wrote, was "ennobled by the stamp of ancient times," like Rembrandt's subjects. Here, "the admirable Paris of yesterday is reborn."[82] Similar nostalgia for the "old Paris" also drew the photographer Eugène Atget and the Carnavalet Museum curator Georges Cain to the Bièvre. But when Huysmans depicted the more industrial *Bièvre morte*, he evoked exotic places like Venice, the Orient, and Florida. Here, his imagery grew graphic, both sexualized and pathologized in a familiar sexist way, as both Culbert and Koos have noted. *Le monde illustré* similarly called the river "the Venice of putrefaction!"[83]

While the Bièvre's exotic qualities attracted Huysmans, the more conservative Georges Lenotre was repulsed by this "flux of ordure . . . debased,

Figure 8.12. Unbranded postcard (postmark illegible), "Old Paris" (Vieux Paris) series, no. 13, with the caption "Ruelle des Gobelins. The Bièvre going under the Boulevard Arago." Author's collection.

abject, muddied, prostituted." After reviewing writings about the Bièvre by Hugo and Huysmans, his essay ended, "Don't go to the Bièvre"; it was a "bad trip" (*mauvais tour*) that was better to read about: "It's very pretty . . . in a book." Adrien Mithouard, a city councilor from the 13th district, also built on Huysmans in his 1906 pamphlet *The Perdition of the Bièvre*, which exemplifies the biblical image of Eden lost. Mithouard argued that Huysmans never foresaw how "the engineers" would rend the river after 1900, creating a "whirlpool of death." It was "given over to the engineers," whom he scorned as "the natural enemies of waters and forests." For him, the river was "dirty and skinny," a "ballerina" contorted and rerouted by engineers who linked her with sewers and pumps. She was "covered" like the Styx, seen only by sewer workers. "It would be better to have never been born than to finish such a dismal life with a death against nature." Engineers were "like all those who amuse the masses professionally. They have always seen Paris as a giant circus where all that needs to be done is to make activities circular." Citing Eugène Hénard's influential roundabout study, he equated infrastructure—electricity, gas, sewers, railways, automobiles, and horsecars—with "a vast spinning drive" (*entrainement giratoire*), and "this stupid and perfectly useless orbit around Nôtre-Dame."[84] Mithouard saw the Bièvre's death as unjustly politicized

Figure 8.13. J. L. & Co. Postcard (postmark illegible), "Historic Paris" series, no. 67, with the caption "La Bièvre at the Gobelins." Author's collection.

and technologized, and thereby unnatural, undermining narratives of progress.

Memory and mourning shrouded the Bièvre long before its burial. In 1903, the popular encyclopedia *Le magasin pittoresque* called it "dying." In 1907, the travel editor Karl Baedeker called the Gobelins the last place to see the river, "now almost completely covered over." In 1909, the American travel writer Edward Lucas quit seeking the "lost river"; he glimpsed it between buildings along the Boulevard Arago but gave up.[85] E. A. Taylor interestingly claimed that the Bièvre was still being covered in 1915, three years after the work had ended. The Ruelle des Gobelins, among the last open *biefs*, became a landmark in both French and anglophone travel writing around 1900, and it appeared on many postcards (figs. 8.12, 8.13, and 8.14). It became an iconic *lieu de mémoire*, an elusive niche for flâneurs. Photographers, including Atget and postcard producers, were common enough there that Georges Cain met two scruffy little girls who shouted, "Shoot [us] for postcards!"[86] Roughly half of these postcards depict children. Visual and textual sources reveal a rich field of memory work around the dying Bièvre, a vestige of the un-Haussmannized urban body awaiting amputation, but until then drawing flâneurs, nostalgia, and photographs like Montmartre's Maquis (chapter 2).

Figure 8.14. Unused C. M. Postcard no. 1425, with the caption "The Old Paris (13th dist.)—The banks of the Bièvre," ca. early 1900s. Author's collection.

Thus, the river's end elicited strong feelings of disgust, fascination, nostalgia, and pity. Both Culbert and Koos noted the river's strong, ambivalent attractive-repulsive force. Its waters were saturated with pollution and symbolically overloaded with meaning and memory. Parisians tapped the European discourse of disgust, articulated in recent cultural and intellectual histories, and applied it to the Bièvre to express the experience of industrialization, Haussmannization, and human-made ecological change.[87] In his decadent phase, Huysmans expressed both disgust and desire, but for most the river was a paragon of filth, "abject," "cloaca," or "grave."

Across the twentieth century, the Bièvre, erased from the cityscape, survived in the urban imaginary, local legend, and history books. For Parisians, debating the river's life and death raised bigger questions about how living nature fitted within the human-made city. Some mourned the Bièvre's end as a "death of nature" of the old city or of functioning riverine infrastructure. Others welcomed its death, favoring biopolitics and sanitation. While engineers worked to invert the river, its defenders worked to invert Haussmannization. The broader question provoked disagreement: whether the life teeming around the Bièvre was healthy. This dispute helps explain the river's slow "death," another plan for infrastructure engineering slowed by contentious politics and hybrid complexity. Burying the Bièvre removed some of Paris's most noxious industry, from laundries to tanneries. The Bièvre's end, like the Seine's sanitation and flooding, demonstrates how the Office of Works struggled to manage hybrid rivers shaped by both human agency (resistance) and nonhuman agency (flooding), in ways that challenged the modern infrastructural ideal and its valorization of the technological mastery of "nature," Georges Bechmann's "artificial life," and republican "moral and material improvement."

As infrastructures, rivers could practically make or break Paris's urban metabolism and symbolically make or break its urban modernity. They could be friend or foe, bringing resources and risks, vitality and vulnerability. As Helen Gibbons wrote of the 1910 flood, "The best friend of Paris had turned against her."[88] Primary and secondary Haussmannization, from the 1850s through the 1910s, marked a new era of envirotechnical transformation. In the republican period, while the polluted Seine needed maintenance lest it become a sewer, the overworked Bièvre "died" and was buried in a sewer. But no matter how envirotechnical or hybrid they became, rivers were never fully human-controlled. While the modern infrastructural ideal promoted the technological control of nature, and the Office of Works inverted the controversial sewers through public sewer tours, lay Parisians inverted infrastructure to stress the fragility and "powerlessness" of engineering. Riverine infrastructures enabled drainage, dumping, fishing, laundries, shipping, swimming, animal rendering, and more. But floods could transform Paris into Venice, unleash cascading infrastructure failures, or bring death, destruction, and disaster instead of resources and services. Floods illustrate the pattern that, when nonhuman agents inverted infrastructure, lay Parisians reacted by inverting Haussmannization. While engineers protected the modern infrastructural ideal by naturalizing disasters, laypeople politicized disasters by blaming engineers, defending un-Haussmannized places as heterotopias, and telling progress counternarratives. Because Paris could not live without the resources of rivers, the question has always been precisely how to live with their risks. Such divergent potentials characterize all infrastructures but are made particularly clear by Paris's rivers.

Conclusion

Inverting Haussmannization

An unbranded, undated, and unused postcard from a series called "Paris—Les Trous" (Paris—The Holes) is easily datable as a depiction of the sinkholes opened by the storm of 15 June 1914, with which this volume opened (fig. 9.1). The series title applied the language of "holes," familiar from popular debates about construction sites and public works, to storm damage, recalling previous comparisons of construction to storms. The caption sets the scene socially: "After a storm—Excavation of the Place St-Philipe-du-Roule—Workers searching for cadavers at the corner of the Rue de la Boëtie." The photograph documents the storm's inversion of underground infrastructure, depicting a large sinkhole filled with workers seeking human victims, as workers, police, and passersby gather on the rim. The image foregrounds infrastructure, the workers who maintain it, its vulnerability to water and weather, its risks for everyday users, and the public's interest in experiencing and inverting infrastructure. Like many of my sources, this postcard inverts infrastructure while documenting how Parisians inverted infrastructure. By bringing together infrastructure, weather, workers, and risk, it makes a tidy bookend for this history of the sociotechnical and envirotechnical systems that Parisians learned to live with from 1870 to 1914.

This book has documented the biological, social, and affective dimensions of living with infrastructure. It has shown how Parisians tried to invert

Figure 9.1. Undated postcard (1914) from the series "Paris—The Holes," with the caption: "After a storm—Excavation of the Place St-Philipe-du-Roule—Workers searching for cadavers at the corner of the Rue de la Boëtie." Author's collection.

Haussmannization in various ways, following recent methodological calls from infrastructure scholars to recognize "different kinds of inversion."[1] Already during primary Haussmannization, Parisians from the poet Charles Baudelaire to the revolutionary Communards of 1871 sought to invert Haussmannization.[2] But the task of inverting Haussmannization became doubly important after 1871, as popular inversion accelerated while elites under the new republican regime attempted their own inversion of Haussmannization from within the halls of power and planning. Secondary Haussmannization after 1871 inspired multiple and competing visions of how to invert Haussmannization.

The first, top-down, inversion was practiced by the Office of Works. As republican officials attempted to make the process of Haussmannization more egalitarian, more sanitary, more utilitarian, and more universal, politicians and engineers claimed that expanding and upgrading the city's roads, railways, reservoirs, and rivers would bring "moral and material improvement" for all. Haussmannization continued, and the debates continued, long after Haussmann left office. The republic continued to practice Haussmannization despite its controversy, hoping to reduce that disagreement with the tangible benefits of infrastructure: comfort, health, mobility, and security. Major strides were made: the provision of water, mobility, and sewerage expanded while cholera receded.

The municipal Métro was a notable success, and a real departure from the departmental government's traditional dominance of infrastructure provision. The public flocked to it, confirmed by ballooning ticket sales as well as foreign observers from the *London Times* to Franz Kafka, who visited in 1911 and memorably called the subway "a frail and hopeful stranger's best chance to think that he has quickly and correctly, at the first attempt, penetrated the essence of Paris."[3] When the 1910 Great Flood hobbled the subway, the engineers Bienvenüe and Hétier found that Parisian dependence on the railway was both "profound" and "unexpected." Unlike infrastructures installed by the Office of Works, the Métro did not pit Parisians against the Prefecture of the Seine. Eugène Poubelle called it "un Métro à tutoyer," a familiar infrastructure that Parisians could call their own (or literally, greet with the personal informal mode of address). The historian Alain Cottereau claimed the Métro as a victory for local control or democracy, and the literary scholar David Pike noted the "populist thrust" of Métro representations.[4] Further, as a platform for creating construction and transportation jobs with generous wages and benefits, the Métro was a victory for local welfare provision. The Métro provides good evidence, therefore, that the municipality succeeded more than the department did in inverting Haussmannization.

But there were notable limits that dragged long-standing infrastructure problems into the 1920s. Water and sanitation problems persisted, evinced by the Great Stink of summer 1926, and a lingering 18 percent of houses without all-to-sewer drainage in 1928.[5] Although it is typical for infrastructures to develop on long, slow timescales comprising planning time, building time, and the long operating life of systems, it is clear from the battles over tramways, sewers, the Métro, and the Bièvre that many Parisians believed that infrastructure development was moving too slowly.[6] Infrastructure was laden with so many dreams, expectations, and promises that both providers and users described feelings of impatience ("waiting"), even shame. The slowness of infrastructure development sometimes seemed a grind; those more concerned about the negative impacts of infrastructure sometimes despaired that Haussmannization might never end.

These various problems of living with infrastructure pushed Parisians toward inversion. If infrastructures worked perfectly and seamlessly everywhere, if their function was homogeneous, uniform, unified, and universal, there would be no need to shift our frame of vision, turning them inside out to reveal their secrets. It is precisely because infrastructures do not live up to the modern infrastructural ideal, because they do not fulfill their promises, that people seek to invert them. Ultimately, therefore, inversion aims to improve the workings of infrastructure in society.

The second type of inversion documented in this book was bottom-up, attempted by the users of infrastructure rather than its providers. Who makes technological history? As historians of technology have turned to studies of users, appropriation, and vernacular innovation, they have shown that technological history is not only made by engineers or only by politicians, entrepreneurs, or workers who build, maintain, and operate technologies. This book has documented a broad range of actors—cartoonists, journalists, photographers, anarchists, bohemians, a baker, a barkeep, and many more—who sought to intervene in infrastructural development. Importantly, this book has also presented a range of civil associations working to steer infrastructure development in social blocs on the terrain of civil society, from civil engineering societies to landlord groups, tenant groups, reform organizations, neighborhood organizations, and historic preservation advocates. Suburban groups, decentralists, Orleanists, and the Academy of Medicine also participated. Therefore, because Paris infrastructure users were diverse, bottom-up inversions of Haussmannization took varied forms, although all such inversions suggested a strong sense of laypeople's right to the city.

Bourgeois critics of Haussmannization often began with aesthetic issues, attachment to the beauty of the cityscape, and suggested that public works—in the form of railways, trolley wires, or construction sites—made the city uglier. They argued that interrupting sight lines and traffic flows raised cultural issues, troubling traditional pedestrian practices of flânerie and the promenade, and making it harder to visit, view, and revere monumental landmarks. Similarly, shopkeepers claimed that Haussmannization, ironically intended to improve traffic flows, made it harder for consumers to see or visit the shops. While officials intended infrastructures as tools for civilizing, users raised various concerns about threats to civilization from stinking cesspits, water rationing, trolley wires, construction sites, and tramways fit for "barbarians." The shrillest bourgeois critics suggested that modernized infrastructures would threaten French culture with unwanted American, Asian, or otherwise foreign influence. Overall, bourgeois inversions focused on cultural issues and were more nationalist and negative than those of other groups.

While the preceding inversions raised more symbolic concerns, a range of more material issues emerged from infrastructure's entanglement with the workings of society and the maintenance of biological life. A range of actors raised health and safety issues. Cyclists worried about tramway safety. Construction workers worried about jobsite safety. Cartoonists worried about electrical safety. Tenants tackled threats from toilets to tuberculosis. Journalists were particularly hungry for stories about dangers of all kinds, which they tarred with infrastructure epithets such as "intersection of the crushed," "murderous tramways," the "ravages of electricity," and "puree of microbes," and richly documented in the republic's booming periodical press. Common concerns about disease, pollution, sanitation, and ventilation indicate that health and safety concerns paralleled issues of environmental quality. Environmental quality was a widespread concern, driving debates about trolley wires and motorization, water pollution and "the odors of Paris," construction sites, subway ventilation, and "unclean lodgings." The French conception of organisms embedded in their milieu or environment suggested strong ties between environmental quality and the health and safety outcomes of individuals.[7]

Infrastructure users also inverted Haussmannization to raise social justice issues. In words and images, they documented and publicized heterotopic places where infrastructure was lacking, whether shoddy furnished apartments, peripheral districts, or the parched Zone. Low-rent housing (chapter 2) and construction sites (chapter 5) inspired Parisians to publicize the filth, disarray, and danger that they witnessed inside these sites to people on the outside. Others defended areas in Paris's periphery that had not yet been Haussmannized, such as Montmartre's Maquis (chapter 2) and the Bièvre Valley (chapter 8). They inverted progress narratives by highlighting the uneven geography of infrastructure development, or even by arguing that parts of Paris would be better left untouched by Haussmannization.

The unequal and uneven provision of infrastructure was particularly problematic given the republic's democratic political culture. Therefore, when Parisians inverted infrastructure, they often did so on behalf of the public, the people, or the poor. The age of secondary Haussmannization raised important questions about the relationships between the public, the republic, public works, and the French legal term "public utility" (*utilité publique*). If infrastructures are the "material culture of liberal government," then they reflect the exclusions and limits of liberalism.[8] This book has thus drawn together the vexed republican universalism of French political culture, a towering topic in French political history, and the misleading infrastructure universalism of the modern infrastructural ideal. Because infrastructures were "public works" provided by the state, critics of the republic could leverage infrastructure issues as evidence of the regime's limits or failings. Once Haussmannization, both primary and secondary, durably coupled infrastructure to politics, technical limits and failings could reflect political limits and failings and vice versa. In other words, if the Third Republic was serious about equality and universalism, its infrastructures should reflect that. Many in Paris, both left and right of the republican regime, questioned whether it did. Critics often used the French political language of rights, whether referencing rights to quality housing, clean water, or unencumbered mobility. Similarly, contemporaries often used the term "responsibilities" in the aftermath of infrastructure failures, accidents, and disasters.

Living with infrastructure taught Parisians that tax-funded large envirotechnical systems were shared or public resources whose ability to meet the needs of a diverse population would depend on careful debate and negotiation, followed by careful management and organization. Official plans from Poubelle's "unity of piping" in the 1880s to Cherioux's "unity of worksites" in the 1900s were intended to address these issues but they never materialized. More durable public solutions to these problems would have to wait until the public housing and municipalization of Paris transit in the 1920s. Critics further to the left tended to invert Haussmannization in order to criticize the French system of concessions, or state contracts for private companies, and the corruption of contractors, politicians, and engineers in the capitalist economy. Leftists called out republican moral and material improvement for the limits of its democracy and equality. It is important to note that Paris's popular political culture also had its limits. Although many critics of Haussmannization were strong advocates of democracy, they were vulnerable to the antidemocratic undercurrents of antisemitism, imperialism, sexism, and knee-jerk populism that have long haunted French republican politics. For all these reasons, throughout the struggles to invert Haussmannization after 1870 that pitted elite providers against public users, both sides were vocal advocates of democracy and both sides carried antidemocratic baggage. Long-running discussions of infrastructure luxuries, necessities, and rights reveal the limits to the universalism of republican politics and the modern infrastructural ideal.

With these human struggles over infrastructure in the foreground, it is important to remember the nonhuman agency that also shapes physical infrastructures as envirotechnical systems, hybrids of nature and artifice. Parisians were often inspired to invert infrastructure, invert Haussmannization, or challenge progress narratives after nonhuman agents (electricity, horses, microbes, rivers, or weather) disrupted infrastructure's smooth functioning, waging an inversion of their own. Thus, the later chapters of this book showed how moments of technological stress, failure, or crisis, often provoked by infrastructure's embeddedness in the environment, invert infrastructures: accidents in chapter 6 and disasters in chapters 7 and 8. Storm damage, flooding, water shortages, typhoid outbreaks, river pollution, and short circuits all represent infrastructural problems caused by technology's embeddedness in the environment, and more specifically the role of water circulating among rain, reservoirs, and rivers. From Georges Bechmann's "artificial life" to making a stink about engineers' "powerlessness" to control nature, Parisians found many ways to express and reflect on the fact that the efficient function of infrastructure always depended on a fragile and incomplete mastery of the environment. It takes continuous work to maintain coherent large envirotechnical systems.

Draft horses that pulled vehicles including omnibuses and trams were components in envirotechnical systems as well as living, feeling bodies that had to live with infrastructure the way people did. The Diatto tramway accidents of 1900–1901, which electrocuted and killed dozens of horses, offer a particularly stark example of why we must have a more-than-human view of infrastructures. Although draft horses were chosen, deliberate parts of horse-drawn transportation infrastructures, their interaction with electric-powered motorized trams was accidental. Something similar applies to water: although water is intended to be part of water supply and sewer systems, it leaked into other systems, disrupting traffic, corroding equipment, causing short circuits, or conducting electricity. Electricity, in turn, sometimes infiltrated the gas network, causing explosions and fires. One consistent reason for inverting infrastructure, both during secondary Haussmannization and in our own time, is the accidental entrance of nonhuman forces or actors into systems designed to steer, exclude, or ignore them. These unintended and unpredictable interactions of physical infrastructures with their environmental context offer a fundamental explanation of the contingency, risk, and uncertainty that, then as now, characterize the human experience of infrastructure and inspire its inversion.

The third and final form of inversion discussed in this book is my own attempt to cast infrastructure in a new light. Building on recent infrastructure studies that place infrastructures durably in their social and environmental contexts, I gathered methodological tools that could reveal the inner workings of infrastructure and its everyday experience. I found these tools in histories of everyday life and technology, urban studies of street life, and histories of embodied, emotional, and sensory experience, as well as cognate research in anthropology, sociology, geography, and science and technology studies. My specific methods included close reading of the everyday language used by Parisians to describe their experience of infrastructure, and an analysis of visual sources including illustrations, caricatures, and photographic postcards. Visual sources not only let me see infrastructure in new ways, but also document and embody the culture of visualizing and inverting infrastructure among the Parisians who made these images. I can use postcards to invert infrastructure today because Parisians used them in the past to do the same.

One way for scholars to question the modern infrastructural ideal has been to shift geographically from northern or western cities to those in the Global South. Another way to do the same is to "provincialize" Paris, to show that even in the flagship city of Haussmannization, the process was contested and incomplete, with contingent and divergent outcomes. Urban modernity and urban infrastructure are more fraught and fragile than we usually believe—even in this capital of modernity, the City of Light. As David Harvey, Patrice Higonnet, and others have argued, Paris is a city especially prone to myth-making. Harvey has further analyzed how myths establish breaks between past and present. The republic's attempted inversion of Haussmannization can be read in this frame, as an attempt to make a founding republican myth, which distinguished secondary from primary Haussmannization, while maintaining the reputation of the City of Light. We can also read the modern infrastructural ideal as the wellspring of a series of myths about what physical infrastructure is and does: that it consists of complete and ubiquitous networks, or that it provides resources and services uniformly and universally. While these remain valuable as aspirations and ideals, to equate them with reality puts the cart before the horse. The dark corners of the City of Light offer concrete evidence of how far a city, even a model city, remained from its goals. Haussmann made Paris a networking city, not a networked one. In part, networking remained unfinished because all large envirotechnical systems require so much funding, labor, materials, and time. Thus, the process Haussmann started in 1853 reached, contested and unfinished, into the 1920s. But Haussmannization also faced resistance from disgruntled Parisians and recalcitrant nonhumans. The efficient development and function of infrastructures requires recognizing their embeddedness in society and the environment. By ignoring these truths, the modern infrastructural ideal inspired dreams, hopes, predictions, and promises that did not come to fruition. I argue that these insights help revise our view of Haussmannization—as not only a model but also a question, both a solution to and a source of urban problems.

The fact that Parisians found varied ways to politicize infrastructure shows that the politics of infrastructure go far beyond narrowly institutional questions. Republican politicians, engineers, and entrepreneurs pursued biopolitics and technopolitics, while infrastructure users activated political economy, cultural politics, and the politics of recognition and redistribution. This book has documented how, outside the halls of power, infrastructural politics took place in the home, the press, civil society, and the streets. Because public works and large technical systems are directed by the state and by corporations, it is easy to forget that, for users, they are everyday technologies experienced in bottom-up, concrete, embodied, and local ways. Though the Third Republic self-identified as democratic and egalitarian, the regime's sense of infrastructural democracy was limited, and citizens consistently called out the limits, failings, exclusions, and injustices of infrastructures. To cultivate democratic, egalitarian, inclusive, and just approaches to infrastructure today, we require both a diverse plurality of participants and a capacious sense of infrastructure's entanglement with and relevance for politics.

Struggles over infrastructure, which so saturated the period of secondary Haussmannization, did not decide *whether* Parisians would live with infrastructure, but *how* they would live with infrastructure. Although most Parisians would rather live with these infrastructures than without them, they have not always been easy to live with. The challenges of living with infrastructure, as decades of infrastructure studies have argued, are ultimately what inspire scholars to invert them. This book demonstrates that the same held true for Parisians in the time of secondary Haussmannization.

Notes

Introduction: Living with Infrastructure

1. On inversion, see Bowker and Star, *Sorting Things Out*, 34–37; on surfacing, see Starosielski, *Undersea Network*.

2. Bijker, *Of Bicycles, Bakelites, and Bulbs*; Pritchard and Zimring, *Technology and Environment*; and Cutcliffe and Reuss, *Illusory Boundary*.

3. Schivelbush, *Disenchanted Night*, 3–4; Clayson, *Illuminated Paris*, 3–8; DeJean, *How Paris Became Paris*; and Delattre, *Les douze heures noires*.

4. Jordan, *Transforming Paris*; Meyer, *Dividing Paris*; and Saalman, *Haussmann*.

5. Scott, *Seeing Like a State*, 60; Graham and Marvin, *Splintering Urbanism*, 53–55, on "modern infrastructural ideal," 49, 52, 62, on "icon," 44, and on "archetype," 53; Harvey, *Paris, Capital of Modernity*, 7–14, 80–89, 107–24; Rabinow, *French Modern*, 73–81; Schivelbusch, *Railway Journey*, 181–88, 195–96; Berman, *All That Is Solid Melts into Air*, 146–64; Tarr and Dupuy, *Technology*; and Benjamin, *Arcades Project*, 120–49.

6. Soppelsa, "How Haussmann Haunted the Early Third Republic," 35; Harvey, *Paris, Capital of Modernity*; Higonnet, *Paris*; Benjamin, *Arcades Project*; and Willms, *Paris*.

7. Fressoz, "Gas Lighting Controversy"; Bressani, "Light into Darkness"; and Bouman, "Luxury and Control."

8. Chakrabarty, *Provincializing Europe*. On Paris myths, see Harvey, *Paris, Capital of Modernity*; Higonnet, *Paris*; and Thompson, "Review Essay."

9. Hughes, *Networks of Power*. On top-down and institutional biases, see Soffer, "Introduction," 4, 5, 8. For social and cultural histories of urban technologies, see Goodman, *European Cities*; and Hård and Misa, *Urban Machinery*.

10. Graham and Marvin, *Splintering Urbanism*; Coutard and Rutherford, *Beyond the Networked City*; and Howe et al., "Paradoxical Infrastructures."

11. On politicizing infrastructure, see Soppelsa, "How Haussmann Haunted the Early Third Republic"; and Soppelsa, "Finding Fragility in Paris." On Paris as a model (or not) abroad, see Rybczynski, *City Life*; and Clayson and Dombrowski, *Is Paris Still the Capital?*

12. Bowker and Star, *Sorting Things Out*; Edwards, *Vast Machine*; and Klinenberg, *Palaces for the People*.

13. Reid, *Paris Sewers*; Berlanstein, *Big Business*; Belenky, *Engine of Modernity*; Papayanis, *Horse-Drawn Cabs*; and Hopkins, *Planning the Greenspaces*.

14. Clayson, *Illuminated Paris*; Beltran, "La difficile conquête d'une capitale"; Beltran, "Du luxe au coeur du système"; and Lopéz Galviz and Dikec, "Modern Atlas."

15. Mukerji, *Impossible Engineering*; Pritchard, *Confluence*; Schneider, *Hybrid Nature*; Soppelsa, "Water and Power"; "Water"; *Cultural Anthropology*, special section; Barnes and Alatout, "Water Worlds"; and Kinkela et al., *Radical History Review*, special issue.

16. Tarr and McShane, *Horse in the City*; Greene, *Horses at Work*; and Soppelsa, "Instrumentalisation of Horses."

17. Law, "Technology"; and Latour, *We Have Never Been Modern*.

18. White, "From Wilderness to Hybrid Landscapes"; Bess, "Artificialization"; Whatmore,

Hybrid Geographies; Schneider, *Hybrid Nature*; Swyngedouw, "Circulations and Metabolisms"; Haraway, "Companion Species Manifesto"; and Latour, *We Have Never Been Modern*.

19. Pritchard, "Envirotechnical Disaster"; and Zeller, "Aiming for Control."

20. Soppelsa, "How Haussmann Haunted the Early Third Republic"; and Carse, "Keyword." On the verb "Haussmannize," see Lanfrey, *Chroniques politiques*, 362–66; "Haussmanniser," in Barrère, *Argot and Slang*, 206; and "Haussmanniser."

21. Park, *Ideals of the Body*; Tresch, *Romantic Machine*; Harvey, *Paris, Capital of Modernity*; Bowie, *La modernité avant Haussmann*; Papayanis, *Planning Paris*; DeJean, *How Paris Became Paris*; Benjamin, *Arcades Project*; and Marcus, "Haussmannization as Anti-Modernity."

22. J. Simon, "Les comptes fantastiques d'Haussmann"; and Haussmann, *Mémoires*, 2:ix–xii.

23. Levin, "Bringing the Future to Earth." For more on the continuities of the Second Empire and Third Republic, see Sutcliffe, *Autumn of Central Paris*; Aisenberg, *Contagion*; Chevallier, *Le Paris moderne*; and Hopkins, *Planning the Greenspaces*.

24. Chevallier, *Le Paris moderne*, 17, 45; Graham and Marvin, *Splintering Urbanism*, 55–57; Latour, *Pasteurization of France*, esp. 16–19; Aisenberg, *Contagion*, ch. 3; Barnes, *Great Stink of Paris*; Rabinow, *French Modern*; Goubert, *Conquest of Water*; R. Nye, *Crime, Madness, and Politics*; and Weber, *Peasants into Frenchmen*.

25. On retrofitting, see Howe et al., "Paradoxical Infrastructures." On baggage and continuity, see Sutcliffe, *Autumn of Central Paris*.

26. Harvey, *Paris, Capital of Modernity*, 1, 10, 16; and Scott, *Seeing Like a State*, 59.

27. Porter, *Trust in Numbers*, ch. 6; Picon, "French Engineers"; and C. Smith, "Longest Run."

28. Evenson, *Paris*, 22; and Schivelbusch, *Railway Journey*, 181–83.

29. Rabinbach, *Human Motor*, 8; and Rabinow, *French Modern*, 9.

30. Levin, "Bringing the Future to Earth"; Levin, "Dynamic Triad": Soppelsa, "Paris's 1900 Universal Exposition"; Geppert, *Fleeting Cities*; Grubbs, "Terminus 1900"; and Benjamin, *Arcades Project*, 3–26, 171–202.

31. Biggs, *Quagmire*, 39–40.

32. Conklin, *Mission to Civilize*, ch. 2.

33. T. Mitchell, "Introduction," 437.

34. Graham and McFarlane, "Introduction," 1.

35. Amin, "Lively Infrastructure."

36. Melosi, "Urban Infrastructure."

37. Berman, *All That Is Solid Melts into Air*; Misa et al., *Modernity and Technology*; Lefebvre, "Right to the City"; and Harvey, "Right to the City."

38. Figuier, *Les nouvelles conquêtes de la science*, 2:519; Neuman, "Infiltrating Infrastructures"; and Ruby and Ruby, "Preface," 5.

39. Carse, "Nature as Infrastructure"; Cardoso da Silva and Wheeler, "Ecosystems as infrastructure"; Barua, "Infrastructure"; and Bélanger, "Landscape as Infrastructure."

40. Bechmann, *Salubrité urbaine*, 3–4.

41. Papayanis, *Planning Paris*; Rabinow, *French Modern*; Schivelbusch, *Railway Journey*; and Rabinbach, *Human Motor*.

42. J. Strauss, *Human Remains*; Rabinow, *French Modern*; Latour, *Pasteurization of France*; R. Nye, *Crime, Madness and Politics*; Chevallier, *Le Paris moderne*; and Joyce, *Rule of Freedom*.

43. Heynen et al., *In the Nature of Cities*; Schott et al., *Resources of the City*; Brantz and Dümpelman, *Greening the City*; Leman-Stefanovic and Scharper, *Natural City*; and Cronon, *Nature's Metropolis*.

44. Haussmann, *Mémoires*; du Camp, *Paris*; Papayanis, *Planning Paris*, 103, 118–19, 253. On surgery, see Jordan, "Haussmann and Haussmannization," 90; Barnes, *Great Stink of Paris*, 50; Miles, *Urban Avant-Gardes*, 65; Olson, *City as a Work of Art*, 142; and Kostof, "His Majesty the Pick," 11. On circulation, see Schivelbusch, *Railway Journey*, 194–95; Pinkney, "Napoleon III's Transformation," 134; and Illich, *H_2O and the Waters of Forgetfulness*, 38–47, 66, 73–74.

45. Du Camp, *Paris, ses organs, ses fonctions, sa vie*; Boucard, *La vie de Paris*; and d'Avenel, *Le mécanisme de la vie moderne*.

46. Barles, "Urban Metabolism"; Tarr, "Metabolism of the Industrial City"; Swyngedouw, "Metabolic Urbanization"; and Gandy, "Rethinking Urban Metabolism."

47. Massard-Guilbaud, "Introduction"; Neuman, "Infiltrating Infrastructures"; Larkin, "Politics and Poetics"; Barua, "Infrastructure"; Bakker, "Water"; Folkers, "Existential Provisions"; and Ten Hoor, "Architecture and Biopolitics."

48. Foucault, *"Society Must Be Defended,"* ch. 11.

49. Chevallier, *Le Paris moderne*; Aisenberg, *Contagion*; Pritchard, *Confluence*; Mukerji, *Impossible Engineering*; Guldi, *Roads to Power*; Joyce, *Rule of Freedom*, esp. ch. 2; Barnes and Alatout, *Social Studies of Science*, "Water Worlds" special issue; and Otter, "Vital City."

50. Foucault, *"Society Must Be Defended,"* 246–47. The verb "regularize" appears in Haussmann, *Mémoires*, 3:69, 81, 91; Choay, *Modern City*, 15; Graham and Marvin, *Splintering Urbanism*, 53–54; Papayanis, *Planning Paris*, 3; Rabinow, *French Modern*, 76; and Schivelbusch, *Railway Journey*, 118.

51. Arnold, *Everyday Technology*; Parr, *Sensing Changes*; Josephson, *Fish Sticks*; Friedel, *Zipper*; Myllyntaus, "Prologue"; and Soppelsa and Rodgers, "Origins of the Flyswatter." On everyday infrastructure, see Graham and McFarlane, *Infrastructural Lives*; and Anand et al., *Promise of Infrastructure*.

52. Oudshoorn and Pinch, *How Users Matter*; Eglash et al., *Appropriating Technology*; Edgerton, *Shock of the Old*; and Russell and Vinsel, "After Innovation."

53. Schivelbusch, *Railway Journey*; Herf, *Reactionary Modernism*; Hård and Jamison, *Intellectual Appropriation*; Marvin, *When Old Technologies Were New*; Rieger, *Technology*; Reid, *Paris Sewers*; and Soppelsa, "Reworking Appropriation."

54. On Paris urbanism, see Hopkins, *Planning the Greenspaces*; and Park, *Ideals of the Body*. For street-level urban histories, see Berman, *All That Is Solid Melts into Air*; Jerram, *Streetlife*; Ladd, *Streets of Europe*; and Wakeman, *Heroic City*.

55. Coutard et al., "History (and Stories)"; and Egyedi and Mehos, *Inverse Infrastructures*.

56. Foucault, "Of Other Spaces."

57. Giedion, *Mechanizaton Takes Command*; Latour, *We Have Never Been Modern*; Coole and Frost, *New Materialisms*; Brown, "Thing Theory"; and Gieryn, "What Buildings Do."

58. For "technopolitics," see Hecht, *Radiance of France*, 15–17, 89, 91; T. Mitchell, *Rule of Experts*; and Folkers, "Existential Provisions." For "congealed social interests," see Pinch and Bijker, "Social Construction of Facts"; and Graham and Marvin, *Splintering Urbanism*, 11.

59. Cottereau, "Les batailles pour la création du Métro"; Cottereau, "Les batailles du métropolitain"; Jacquemet, "Urbanisme Parisien"; and Guerrand, "La bataille du tout-a-l'égout."

60. Chatzis and Coutard, "Water and Gas"; Beltran, "Du luxe au coeur du système"; Weber, *France, Fin de Siècle*, 53, 58, 71, 186, 197; Marvin, *When Old Technologies Were New*, 101, 123, 162, 177; Ladd, *Streets of Europe*, 171; and Benjamin, *Arcades Project*, 152, 184, 885.

61. C. Smith, "Longest Run"; Aisenberg, *Contagion*; Cole, *Power of Large Numbers*; and Fox, *Savant and the State*.

62. Soppelsa, "Finding Fragility in Paris"; Soppelsa, "Paris's 1900 Universal Exposition"; Fraser, "Rethinking Recognition"; Fraser, "From Redistribution to Recognition?"; and M. Shapiro, *Politics of Representation*.

63. Anand et al., "Introduction," 3; and Graham and Marvin, *Splintering Urbanism*, 47.

64. Graham and Marvin, *Splintering Urbanism*, 40, 49–65.

65. Rabinow, *French Modern*, 76–78; on lustrous, see Graham and Marvin, *Splintering Urbanism*, 47, see also 53–59; and Gandy, *Fabric of Space*, 27–54. On legibility, see Scott, *Seeing Like a State*, 59.

66. Jasanoff and Kim, *Dreamscapes of Modernity*.

67. Barnes, *Great Stink of Paris*, 60.

68. Soppelsa, "Theorizing Infrastructure." For influential sources of affect theory, see Ahmed, *Cultural Politics of Emotion*; Sedgwick, *Touching Feeling*; and Berlant, *Cruel Optimism*.

69. Hughes, *American Genesis*; D. Nye, *American Technological Sublime*; Rieger, *Technology*; Marvin, *When Old Technologies Were New*; Schivelbusch, *Railway Journey*; Barnes, *Great Stink of Paris*; and J. Strauss, *Human Remains*.

70. Otter, *Victorian Eye*; Kenny, *Feel of the City*; Clayson, *Illuminated Paris*; D. Nye, *When the Lights Went Out*; Marvin, *When Old Technologies Were New*; and Saraiva and Cardoso De Matos, "Technological Nocturne."

71. Kenny, *Feel of the City*; Ladd, *Streets of Europe*; Parr, *Sensing Changes*; Graham and McFarlane, *Infrastructural Lives*; and Boutin, *City of Noise*.

72. J. Engels, "Rhythm Analysis"; Jalas et al., " Rhythms of Infrastructure"; Edwards, "Infrastructure and Modernity"; and Augé, *In the Metro*, xv, 69.

73. Bowker and Star, *Sorting Things Out*, 35.

74. Graham, "When Infrastructures Fail"; Starosielski, "'Warning"; Edwards, "Mechanics of Invisibility"; Larkin, "Politics and Poetics"; Cushman and Risteen, "Efficient Ruin of Infrastructure," 1; Appadurai, "Foreword"; and P. Harvey et al., "Introduction."

75. Soppelsa, "Urban Railways"; and Soppelsa, "Visualizing Viaducts."

76. Barnes, *Great Stink of Paris*; Reid, *Paris Sewers*; Parr, *Sensing Changes*; Robbins, "Smell of Infrastructure"; Mavhunga, "Cidades Esfumaçadas"; and Suominen et al., "Smelling Machine History."

77. Knowles, "Learning from Disaster?"; Pritchard, "Envirotechnical Disaster"; Zeller, "Aiming for Control"; Jenks, *Perils of Progress*; D. Nye, *When the Lights Went Out*; and Graham, *Disrupted Cities*. On Paris disasters, see Keller, *Fatal Isolation*; and Jackson, *Paris Under Water*.

78. On scripts, see Akrich, "De-Scription of Technical Objects"; Oudshoorn and Pinch, *How Users Matter*; Fallan, "De-scribing Design"; Ferreira da Silva and Sousa, "'Script' of a New Urban Layout"; and Certeau, *Practice of Everyday Life*, ch. 7 and 9.

79. Norton, *Fighting Traffic*; and Soppelsa, "Reworking Appropriation."

Chapter 1. Streets, or the Politics of Infrastructure

1. On roads, see Freed, "Networks"; Guldi, *Roads to Power*; Christopher Jones, *Routes of Power*; Denning, "Infrastructural Propaganda"; and Denning, "Mobilizing Empire." On streets, see Esbester and Wetmore "(Auto)mobility"; Norton, *Fighting Traffic*; Ladd, *Autophobia*; Ladd, *Streets of Europe*; Josephson, *Traffic*; and M. Flonneau, *Paris et l'automobile*.

2. Benjamin, *Arcades Project*, stone pavers 131, 139, 359, 737; wood pavers 89, 121, 913.

3. "À qui la rue? . . . Au populo!" *Le père peinard: Un journal "espatrouillant,"* 9 October 1898, 326.

4. Benjamin, *Arcades Project*, flâneur 416–55, streets 516–26, quotes on 421–22. See also Meyer, *Dividing Paris*, ch. 3.

5. Barles and Guillerme, "La rue Parisienne," 28; and colonizing metaphor in M. Flonneau, *Paris et l'automobile*, Part 1.

6. Hecht, *Radiance of France*, 15.

7. Levin, "Bringing the Future to Earth"; Friedel, *Culture of Improvement*; Fox, *The Savant and the State*; and Tresch, *Romantic Machine*.

8. "Faits-Paris," *Le Figaro*, no. 2–3 (2–3 January 1872), 2; Schivelbusch, *Disenchanted Night*, 97, 120, 142; and Clayson, *Illuminated Paris*, Opéra on 8, 1–13, 22–23, 41, 62–66, 87, 125, 136.

9. "Chronique de Paris," *Le temps*, 4 January 1872, 3.

10. Weiss, "'Dravert Affair'"; dossier "Affair Cabrin/Courcier" in AP V.ONC 227.

11. "Chronique de Paris," *Le temps*, 16 February 1872, 2.

12. "Chronique de Paris," *Le temps*, 8 March 1872, 2.

13. "Chronique de Paris," *Le temps*, 8 March 1872, 2.

14. Sante, *Other Paris*, 40; Giedion, *Space, Time, and Architecture*, 770, 772; Benjamin, *Arcades Project*, 24, 132, 138, 146; Soppelsa, "How Haussmann Haunted the Early Third Republic"; and Lee, *Tyranny of the Straight Line*.

15. Merriman, *Margins of City Life*, 80; Colin Jones, *Paris*, 325; Marchand, *Paris*, 114; and Gould, *Insurgent Identities*.

16. Bachelle, "Une cauchemar," *Le Gaulois*, 16 July 1872, 2–3; and Dickens, "At Bordeaux."

17. Built from 1668 to1705 under Louis XIV as the Nouveau Cours. See "Dernière heure: l'inauguration de l'avenue de la République," *Le temps*, 14 July 1891; and *Bulletin municipal officiel de la ville de Paris*, 14–16 July 1891, 1611–14.

18. Defrance, *Histoire de l'éclairage des rues*, 118; and Clayson, *Illuminated Paris*, 193n17, 194n39.

19. In 1872, the most industrialized arrondissements were the 2nd, 3rd, 9th, 10th, and 11th. See Dickens, *Dickens's Dictionary of Paris*, 193; and Gould, *Insurgent Identities*, map on 90.

20. Galignani, *Galignani's Illustrated Paris Guide*, "Boulevards," 21.

21. The original was longer than today's Avenue de la République, which connects the Place de la République and the Père Lachaise cemetery, crossing the 11th. Today the street changes name thereafter: across the 20th, it is called Avenue Gambetta and then Rue Belgrand. Notable exceptions to Haussmann's relative neglect of the northeast are the Parc des Buttes Chaumont and the Rue de Puebla; see Harvey, *Paris, Capital of Modernity*, map on 307; Jordan, *Transforming Paris*, 185–95; and Hopkins, *Planning the Greenspaces*, 53.

22. "Dernière heure: l'inauguration de l'avenue de la République," *Le temps*, 14 July 1891.

23. "Dernière heure: l'inauguration de l'avenue de la République," *Le temps*, 14 July 1891.

24. Van Zanten, "Paris Space"; and Harvey, *Paris, Capital of Modernity*, 210.

25. Poubelle's speech is transcribed in Duprey, *L'Oeuvre sociale*, ix–x; and *Bulletin municipal officiel de la ville de Paris*, 14–16 July 1891, 1611–13.

26. Chevallier, *Le Paris moderne*, 249–51; Higonnet, *Paris*, 224; Sante, *Other Paris*, 59; and Weber, *France, Fin de Siècle*, 58.

27. Stone, *Search for Social Peace*; and Jackson, "Solidarism in the Paris Streets."

28. Shaw, *Municipal Government*, 12–13; the Poubelle quote is in *Bulletin municipal officiel de la ville de Paris*, 14–16 July 1891, 1613; Haussmann's various motives are discussed in Benjamin, "Paris, Capital of the Nineteenth Century"; Jordan, *Transforming Paris*; Harvey, *Paris, Capital of Modernity*; Evenson, *Paris*; and Schivelbusch, *Railway Journey*. On hygiene as a "master concern," see Pick, *Faces of Degeneration*; R. Nye, *Crime, Madness, and Politics*; and Latour, *Pasteurization of*

France. On the republic's deeper commitment to hygiene, see Chevallier, *Le Paris moderne*, 54, 58, 333.

29. Gandy, *Fabric of Space*, 43; and Reid, *Paris Sewers*, 80.

30. Schivelbusch, *Railway Journey*, 178; Rabinbach, *Human Motor*; "ecological slippage" in Barnes, *Making of a Social Disease*, 113; A-L. Shapiro, *Housing the Poor*; Horne, *Social Laboratory*; and Canguilhem, *Knowledge of Life*.

31. On zoning, see Marcus, *Apartment Stories*, 142; and Higonnet, *Paris*, 185.

32. Rabinow, *French Modern*, analyzes French garden cities thinking in the work of architects Prost (140–41), Hénard (212–13), Garnier (225), Howard (257–60), and Sellier (334–35, 338–42, 348–49); see also Horne, *Social Laboratory*, 246–47, 256; Josephson, *Traffic*, 22; and Charvet, *Les fortifications de Paris*.

33. Levin, *When the Eiffel Tower Was New*; and Levin, *Republican Art and Ideology*.

34. "Dernière heure: l'inauguration de l'avenue de la République," *Le temps*, 14 July 1891; *Bulletin municipal officiel de la ville de Paris*, 14–16 July 1891; and Duprey, *L'oeuvre sociale*, x.

35. Duprey, *L'oeuvre sociale*, xii.

36. Conklin, *Mission to Civilize*, 73, 184, 240, 313; Weber, *Peasants into Frenchmen*, 485–96; Pritchard, *Confluence*, 185; and Latour, *Pasteurization of France*.

37. Guizot, *Cours d'histoire moderne*, 1:3, 19; Cabanis in Aisenberg, *Contagion*, 8; A-L. Shapiro, *Housing the Poor*, 4; and Rabinow, *French Modern*, 20.

38. Hugo, *Oeuvres complètes: Philosophie*, 1:269; and Hugo, *Histoire d'un crime*, 259. For other early examples, see Lafarelle, *Du progrès social au profit*; Lajoulet, *Amélioration morale*; and Vanderslaghmolen, *De l'amélioration*.

39. Haussmann, *Mémoires*, 2:257.

40. Baron, *Le paupérisme*, art. 7, 233–56; Hautière, *La constitution*, 115, 199–200; *Le radical*, 19 July 1900, quoted in Hausser, *Paris*, 37.

41. "Le tramway électrique de Romainville à la place de la République," *Le temps*, 2 June 1896, 3; and Jacquin, "Le tramway électrique."

42. "Compagnie du Tramway Électrique."

43. A. Dumas, "Le tramway électrique," 294; Jacquin, "Le tramway électrique"; Parville, *Les tramways électriques*, 121–28, 279; and Robert, *Les tramways Parisiens*, 13.

44. Say, *Traité d'économie politique*, 463; and Figuier and Gautier, *L'année scientifique*, 111.

45. Cheysson, "La circulation ," 152–54.

46. Dumarest, "Quelques notes recueillies"; Rigaud, *Dictionnaire du jargon*, 71; and Rigaud, *Dictionnaire d'argot*, 79. On the term's coinage, see Rochegude, *Promenades*, 32; Nord, *Politics of Resentment*, 125; Verne, *Journey*, 172; and "Maelstrom" in M. Flonneau, *Paris et l'automobile*, 67.

47. Benjamin, *Arcades Project*, 32; Colin Jones, *Paris*, 288; Hazan, *Invention of Paris*, 40; Walz, *Pulp Surrealism*, 24; and C. Clark, *Paris*, 43, 54.

48. France, Assemblée nationale, *Annales du Sénat*, 45.

49. Claux, "Le pavage en bois"; "Engineering"; and Rochard, *Traité d'hygiène publique*, 239–40.

50. "Charivaria"; tunnels in Lapparent, "La question du métropolitain" (part 2), 409; "Les passerelles," 128–29; and Vèze, "Les embarras de Paris," 347.

51. André, *Note sur les variations*, 17–18, quote 17, 1887 on 47. AP V.ONC 239.

52. Figuier, "Le chemin de fer métropolitain," 216.

53. M. Flonneau, *Paris et l'automobile*, 21–79; Ladd, *Autophobia*, 15; and Black, *Crude Reality*, 99–100.

54. Jules Claretie, "La vie à Paris," *Le temps*, 9 June 1898, 2.

55. "Résultat de concours."

56. Simond, *La vie Parisienne*, 3:14.

57. Croix-Rouge in "Le carrefour des écrasés"; "Carrefour des écrasés"; and Rue Saint Lazare in A. Martin, *Une visite à Paris*, 35.

58. New York in "De tout un peu," *Le petit Parisien: Supplément littéraire illustré*, 19 April 1891, 7–8; and Marseilles and Algiers in Baudel, *Un an à Alger*, 39. Several intersections in Lapparent, "La question du métropolitain" (part 2), 405.

59. Hénard, *Études sur les transformations*, fascicule 5, 172–78, 276–300, quote 277–78.

60. Veron quoted in Vèze, "Les embarras de Paris," 346.

61. Chrysale, "Notes de la semaine"; and Bournon, *La voie publique*, 71.

62. M. Flonneau, *Paris et l'automobile*, 61.

63. M. Flonneau, *Paris et l'automobile*, 56; and Deluermoz, "Rouler à la baguette?" 9–10.

64. Speed limit, in Jackson, "Solidarisme," 247; and M. Flonneau, *Paris et l'automobile*, coupe-files 55–57, marauding 64.

65. M. Flonneau, *Paris et l'automobile*, 66–69; and geometric and symbolic, in Josephson, *Traffic*, 25.

66. Jackson, "Solidarisme," Mirabeau on 243, Hénard on 253; Guillerme, "Autophiles et autophobes," 2; and Ladd, *Streets of Europe*, 205.

67. Conseil municipal de Paris, *Rapport no. 17* (30 May 1910), Lepine on 11, Massard on 12.

68. Chrysale, "Notes de la semaine," 590.

69. *Bulletin municipal officiel de la ville de Paris*, 30 April 1911, 1869; and Conseil municipal de Paris, *Procès-verbaux* (1911), 365.

70. "Le kiosque-signal."

71. "Nouvelles du jour: l'inauration du kiosque à signal," *Journal des débats politiques et litteraires*, 12 April 1912, 3; and "Le kiosque signal ne fonctionne plus!" *L'intransigeant*, 12 April 1912, 1.

72. Le Bachelier, "Le kiosque de l'entente cordiale," *Gil Blas*, 12 April 1912, 2.

73. Lecoq, "Pour la beauté de Paris," *Le petit journal*, 22 May 1912, 1; "Échos," *La petite presse*, 29 April 1912, quote 2; and "Échos," *La petite presse*, 20 May 1912, 2. See also M. Flonneau, *L'automobile*, 114; and M. Flonneau, *Paris et l'automobile*, 78.

Chapter 2. At Home with Infrastructure

1. Adams et al., "Collapse and Expand"; Gieryn, "What Buildings Do"; and Bray, "Technics and Civilization."

2. Evenson, *Paris*, 212; Chevallier, *Le Paris moderne*, 163–85; and Levin, "Bringing the Future to Earth," 48.

3. Kamoun, *V'là Cochon qui démenage*, 14.

4. Wagenaar, "Conquest of the Center?"; R. Nye, *Crime, Madness and Politics*, 54–56; Jordan, *Transforming Paris*, ch. 4; and Chevalier, *Laboring Classes*.

5. Evenson, *Paris*; A-L. Shapiro, *Housing the Poor*; Bullock and Read, *Movement for Housing Reform*; Loyer, *Paris XIXe siècle*; Rabinow, *French Modern*; Marcus, *Apartment Stories*; Harvey, *Paris, Capital of Modernity*; Aisenberg, *Contagion*; Stone, *Search for Social Peace*; Elwitt, *Third Republic Defended*; Barnes, *Making of a Social Disease*; Horne, *Social Laboratory*; and Yates, *Selling Paris*.

6. Harvey, *Paris, Capital of Modernity*; Yates, *Selling Paris*; Benjamin, *Arcades Project*, 145; and Soppelsa, "How Haussmann Haunted the Early Third Republic," 26.

7. Bullock and Read, *Movement for Housing Reform*, 301; Harvey, *Paris, Capital of Modernity*, 125–40; and Evenson, *Paris*, 200–204.

8. For encyclopedias, see "Les loyers à Paris," 113–14; Bertillon, "La crise des loyers"; for savants, see d'Haussonville, *La vie et les salaires*, 16; and Biget, *Le logement de l'ouvrier*, 23.

9. Tourdonnet, "De la crise des loyers"; Cochut, "De l'enchérissement des marchandises"; "Le prix de la vie," *L'humanité*, 3 September 1910; *L'assiette au beurre* special issue on "la vie chère," no. 496 (1 October 1910); Appelbaum, *French Satirical Drawings*, 69; J-M. Flonneau, "Crise de vie"; and Stovall, "Consumers' War."

10. Chevalier, *Laboring Classes*; and Marchand, *Paris*.

11. Amin and Thrift, *Cities*, 2; on Berlin, see McElligott, *German Urban Experience*, 8–16; Ladd, *Ghosts of Berlin*, 96; on London and Paris, see *Commission d'extension de Paris*, vol. 2: aperçu technique, plate 1, and on "fortress," see vol. 1: historique, 1; and Hénard, *Études sur les transformations*, 55–88.

12. Verne, *La tour du monde*, 6; Claretie, *La vie à Paris* (1895), 176; Sancha, "La plus Parisienne"; Colin Jones, *Paris*, xviii; Cobb, *Paris and Elsewhere*, 71, 223; and Sante, *Other Paris*, 35, 57, 65.

13. Marcus, *Apartment Stories*; Periton, "'Coupe Anatomique'"; and Hazan, *Invention of Paris*, 3–16.

14. "Dernière heure: l'inauguration de l'avenue de la République," *Le temps*, 14 July 1891, 1; *Bulletin municipal officiel de la ville de Paris*, 15 July 1891; Petitjean, *Les grands travaux*, 9–10; Corday, "Science et moeurs," 771; *Commission d'extension de Paris*, 1:166 and 2:45; and Benjamin, *Arcades Project*, 304.

15. A-L. Shapiro, *Housing the Poor*; Aisenberg, *Contagion*; and Keller, *Fatal Isolation*.

16. Veuillot, *Les odeurs de Paris*; Brunfaut, *Les odeurs de Paris*; Chrétien, *Les odeurs de Paris*; Sarcey, *Les odeurs de Paris*; Baer, "Revue comique du mois de juin," *La lanterne*, 4 July 1895; "Les odeurs de la Seine," *Le petit journal*, 8 September 1895; "Les mauvaises odeurs," *Le radical*, 9 September 1895; *Le petit Parisien*, 30 July 1896 and 12 July 1897; "Les mauvaises odeurs de Paris," *Le petit Parisien*, 10 July 1908; Trillat, *Commission des odeurs*; and Barnes, *Great Stink of Paris*.

17. In addition to the discussion of Floquet in chapter 1, see "Échos de l'escalier," *L'anti-concièrge*, year 2, no. 5 (April–July 1882), 2; "Moyens de secher"; Deuzères, "À travers la science: les murs humides," *Le petit Parisien*, 11 July 1904; DuMesnil, *L'hygiène à Paris*, 162–64; and Aisenberg, *Contagion*, 21, 54.

18. R. Nye, *Crime, Madness, and Politics*; Pick, *Faces of Degeneration*; Latour, *Pasteurization of France*; Strauss, *Human Remains*; and M. Anderson, *Regeneration Through Empire*.

19. Barles, *La ville deletère*; Strauss, *Human Remains*, ch. 3; and Aisenberg, *Contagion*.

20. A-L. Shapiro, *Housing the Poor*, 135; and Aisenberg, *Contagion*.

21. Rabinbach, *Human Motor*, 8; and Rabinow, *French Modern*, 9.

22. Fox, *The Savant and the State*; Nord, *Republican Moment*, ch. 2; and A-L. Shapiro, *Housing the Poor*, ch. 6.

23. Weisz, *Medical Mandarins*; and Ellis, *Physician-Legislators*.

24. Chevallier, *Le Paris moderne*; Rabinow, *French Modern*; Horne, *Social Laboratory*; Aisenberg, *Contagion*; and Strauss, *Human Remains*.

25. Horne, *Social Laboratory*; Elwitt, *Third Republic Defended*; Stone, *Search for Social Peace*; and Latour, *Pasteurization of France*.

26. Rabinow, *French Modern*, 36–39, 126–38; Canguilhem, *Knowledge of Life*; Schivelbusch, *Railway Journey*; and Rabinbach, *Human Motor*.

27. Cole, *Power of Large Numbers*, 7; Chevallier, *Le Paris moderne*; and Rabinow, *French Modern*.

28. The catchphrase recurred in advertisements: 45 in *Le Figaro* (1873–1914); 27 in *Le monde illustré* (1880–1896); 23 in *Gil Blas* (1883–1914); 15 in *Le petit Parisien* (1880–1906); 14 in *Le*

Gaulois (1882–1914); 14 in *Le temps* (1878–1936); 13 in *Le radical* (1884–1906); 7 in *La vie Parisienne* (1885–1933); 7 in *Le XIXᵉ siècle* (1884–1895); 7 in *Le matin* (1890–1925) (all results from Gallica.bnf.fr). See also "Revue foncière et bulletin de mutations immobilières," *La vie Parisienne* (2 April 1892): 194; *Journal des débats politiques et littéraires* (31 October 1878); *La revue mondiale* 75 (15 August 1908): 12; *La revue mondiale* 51 (15 July 1904): 12; *La revue bleue* 18, no. 2 (12 July 1902), back cover.

29. Figuier, *Les nouvelles conquêtes*, 2:519; Huret, "En Allemagne," *Le Figaro* (25 December 1906), 3; Grand-Carteret, *XIXᵉ siècle*, 730.

30. Neuman, "Infiltrating Infrastructures"; and Barles, "Urban Metabolism."

31. Figuier, *Le téléphone*, 282; Figuier, *Les merveilles*; "Bibliographie," 224; and Graffigny, *L'électricité pour tous*, 271.

32. Rieger, *People's Car*, 39–41; Rieger, *Technology*; and L. Simon, *Dark Light*.

33. Havard, *La maison salubre*, 24–25. For luxury vs. needs, see A-L. Shapiro, *Housing the Poor*, 152; Evenson, *Paris*, 209; and Beltran, "Du luxe au coeur du système."

34. P. L., "La maison salubre," 14–16.

35. Evenson, *Paris*, 149–50; Sutcliffe, *Paris*, 114–15; Jordan, "Haussmann and Haussmannisation"; Rabinow, *French Modern*; and Chevallier, *Le Paris moderne*.

36. Commission des logements insalubres, *Rapport général sur les travaux de la commission pendant les années 1877 à 1883*, 41–43, esp. art. 8 (AP V.O3 63).

37. Commission des logements insalubres, *Rapport général sur les travaux de la commission pendant les années 1877 à 1883*, 37; *Rapport général sur les travaux de la commission pendant les années 1870 à 1876*, urinals 40–44 (both sources AP V.O3 63); Aisenberg, *Contagion*, 147; A-L. Shapiro, *Housing the Poor*, 55, 137–39; Levin, "Bringing the Future to Earth"; Harvey, *Paris, Capital of Modernity*, 180; Bullock and Reid, *Movement for Housing Reform*, 299, 302–4; Benjamin, *Arcades Project*, 133–35. On floating population, see "Vital Statistics of Paris and Vienna"; Chevalier, *Laboring Classes*, 227–31; and Merriman, *Margins of City Life*. On the male lodging house majority, see Corbin, *Women for Hire*, 188.

38. Commission des logements insalubres, *Rapport général sur les travaux de la commission pendant les années 1870 à 1876*, 18–21; Rabinow, *French Modern*, 82; Callon, "Society in the Making"; and Levin, "Eiffel Tower Revisited."

39. Zola, *"Assommoir,"* 46–47; and Marcus, *Apartment Stories*, 166–98, quotes 169 and 180, respectively.

40. Zola, *Ladies' Paradise*, 12–13, 182–83, and 282, respectively; for a hygienist view, see Graux, *La tuberculose*, 11–12.

41. Deuzères, "A travers la science," *Le petit Parisien*, 4 July 1904. On abjection, see Strauss, *Human Remains*; and on Yersin, see Mendelssohn, "Microscopist of Modern Life."

42. DuMesnil, *L'hygiène à Paris*, 23; Colin, in Aisenberg, *Contagion*, 113–14; and Benjamin, *Arcades Project*, 123, 137.

43. DuMesnil, *L'hygiène à Paris*, 14, 34, and 29–30, respectively; Fillassier, *Les casiers sanitaires*, 1; on "microbe factories," see A-L. Shapiro, *Housing the Poor*, ch. 6.

44. A-L. Shapiro, *Housing the Poor*, 75–78; Horne, *Social Laboratory*, 226–28; Bullock and Reid, *Movement for Housing Reform*, 307; Evenson, *Paris*, 208; and Aisenberg, *Contagion*.

45. The word "casier" translates as rack, filing cabinet, compartment, or pigeonhole, conveying a grid of cells; see Barnes, *Making of a Social Disease*, 117–28; Fijalkow, *La construction des îlots insalubres*, 127; and Biggs, *Quagmire*, 106–24. On medical geography, see Osborne, *Emergence of Tropical Medicine*, 50–53; A-L. Shapiro, *Housing the Poor*, 144.

46. Chevallier, *Le Paris moderne*, 154.

47. Regnier, "La lutte contre la tuberculose," 52; Acheray, *La lutte contre la tuberculose*; Brouardel, *La lutte contre la tuberculose*; Thoinot, "La vie et l'oeuvre," 229. David Barnes uses the phrase "war on tuberculosis" in *Making of a Social Disease*, 13–18; and Rendu in "La lutte contre la tuberculose."

48. *Transactions of the British Congress*, 1:200–21; Robin, "La lutte contre la tuberculose"; Savary and Collet, "La lutte contre la tuberculose"; and Patrikios, "La lutte contre la tuberculose," 28.

49. *Congrès international de la tuberculose*, 8–9, 13–14, 334–36, 462–69.

50. *Congrès international de la tuberculose*, 8–9.

51. *Congrès international de la tuberculose*, 334–36, 462–69; and Benjamin, *Arcades Project*, 103.

52. "L'exposition," *Le petit Parisien*, 2 July 1900; and Auslander, *Taste and Power*, 219.

53. On the decline, see Barnes, *Making of a Social Disease*, 18; Juillerat, *Rapport à Monsier le Directeur des affaires départementales*, 20 December 1904 (AP V.ONC 1342).

54. Rendu, *Rapport au nom de la 6e Commission, sur l'assainissement des îlots insalubres à Paris* (Conseil municipal de Paris, no. 69, 1 July 1909), AP V.ONC 1342; AP Perotin 10653 150.

55. Sonn, "Marginality and Transgression," 121, 123; see also Sonn, *Anarchism and Cultural Politics*, 52; Commission des logements insalubres, *Rapport 1870–76*, 89; and Commission des logements insalubres, *Rapport 1877–83*, 6–12.

56. Conseil municipal de Paris, no. 16 (20 April 1891), *Contre-projet de délibération*, annexes (AP V.O3 220); *Annuaire statistique de la ville de Paris 1906*, year 27 (1908), disease 179, crowding 615 (AP V.O3 63); and Hewitt, *Montmartre*, 115–16.

57. On 1898 visitor, see Merriman, *Margins of City Life*, 12; Emanuel, *Illustrators of Montmartre*, 4–5; on cemeteries, see Benjamin, *Arcades Project*, 99; M. Flonneau, "Du profil de la rue"; Jackson, "Artistic Community"; Cate, "Spirit of Montmartre"; and Kenny, "Je Cherche Fortune."

58. Hewitt, *Montmartre*, 6, 16, 116–17, 136, 151; and Jackson, "Artistic Community," 5–6.

59. M. Flonneau, "Du profil de la rue," 69.

60. Derville, "Montmartre pittoresque," 372.

61. Charpentier, *À travers Montmartre*, 34–43, quote 34.

62. Benjamin, *Arcades Project*, 289.

63. Lévy cited in Cate, "Spirit of Montmartre," 28; Jerrold Seigel chose Émile Goudeau as the scene's originator in *Bohemian Paris*, 216–41; and Sante, *Other Paris*, 135–58, esp. after 153.

64. They were called *Hydropathes* for drinking absinthe without water, perhaps to separate themselves from the 1840s "water-drinkers" bohemian clique. *Fumistes* (literally "smokers") meant roughly "pranksters." Their humor borrowed the *blague* (prank or put-on) from Latin Quarter student culture, via Jules Lévy and Émile Goudeau. See Grojnowski, "Hydropathes" 96; Cate, "Spirit of Montmartre," 23; Weisberg, *Montmartre*; and Dousteyssier-Khoze, "Fumisme." On minor literature, see Deleuze and Guattari, *Franz Kafka*; and Spector, *Prague Territories*.

65. Jarry and Apollinaire cited in Shattuck, *Banquet Years*, 213–14.

66. Cartoon in "La question des loyers"; for more on vulgar caricatures, see Clayson, *Illuminated Paris*; and Brigstocke, *Life of the City*.

67. The historian Ann-Louise Shapiro found "heightened political activity among urban workers" around housing from 1879 to 1883, including petitions, community meetings, and rent strikes. See A-L. Shapiro, *Housing the Poor* , 113.

68. *L'anti-concièrge*, year 1, no. 1 (1 December 1881), 2; *L'anti-concièrge*, year 2, no. 7 (October 1882–January 1883), 1; Jules Jouy, "Fantasie concièrgicide," *L'anti-concièrge*, year 1, no. 2 (15

December 1881), 1; for Sapeck, see "Locataires et proprios," *L'anti-concièrge*, year 2, no. 6 (July–October 1882), 1. For two Daumier landlord cartoons, see Harvey, *Paris, Capital of Modernity*, 126; and Lafargue, *Monsieur Vautour.*

69. "La dernière quinzaine," *L'anti-concièrge*, year 2, no. 3 (1 January 1882), 1–2.

70. The Allais quote is from *L'anti-concièrge*, year 2, no. 3 (1 January 1882), 4; "Échos de l'escalier," *L'anti-concièrge*, year 2, no. 5 (April–July 1882), 2, 4; and Certeau, *Practice of Everyday Life*, 91–130.

71. "Locataires et proprios," *L'anti-concièrge*, year 2, no. 6 (July–October 1882), 1; Jules Simon, preface to Du Mesnil's *L'hygiène à Paris*, 6; and Pouget, "Mon drapeau rouge," *Le père peinard*, 12 May 1889, 40.

72. Sapeck, "Le nommé 'Cerbère,'" *L'anti-concièrge*, year 1, no. 1 (December 1881), 3; the Marxist paper *L'égalité* often covered housing from summer 1882 to fall 1883. See also Harvey, *Paris, Capital of Modernity*, 196; Ross, *Emergence of Social Space*; Gould, *Insurgent Identities*; J-M. Flonneau "Crise de vie," 65; and Péchu, "Entre résistance et contestation," 12–15.

73. See Péchu, "Entre résistance et contestation," 12–15; Seigel, *Bohemian Paris*, 234–39; Cate, "Spirit of Montmartre," 38; Grojnowski, "Hydropathes and Company," 107; and Dousteyssier-Khoze, "Fumisme," 153. On *prudhommie*, see Nord, *Republican Moment*; and Tucker, *French Revolutionary Syndicalism.*

74. In Bruant and de Bercy, *L'argot au XXe siècle*, "clochard" is a synonym for "Miserable sans domicile" (319), "Noctambule" (332), and "Vagabond," (435); and Sante, *Other Paris*, 144–48.

75. *Le père peinard*; *Le Figaro*, supplement literaire de dimanche, 13 January 1894, 6; and Merriman, *Dynamite Club.*

76. Carnot, in Pouget, "Mon drapeau rouge," *Le père peinard*, 12 May 1889, 40; and Pouget, "Les bouffe-gallette socialos," *Le père peinard*, 22 December 1889, 81–83.

77. Wileman, "Not the Radical Republic," 595; Rabinow, *French Modern*, 185; A-L. Shapiro, *Housing the Poor*, 146; and Pouget, "Barbarie française," *Le père peinard*, 12 January 1890, 89.

78. Jouy, "À la cloche de bois."

79. "À propos des chiottes," *Le père peinard*, 16 February 1890, 89–91.

80. Pouget, "Tous kif-kif!"

81. On police, see Pouget, "La mistoufle"; "Toujours la dèche," *Le père peinard*, 29 December 1889, 85–87; on murder, see Pouget, "La fête aux proprios"; and on water taps, see Pouget, "À la cloche de bois."

82. Pouget, "Garce d'influenza."

83. Pouget, "Le choléra"; more on "choléra bourgeois" in Pouget, "La mistoufle"; and Kudlick, *Cholera*, 183–85.

84. On Ferry-Choléra, see *Le père peinard*, 5 March 1893, 176; "V'la le choléra," *Le père peinard*, year 4, no. 182, 11–18 September 1892, 1–2; on original rat, see *Le père peinard*, year 4, no. 183, 18 September 1892, back page, reprinted in Dubois, *Le peril anarchiste*, 54.

85. Cate and Shaw, *Spirit of Montmartre*, biography appendix; Applebaum, *French Satirical Drawings*, xiii–xiv, vagabonds on 4, rent on 116; on concierges, *L'assiette au beurre*, no. 200 (28 January 1905), see Kamoun, *V'là Cochon qui déménage*, 28; and on cholera, see Ragon, *L'assiette au beurre*, 271.

86. Ragon, *L'assiette au beurre*, Steinlen 14, Willette 47; mail 82; Applebaum, *French Satirical Drawings*, suicide 153; and Willette "Annoyers" cover, 155.

87. Weber, *France, Fin de Siècle*, 54.

88. Ragon, *L'assiette au beurre*, 288–89.

89. Kamoun, *V'là Cochon qui déménage*; and J-M. Flonneau, "Crise de vie."

90. Péchu, "Entre résistance et contestation," 19–25.

91. Kamoun, *V'là Cochon qui déménage*, 45–46, 81–84.

92. Kamoun, *V'là Cochon qui déménage*, 60.

Chapter 3. The Street Life of the Tramways

1. Crew, *Germans on Welfare*, 8–10; Lüdtke, *History of Everyday Life*; Certeau, *Practice of Everyday Life*; and Eley, "Labor History."

2. Schmucki, "Toward New Routes"; Schmucki, "On the Trams"; Schmucki, "Machine in the City"; Schatzberg, "Mechanization of Urban Transit"; Schatzberg, "Culture and Technology"; Davis and Lorenzkowski, "Platform for Gender Tensions"; Zylstra, "Whiteness"; De Block and Polasky, "Light Railways"; Passalacqua, "Resident Alien"; and Faure, "Les invasions barbares." On everyday technology, see Arnold, *Everyday Technology*, 10–11; Myllyntaus, "Prologue"; Parr, *Sensing Changes*; Josephson, *Fish Sticks*; and Friedel, *Zipper*.

3. Hussey, *Paris*; Wakeman, *Heroic City*; Jerram, *Streetlife*; Sante, *Other Paris*; and Ladd, *Streets of Europe*.

4. Jones-Imhotep, *Unreliable Nation*, 10–13; and Passalacqua, "Reluctant Capitals."

5. Edgerton, *Shock of the Old*, i–xviii; Knowles, "Learning from Disaster?"; and Russell and Vinsel, "After Innovation."

6. Augé, *In the Metro*, xv, 69.

7. Michel, *Léon Say*, 177–81.

8. Société des ingénieurs civils, *Mémoires et compte rendu*, 25:268; Michel, *Léon Say*, 177–81; Parville, *Les causeries scientifiques* (1873), 377–405; and "Communications et avis divers," *Le temps*, 6 September 1872.

9. Schivelbusch, *Railway Journey*, 178–88; and Sérafon, *Étude sur les chemins*, v.

10. *Le temps*, 29 July 1872, 4.

11. Préfecture de la Seine, *Tramways du Département de la Seine: Cahier des charges* (Paris: Imprimerie Charles de Morgues Frères, 1873), AN F[14] 9189.

12. McKay, *Tramways and Trolleys*, 40–51.

13. "Communications et avis divers" *Le temps*, 6 September 1872.

14. Quotes from "impossible" to "instantly" in Michel, *Léon Say*, 179; see also Passalacqua, "Resident Alien" and "Reluctant Capitals."

15. "Locomotive sans fumée et sans feu"; "Études sur les travaux publics: routes, chemins et tramways"; "Les tramways de Paris"; and Parville, *Les causeries scientifiques* (1872), 219–33; (1873), 224–27.

16. "Correspondance: Les tramways dans les promenades"; "Correspondance: Les tramways et la circulation"; and "Correspondance: Les chemins de fer."

17. Krantz, *Ministère de l'agriculture et du commerce*.

18. Lanthier, "Relationship between State and Private Electric Industry," 593; 1894 tramway statistics in AN F[14] 8588, hereafter "Guyot Tramway Stats 1894."

19. Figuier and Gautier, *L'année scientifique*, 310; Backer, *Tramways*; McKay, *Tramways and Trolleys*; Beltran and Carré, *La fée et la servante*, 77; Bouchet, *Le cheval à Paris*; Weil, "Purebreds and Amazons"; and Soppelsa, "Instrumentalisation of Horses."

20. Quote from Société des ingénieurs civils, *Discussion sur l'emploi*, 3; Commission on Mechanical Motors, "La machine à air comprimé (système Mékarski)," second dossier, 1876 (both sources AN F[14] 9198); Guyot Tramway Stats 1894; and Baedeker, *Paris and Environs* (1888), 21.

21. Letter from Inspecteur Général des Ponts et Chaussées Graeff to the Minister, Paris, 10 July 1876; First Dossier from the Commission on Mechanical Motors, 1876: Locomotive avec foyer (Harding) (both sources AN F^{14} 9189); and Sérafon, *La verité*, 17.

22. Quote from Compagnie continentale de locomotion par la machine thermo-spécifique, *Précis sur la locomotive sans foyer*, 7; *Rapport*, 3rd subcommittee of the Minister's Commission, 18 July 1876, in third dossier from Commission on Mechanical Motors, 1876: Locomotive sans foyer (Francq) (both sources AN F^{14} 9189).

23. "Chronique de Paris," *Le temps*, 8 March 1872; and "Paris," *La lanterne*, 31 May 1879, 3.

24. Plans for Tronchon's 1876 "Frein protecteur tronchon," in AN F^{14} 14999; and Folacci (1879) and Marsillon (1880) body catcher plans, in AN F^{14} 9189. Popular image in *La clé des omnibus et tramways*; and "Les tramways de Paris."

25. Letter, President of the South Tramway Company to Rouselle, Chief Engineer for the Dept. of the Seine, 27 October 1877; Letter, Ruthier of the CGO to Rouselle, 2 November 1877; Letter, President of the North Tramway Company to Rouselle, 7 November 1877; Letters, Melon to the Minister of Public Works, 13 February 1878 and 20 October 1879; and Directeur des routes et de la navigation to Melon, 19 November 1879 (all sources AN F^{14} 9189).

26. Letter, Rousseau to the Prefect of Police, 29 October 1878 (AN F^{14} 15000).

27. Champsaur, *Entrée de clowns*, 89–102.

28. Blerzy, "Études sur les travaux publics," 657.

29. Larivière, "Le climat?" 110.

30. Tracy, "Problem of Municipal Nuisances"; and Wagner, "Décembre 1879 à Paris."

31. "Conseil municipal," *Le Figaro*, 12 December 1879, 2–3.

32. "Un chasse-neige pneumatique."

33. Tainturier, *La traction électrique*, 8; and "Electric Tramway Collision."

34. Lopez, "Belgian Investment."

35. Sérafon, *La vérité*, 17.

36. *Compagnie des Omnibus: Tramways dans Paris. Cahier des charges* (Paris: Ch. De Mourgues Freres, 1873), article 22, 12–13 (AN F^{14} 14999).

37. Sérafon, *La vérité*; and Colin, *Paris*, 126.

38. Quote from Sérafon, *La verité*, 22; Robert, *Les tramways Parisiens*, 29–30; Marsoulan, *Vérités nécessaires . . . !*; and Em. Lemoine, *Étude sur la formation*. Administrators from the two tramway companies coauthored the pamphlet *Observations présentées au Conseil général de la Seine*. See also Mitchell, "Private Enterprise or Public Service?"

39. Minister of Public Works, *Circulaire*, 26 September 1887, in Doniol, *La reglementation des chemins de fer . . .* (1900), 257–58; and *Bulletin municipal officiel de la ville de Paris*, 2 July 1885, 1311.

40. Leroy-Beaulieu, "Les chemins de fer"; and Weber, *France, Fin de Siècle*, 70–71 (trams), 195–212 (bikes).

41. Baedeker, *Paris and Environs* (1891), 20–22; P. Vibert, *La concurrence étrangère*, vol. 2; Francq, *Chemin de fer métropolitain*; Weber, *France, Fin de Siècle*, 204–5; and McKay, *Tramways and Trolleys*.

42. Williams, *Dream Worlds*; Miller, *Bon Marché*; Schwartz, *Spectacular Realities*; and Shaya, "Flâneur."

43. Robert, *Les tramways Parisiens*, 1890–95, figures on 36, and appendixes "Évolution du trafic," comparing long-term omnibus, tramway, and Métro traffic. The 1879–1883 tramway and omnibus figures are from *Variations dans le nombre des voyageurs et la recette brute* (AN F^{14} 9154, dossier "Chemin de fer métropolitain: Notes diverses relatives à l'évalutation de la dépense," mid to late 1880s); and Métro figures are from AP V.O8 13.

44. Ministère des travaux publics, *Réseau complémentaire de tramays du Département de la Seine, Rapport de l'inspecteur général*, 8 December 1898 (AN F[14] 15024).

45. Quote from *Le rappel*, 12 November 1892; and Lopin, "Omnibus et tramways," *Le radical*, 7 December 1892. On "cold, wet feet," see "Le chauffage des tramways," *Le parti national*, 5 December 1892; Guyot, *Trois ans*, 96–97; and Armegnaud, *Nettoyons Paris*.

46. Anatole France, *Histoire contemporaine*, 319. On politicians, see Pouget, "Socialo à la Manque"; and on strikers, see Pouget, "Les tramways de Marseille."

47. For a British account, see "Taxes on Transport," 139–40: "Every railway and tramway that is constructed in France has to obtain its sanction from Paris. In cases of trifling importance that sanction may be given by the Council of State, but in all ordinary cases a *loi*—or, as we should say, an Act of Parliament—is required. But, before that sanction is finally given, no less than three separate local inquiries must have been held. The first, known as the *enquête d'utilité publique*, is of a general character, and is held in the chief town or towns, as the case may be, of the district concerned, by a commission 'composed of members selected by the prefect from amongst the landowners, merchants, and representatives of the local authorities.' Notice of the commission's time and place of meeting is widely published. A summary sketch of the proposed undertaking, with an estimate of expense, and a schedule of the rates and fares proposed to be charged, is made accessible beforehand to every citizen. Chambers of commerce, and even individual private persons, are invited to criticise both verbally and in writing."

48. For fares, see Baedeker, *Paris and Its Environs* (1878), 28; and on women, see Hubault, *Omnibus*, 106–7. For enquête, see Gauthier, *Rapport de l'ingénieur ordinaire*, Paris, 18 November 1896 (AN F[14] 14999). See also Conseil général de la Seine, *Rapport* no. 13, 9 December 1896, and Conseil général de la Seine, *Rapport* no. 5, 2 April 1897 (both sources AP 25W 100).

49. Gauthier, *Rapport de l'ingénieur ordinaire*, Paris, 18 November 1896.

50. "Les tramways à dix centimes," *L'éclair*, 24 October 1897; "Les tramways en Amerique," *La rive gauche*, 10 April 1898; "La question des tramways," *La rive gauche*, 15 July 1898; "La question des tramways," *Paris ouest*, 4 September 1898 (all 1898 papers, AP 25W 100). I thank Gabrielle Hecht for pointing out that the absence of comments from the east side suggests that working-class Parisians did not know how or feel empowered to comment.

51. Conseil général de la Seine, *Rapport* no. 5 (2 April 1897), presented by Councilor Gibert of Saint Mandé (AP 25W 100).

52. *Tramways concédés dans la département de la Seine de juin 1899 à juin 1902* (AN F[14] 15024); "Les tramways à 0.10 centimes," *L'éclair*, 19 June 1896; Conseil général de la Seine, *Rapport* no. 5 (2 April 1897), presented by Gibert, councilor from Saint Mandé; "Les tramways à dix centimes," *L'éclair*, 24 October 1897; Laere, "La question des tramways"; Ferrand, "Les tramways de la Seine"; Conseil général de la Seine, *Rapport* no. 9 (2 July 1898), presented by Gibert (AP 25W 100); *Journal officiel de la République française*, year 34, no. 317, 21 November 1902 (Séance du 20 November 1902), 2679–893; and Berlanstein, *Big Business*.

53. P. Vibert, *La concurrence* étrangère, 2:83–84; Armegnaud, *Nettoyons Paris*, xiii, 67; *Le Figaro*, in Uzane, *La locomotion à travers*, 208; Robida, in Willems, "Stereoscopic Vision," 374; and Robert, *Les tramways Parisiens*, 31–63, 115–60.

54. Laere, "La question des tramways"; and Ferrand, "Les tramways de la Seine." On Grébauval, see *Dictionnaire national des contemporains*, 5:75–76.

55. AN F[14], cartons 15024–15030: Réorganisation des tramways de Paris et du département de la Seine, 1899–1920.

56. Union vélocipédique de France, "Faux départ"; "Le tramway de la mort," *Le Figaro*, 1

January 1900, 4; "Les oeuvres et les hommes"; "Les tramways meurtriers," *Le petit Parisien*, 22 August 1900; "Les tramways meurtriers," *Le petit Parisien*, 5 October 1900, 1; and "Mésaventures électriques," *Le temps*, 19 October 1900, 2–3. Similar concerns attended early automobiles: M. Flonneau, *Paris et l'automobile*; Jackson, "Solidarism"; and Ladd, *Autophobia*.

57. "Tramways empoisonneurs," *Le radical*, 11 October 1901, 2–3.

58. *Mémoires de M. le Préfet de la Seine & de M. le Préfet de police* (1905): 286–309, *avorteurs* on 306; *Bulletin municipal officiel de la ville de Paris*, 29 June 1905, 2182.

59. Sibey, "Jurisprudence"; "Les tramways écraseurs," *La presse*, 7 June 1922, 1; and "Les tramways écraseurs," *Le rappel*, 8 June 1922, 4.

60. Meyan, "Les accidents de tramways," *Le Figaro*, 15 October 1900, 2. Brief reports (this list is not exhaustive!): "Tramways écraseurs," *La lanterne*, 4 April 1901, 2; "Les tramways écraseurs," *L'intransigeant*, 9 June 1901, 4; "Les tramways écraseurs," *Le Figaro*, 8 July 1901, 4; "Les tramways écraseurs," *Le Figaro*, 26 August 1901, 4; "Les tramways écraseurs," *Le petit Parisien*, 26 August 1901, 3; "Tramways écraseurs," *L'aurore*, 14 September 1901, 3; "Les tramways écraseurs," *Le matin*, 21 September 1901, 4; "Les tramways écraseurs," *Le Figaro*, 8 October 1901, 4; "Les tramways écraseurs," *Le matin*, 8 October 1901, 4; "Les tramways écraseurs," *Le radical*, 14 October 1901, 3; "Les tramways écraseurs," *L'intransigeant*, 24 December 1901, 4; "Saint Mandé—tramways écraseurs," *Le petit Parisien*, 25 March 1902, 4; "Tramways écraseurs," *Le petit Parisien*, 20 July 1902, 5; "Les tramways écraseurs," *Gil Blas*, 24 August 1902, 3; "Les tramways écraseurs," *La lanterne*, 19 November 1902, 3; "Les tramways écraseurs," *La petite presse*, 21 December 1902, 3; "Les tramways écraseurs," *La liberté*, 25 December 1902, 2; "Toujours les tramways écraseurs," *Le rappel*, 30 December 1902, 3; "Toujours les tramways écraseurs," *Le rappel*, 6 January 1903, 3; "Les tramways écraseurs," *Le rappel*, 11 January 1903, 3; "Les tramways écraseurs," *Le radical*, 16 August 1903, 3; "Les tramways écraseurs," *L'aurore*, 26 August 1903, 2; "Les tramways écraseurs," *La petite presse*, 12 October 1903, 3; "Les tramways écraseurs," *L'aurore*, 21 December 1903, 3; "Les tramways écraseurs," *La petite presse*, 27 May 1904, 3; "Les tramways écraseurs," *La presse*, 28 May 1904, 4; "Tramways écraseurs," *La petite presse*, 29 May 1904, 3; "Les tramways écraseurs," *La croix*, 6 January 1905, 4; "Les tramways écraseurs," *La lanterne*, 20 June 1905, 3; "Les tramways écraseurs," *Gil Blas*, 4 July 1905, 3; "Les tramways écraseurs," *Le radical*, 3 August 1905, 2; "Les tramways écraseurs," *Le matin*, 15 September 1905, 4; "Les tramways écraseurs," *Le radical*, 9 October 1905, 2; "Les tramways écraseurs," *La presse*, 5 November 1905, 1; "Les tramways écraseurs," *La petite presse*, 30 November 1905, 3; "Les tramways écraseurs," *L'aurore*, 16 March 1906, 3; "Les tramways écraseurs," *L'aurore*, 16 December 1906, 3; "Les tramways écraseurs," *L'aurore*, 12 August 1907, 3; "Les tramways écraseurs," *Gil Blas*, 24 December 1907, 2; "Les tramways écraseurs," *Le rappel*, 5 March 1908, 3; "Les tramways écraseurs," *Le rappel*, 9 March 1908, 3; "Tramways écraseurs," *Le rappel*, 12 March 1908, 3; "Les tramways écraseurs," *Le rappel*, 10 April 1908, 3; and "Tramways écraseurs," *Le rappel*, 16 October 1908, 3.

61. "Toujours les tramways écraseurs," *Le rappel*, 30 December 1902, 3; and "Toujours les tramways écraseurs," *Le rappel*, 6 January 1903, 3. "Crusher" trams appeared regularly in papers from *L'humanité* (far left) to *L'action française* (far right).

62. Claretie, *La vie à Paris* (1897), 110. Quote from "Échos: À travers Paris," *Le Figaro*, 22 June 1897, 1; and "Chronique Parisienne," *La croix*, 22 June 1897, 3. The populist paper *Le matin* added familiar safety and traffic concerns: see "Le tramway des barbares," 9 May 1897, 2; "À l'Hôtel de ville," 21 May 1897, 2; "Les journaux de ce matin," 24 May 1897, 2; "À l'Hôtel de ville," 14 June 1897, 3; and "Les journaux de ce matin," 21 June 1897, 2. On the Champs-Élysées, see Walton, *Paris, from the Earliest Period*, 8:83.

63. "Les interpellations," *Le matin*, 4 April 1897, 1; quote on horses, from "Le tramway des barbares," *Le matin*, 9 May 1897, 2; and "Le tramway des barbares," *Le matin*, 25 May 1897, 2.

64. Quote on descent, from Walton, *Paris, from the Earliest Period*, 8:84; "Chronique Parisienne," 1897; Martial, "Le mouvement de population"; Robert, *Les tramways Parisiens*, 57, 79, 102–3; Warner, *Streetcar Suburbs*; Stovall, *Rise of the Paris Red Belt*; Merriman, *Margins of City Life*; and Chevalier, *Laboring Classes*.

65. "Le tramway des barbares," *Le matin*, 29 May 1897, 3 (all quotes are from this source); Walton, *Paris*, 8:83–84; "Le tramway des barbares," *La croix* (supplement), 30 May 1897, 3.

66. Quote on Montmartre is from Walton, *Paris, from the Earliest Period*, 8:84; "Tablettes théatrales," *Le matin*, 26 June, 27 June, 2 July, and especially 31 July 1897, 4; and "A Paris impromptu."

67. "Échos: À travers Paris," *Le Figaro*, 24 April 1898, 1; 1 January1899, 1; and 1 January 1900, 1; "Le tramway de Romainville," *Le temps*, 11 January 1900, 2; "Le tramway de la rue du Quatre-Septembre," *Le temps*, 18 January 1900, 1; "Nouveaux tramways," *La presse*, 31 March 1900, 1; "Échos: À travers Paris" and "Le tramway de la mort," *Le Figaro*, 1 January 1900, 1 and 4, respectively; Jacquemet, "Voirie," 249n25; and Hausser, *Paris*, 15.

68. Hausser, *Paris*, 15; "Le tramway de Romainville," *Le temps*, 11 January 1900, 2; and "Le tramway de la rue du Quatre-Septembre," *Le temps*, 18 January 1900, 1.

69. Faure, "Les invasions barbares"; Jacquemet, "Voirie," 249; Martial, "Le mouvement de population," 562; Guérard, *L'avenir de Paris*, 211; "Les tramways de Paris au Conseil municipal," 350; Conseil municipal de Paris, *Procès-verbaux* (1901): 2078; and "Les miettes de la vie," 30.

70. "Les oeuvres et les hommes," 416; "Destins," *Le Figaro*, 20 June 1907, 1; and "Actualités."

71. Levallois, *Paris Propre!* 8–9, 18; and pretensions, in Terrail, *L'Angleterre*, 8.

72. "Les tramways à 0.10 centimes," *L'éclair*, 19 June 1896; "Les tramways à dix centimes," *L'éclair*, 24 October 1897.

73. *Bulletin municipal officiel de la ville de Paris*, 6 December 1892, 2809.

74. *Bulletin municipal officiel de la ville de Paris*, 30 November 1893, 2501.

75. "Le tramway fântome," *La petite presse*, 29 June 1897, 3; Parisot, in *Bulletin municipal officiel de la ville de Paris*, 22 June 1899, 2096, and *Bulletin municipal officiel de la ville de Paris*, 22 March 1900, 1069; Cherest, in *Bulletin municipal officiel de la ville de Paris*, 5 July 1900, 2330; and Cherioux, in *Bulletin municipal officiel de la ville de Paris*, 16 December 1906, 4390. See also Mathieu, "Le tramway fântome," *L'avenir républicain*, 1 June 1905, 2.

76. "Tramway fântome," *Le rappel*, 8 October 1903, 3; and "Tramway fântome," *Le XIXe siècle*, 8 October 1903, 3.

77. "Tramway fântome," *Le Figaro*, 2 June 1905, 3.

78. *Bulletin municipal officiel de la ville de Paris*, 21 November 1907, 4387.

Chapter 4. The Dream Life of the Métropolitain

Epigraph: Letter, engineers Vauthier and Deligny to the Minister of Public Works, 22 May 1894 (AN F^{14} 9154): "À partir de 1872, la question métropolitaine dormit longtemps."

1. Jasanoff and Kim, *Dreamscapes of Modernity*, 4; and Anand et al., *Promise of Infrastructure*.

2. On Hausmmann's subway and light-rail neglect, see Higonnet, *Paris*, 191; Jordan, "Haussmann and Haussmannization," 96; Schivelbusch, *Railway Journey*; and Papayanis, *Planning Paris*.

3. Schoeller, *Les chemins de fer*, 318. In "Essai de bibliographie," Frédéric Lions counted 106 plans.

4. Biette, *Le métropolitain de Paris*; Hervieu, *Le chemin de fer*, vol. 1; Evenson, *Paris*, 91–105; Hall-

sted-Baumert et al., *Métro-cité*; Desabres, "Parisian Subway"; Desabres, "Decision Network"; Cottereau, "Les batailles du métropolitain" and "Les batailles pour la création du Métro"; Papayanis, *Planning Paris*; and Mitchell, *Rêves Parisiens*.

5. Ossadzow, "Les pères du métropolitain," esp. 63; and Mitchell, *Rêves Parisiens*.

6. Bobrick, *Labyrinths of Iron*, 14; and Pike, *Subterranean Cities*, 47.

7. Lopéz Galviz, *Cites, Railways, Modernities*; Cottereau, "Les batailles du métropolitain," quote on 77; Cottereau, "Les batailles pour la création du Métro."

8. Bobrick, *Labyrinths of Iron*, 133–68; Pike, *Subterranean Cities*, 47–68; Cottereau, "Les batailles pour la création du Métro"; Lopéz Galviz, *Cites, Railways, Modernities*; Grubbs, "Terminus 1900"; Soppelsa, "Urban Railways"; Soppelsa, "Visualizing Viaducts"; Soppelsa, "Reworking Appropriation"; on retrofit, see Howe et al., "Paradoxical Infrastructures."

9. Williams, *Notes on the Underground*, ch. 1; Lopéz Galviz, *Cites, Railways, Modernities*, ch. 1; and Levin, "Bringing the Future to Earth," esp. 41–42.

10. Lopéz Galviz, *Cites, Railways, Modernities*, 3; Mitchell, *Rêves Parisiens*, 8, 86; Grubbs, "Terminus 1900"; and Schmucki, "Toward New Routes."

11. Bijker, *Of Bicycles, Bakelites, and Bulbs*, 84–88; and Soppelsa, "Reworking Appropriation."

12. Taine, "Lettres de Londres," *Le temps*, 8 January 1872; Spencer, "Fresh Look"; Sérafon, *Etude sur les chemins*, v. On London as a model, see Tomory, "London's Water Supply." On transnational comparisons, see Lewis, *Early Encounter*; Nolan, *Visions of Modernity*; B. Anderson, *Spectre of Comparisons*; Reynolds, *Paris-Edinburgh*; and Çelik, *Empire*.

13. "Lettres de Londres," *Le temps*, 8 January 1872, 2; *Tables du journal Le temps*, 643; and Graham and Marvin, *Splintering Urbanism*, 49, 52, 62.

14. Hamel, *Histoire illustrée*, 386; Simond, *La vie Parisienne*, 2:431; and Colin Jones, *Paris*, 323.

15. Le Hir, *Réseau des voies ferrées*; and Papayanis, *Planning Paris*, 220–24.

16. Pike, *Subterranean Cities*; Bobrick, *Labyrinths of Iron*; Williams, *Notes on the Underground*; J. Strauss, *Human Remains*; Zaretsky, "Sewers of Paris"; Reid, *Paris Sewers*, 1–5; "sewer train," Evenson, *Paris*, 93; and Benjamin, *Arcades Project*, 84–85, 89, 91, 93, underworld on 84. Métro as sewer again, Larroque, "Un aspect residuel," 66.

17. Huet, *Chemin de fer*, 5–6, 27; Letter, Minister of Public Works to the Prefect of the Seine, Versailles, 12 July 1876 (AN F^{14} 9153); Bowie, "Paris"; and Ossadzow, "Les pères du métropolitain."

18. Huet, *Les chemins de fer*.

19. Nansouty, *La question du réseau*, 10–11; and Biette, *Le métropolitain de Paris*, 4.

20. Heuzé, *Chemin de fer*, 5; Olivier, *Chemins de fer*, 8; Evenson, *Paris*, 93–95; Bobrick, *Labyrinths of Iron*, 143; and Pike, *Subterranean Cities*, 47–49.

21. Rammell, *Tramway souterrain*; "Chemin de fer atmosphérique"; Lopéz Galviz and Dikec, "Modern Atlas."

22. Figuier, *Les merveilles*, vol. 1; Fevre, *La locomotive*; Delmer, *Les chemins de fer*, 5; and Weber, *France, Fin de Siècle*, 70.

23. Société des ingénieurs civils de France, *Mémoires et compte rendu*, 38:30; and Haag, *Le métropolitain de Paris*, 16.

24. Flachat, in Société des ingénieurs civils, *Mémoires et compte rendu*, 25:268; Michel, *Léon Say*, 177–81; and Parville, *Les causeries scientifiques* (1879), 377–405.

25. Théry, *Les chemins métropolitains*.

26. Quotes, respectively, from Anonymous, *Deux métropolitains*, 4; and Heuzé, *Chemin de fer*, 1–2.

27. Frémaux, *Chemin de fer métropolitain de Paris. Avant-projet. Résultats de l'enquête d'utilité publique. Rapport de l'inspecteur général*, 12 September 1883, 23; The Milinaire Brothers, Brochure (1885), 14 (both sources AN F[14] 9154).

28. Cottereau, "Les batailles du métropolitain," 82; Desabres, "Parisian Subway," 3; and Polasky, "Transplanting and Rooting Workers."

29. For the 1880 law, see Doniol, *La règlementation des chemins*, 241; and council, in Ossadzow, "Les pères du métropolitain," 61. See also Mitchell, *Rêves Parisiens*; Lanthier, "Relationship Between State and Private Electric Industry"; and Desabres, "Parisian Subway."

30. Frémaux, *Chemin de fer métropolitain*, 25 (AN F[14] 9154).

31. Marisy, "Le déficit communal," 625.

32. Guyot, *Trois ans*, 86.

33. The terms "closure," "stabilization," and "interpretive flexibility" are from Bijker, *Of Bicycles, Bakelites and Bulbs.*

34. Nord, *Republican Moment*; and Tucker, *French Revolutionary Syndicalism.*

35. Haag, "Le chemin de fer."

36. Société des ingénieurs civils de France, *Mémoires et compte rendu* 39:265–69. Studies: Berlin (March to June 1883; vol. 39, 313–30, 614–17, 635–36, 775–81, respectively), New York (vol. 40, 6 July, 9–27), and San Francisco (vol. 40, 20 July, 33–42). The final decision (vol. 40, 3 August, 166–96) has been cited often, see: Jules Garnier (1884), the Milinaire Brothers (1885, 1–2), and Pike, *Subterranean Cities*. Evenson, *Paris*, 95–97, 104; "Souvenirs de l'exposition d'électricité," 59–62; Motorization experiments in AN F[14] 9189 and F[14] 14999; and Figuier, *Les nouvelles conquêtes*, 503–640.

37. Eugène Chardon, Letter to M. Charles Garnier, 30 April 1887; Milinaire Brothers, Brochure (1885), 1 (AN F[14] 9154).

38. Quotes from Frémaux, *Chemin de fer métropolitain*, 25 (AN F[14] 9154). See also Chrétien (1882), Mazet (1884), and Haag (1887). For trees, see Jules Garnier (1884), 11, 14, 28, 39, Milinaire Brothers (1885), 3–7; all sources AN F[14] 9154; and Figuier, *Les nouvelles conquêtes.*

39. Soppelsa, "Visualizing Viaducts."

40. In 1884 Charles Normand published a manifesto, *Société* des Amis des Monuments Parisiens, *constituée dans le but de veiller sur les monuments d'art et la physionomie monumentale de Paris (architecture, peinture, sculpture, curiosités et souvenirs historiques)*. Evenson, *Paris*, 103–4, gives the founding date as 1885, when the society first published a bulletin. But Elisée Reclus addressed a letter to Normand on 24 September 1879, calling him "secrétaire de la Société des Amis des Monuments Parisiens," smudging the chronology. See Reclus, *Correspondance*, 216; "Anti" quote from Evenson, *Paris*, 93 and Pike, *Subterannean Cities*, 47.

41. *Projet de chemin de fer.*

42. Boudenoot, "Le métropolitain de Paris," 156.

43. *La question du métropolitain*, 9.

44. Only three new trams opened from 1887 to 1889, see Tramway Stats 1894 (AN F[14] 8588). On embarrassment, see Papayanis, "Development of the Paris Cab Trade," 55; Mitchell, *Rêves Parisiens*; and Grubbs, "Terminus 1900."

45. Baedeker, *Paris and Environs* (1888), 25.

46. Conseil général de la Seine, session de 1890, Routes nationales et départementales, *Chemins de fer métropolitains. Tramways. Rapport de l'ingénieur en chef du département*, quotes, 57 and 78, respectively (AN F[14] 15000); Villain, *Le Métro*; Villain and Mauger, *Un métropolitain* (both AN F[14] 9154); Guyot, *Trois ans*, 84; Nansouty, *La question du réseau*, 4.

47. *Enquête* in *Le temps*, 18 July 1890, 3; map in the illustrated supplement of 21 July 1890.

48. Petitions in Mitchell, *Rêves Parisiens*, 104–5.

49. Quotes from Nansouty, "Le réseau central métropolitain." See also Nansouty, *La tour Eiffel* and Nansouty, *La question du réseau.*

50. Great Depression, in Hobsbawm, *Age of Empire*, 34–55; Biette, *Le métropolitain de Paris*, 4, where he highlights 1877–1895; and Nansouty, "Le réseau central métropolitain."

51. Guyot, *Trois ans*, quotes 13 and 88, respectively; London on 96.

52. P. Vibert, *La concurrence étrangère*, 1:preface, 231–308. Métro globalization, in Desabres, "Parisian Subway"; and Paris globalization, in Harvey, *Paris, Capital of Modernity*, 13, 47–50, 107–16. See also Adas, *Machines*; Conklin, *Mission to Civilize*; and Hobsbawm, *Age of Empire.* On techno-nationalism, see Hughes, *Networks of Power*; Herf, *Reactionary Modernism*; D. Nye, *American Technological Sublime*; Hecht, *Radiance of France*; Hård and Jamison, *Intellectual Appropriation*; and Hård and Knie, "Grammar of Technology."

53. Pouget, "Toujours les grinches."

54. Odelin, *Métropolitain.*

55. Mollier, *Le scandale*; *L'intransigeant*, in Lapparent, "La question du métropolitain" (part 1), 191.

56. See also P. Vibert, *La concurrence étrangère*, 1:235; and Villain and Mauger, *Un métropolitain*, 43. On Eiffel, antisemitism, and Panama, see Jonnes, *Eiffel's Tower*, 18; Loyrette, *Gustave Eiffel*, 21, 193–200; Lindemann, *Anti-Semitism before the Holocaust*, 65–70.

57. Lapparent, "La question du métropolitain," quotes 189 and 191; see also Lapparent, "La question du métropolitain" (part 2); and Lapparent, "À propos du métropolitain."

58. Burgraff, "Tramway funiculaire"; Figuier, *L'année scientifique* (1893), 109–10; Baedeker, *Paris and Environs* (1884), 25; and Baedeker, *Paris and Environs* (1891), 24.

59. T. Clark, "Rapid Transit in Cities," map on 573; Gray, *Proposed Plan*, 27–30; and Rousselle, "Compte rendu du Sécretariat." On the tubular tramway, see Figuier, *L'année scientifique* (1892), 211–20; Figuier, *L'année scientifique* (1894), 171–74; C. Carré, "L'enquête sur le tramway"; "Chronique et revue de la presse industrielle," 276–79; and "nightmare," in Mitchell, *Rêves Parisiens*, 105.

60. Conseil général du département de la Seine, Première Session de 1892. *Mémoires de M. le Préfet de la Seine & de M. le Préfet de police et procès-verbaux des délibérations*, 81.

61. P. Vibert, *La concurrence étrangère*, 1:231–42 and 254–82, first quote on 266, slurs on 255, 261, and 265.

62. Conseil général du département de la Seine, *Mémoires de M. le Préfet*, 298–99; "Conseil Municipal," *Le temps*, 12 January 1895, 3; Drumont, *Automobiles sur rails*, 167; and Guyot, *Trois ans*, 96–97.

63. "Le métropolitain," *Le temps*, 21 November 1896, 3; Guyot, *Trois ans*, 85; Baedeker, *Paris and Environs* (1896), 23–25; and Baedeker, *Paris and Environs* (1898), 24–26.

64. Quote from Maubry, "Le métropolitain"; Margairaz, "Le réseau métropolitain"; Tricoire, ed. *Le Métro de Paris*; Tricoire, ed. *Métropolitain*; Berton and Ossadzow, *Fulgence Bienvenüe*; Hovey, *Paris Underground*; and Ovenden, *Paris Underground.*

Chapter 5. Under Construction

1. On French public-private contracts, see Bocquet et al., "From Free Good to Commodity."

2. Foucault, "Of Other Spaces"; Harvey, *Paris, Capital of Modernity*; Hommels, *Unbuilding Cities*; "inverting," in Bowker and Star, *Sorting Things Out*, 34; "upside-down," in "Les embarras

de Paris," *Le radical*, 29 August 1908, 1–2; and "Petits paradoxes de Paris," *Le matin*, 18 August 1909.

3. Schipper and Schot, "Infrastructural Europeanism," 247; and Carse, "Keyword," 29–31.

4. Klinenberg, *Palaces for the People*.

5. "Socialistes et monopoles," *Le temps*, 28 April 1895, 1.

6. Bon, *La psychologie des foules*, 74; on the CGT, see Tucker, *French Revolutionary Syndicalism*; Lafargue, "L'idéalisme"; and Pelloutier, "L'anarchisme" and "La situation actuelle."

7. "La grève des omnibus," *Le Figaro*, 23 April 1895, 1–2; "La grève des omnibus," *Le Figaro*, 24 April 1895, 1–2; "Nouvelles à la main," *Le Figaro*, 25 April 1895, 1; Burke, "Foreign and Colonial History," 229–30; and Shorter and Tilly, *Strikes in France*, 115, 150.

8. Paul Vibert's speeches in *La concurrence étrangère*.

9. Ross, *Communal Luxury*.

10. Biette, *Le métropolitain de Paris*, 9; and James, "Conditions Relating to the Treatment of Employees."

11. Mazoyer, *Les conditions du travail*; labor inspections, in Reid, "Putting Social Reform into Practice"; health systems, in Tissanie-Noir, "Le système de santé."

12. Nord, "Welfare State in France"; on scale, see also Carroll, "Water and Technoscientific State Formation."

13. "Chronique politique," *La quinzaine* (1898).

14. Soppelsa, "Paris's 1900 Universal Exposition."

15. "La nonchalence de M. de Selves," *La lanterne*, 24 November 1905; "La grève des terrassiers," *La petite république*, 24 November 1905; "La grève des terrassiers," *L'humanité*, 25 November 1905; and "La grève des terrassiers," *Le petit Parisien*, 25 November 1905. See also "La grève des terrassiers," *Le journal*, 11 January 1906.

16. See "Le caisson du métropolitain"; Tricoire, *Le Métro de Paris*, 124–41; and Lopéz Galviz and Dikec, "Modern Atlas," 17.

17. "Accident dans un chantier," *La lanterne*, 2 July 1906, 2.

18. Conseil municipal de Paris, *Procès-verbaux* (1907), 1458; Fulgence Bienvenüe, *Note* from the *Service technique du métropolitain*, 24 December 1907 (AP V.ONC 129); "À Paris: au chantiers du Métro," *Le rappel*, 25 December 1907, 2; and "La catastrophe du Métro," *Le petit Parisien*, 27 December 1907, 1–2.

19. "À Paris: au Métro," *Le rappel*, 2 March 1908, 2; and "À Paris: au Métro," *Le XIXe siècle*, 2 March 1908, 2.

20. "Situation critique: gares inondées," *Le radical*, 2 April 1909, 1; and "Sabotage criminel," *Gil Blas*, 22 May 1909, 2.

21. "Un caisson à air comprimé explose à Lorient," *Gil Blas*, 28 September 1913, 2; "Un caisson explose," *Le radical*, 28 September 1913, 1–2; "Explosion d'un caisson," *Le rappel*, 29 September 1913, 1; and "Explosion d'un caisson," *Le XIXe siècle*, 29 September 1913, 1.

22. See "Les malfaçons," *Le matin*, 16 January 1906; "Les malfaçons du métropolitain," *Le matin*, 26 January 1906; and sewerman quoted in "Les malfaçons du métropolitain," *Le matin*, 6 February 1906.

23. Quoted in Bogaert, "Les documents d'un conflit," 124.

24. Bogaert, "Les documents d'un conflit," 8; and Bogaert, "Les documents d'un conflit: le lock-out."

25. France, Direction du travail, *Statistique des grèves* (1907), v–vi, ix, 324, 583, 609–10; *Statistique des grèves* (1909), v–vi; and *Bulletin municipal officiel de la ville de Paris*, 9 July 1907, 2918.

26. Bogaert, "Les documents d'un conflit," 18.

27. Quote from "La désorganisation des services publics," *L'intransigeant*, 18 August 1908. See also "Entrepreneurs et terrassiers," *Le siècle*, 19 August 1908; "La crise du bâtiment," *L'action*, 21 August 1908; "Entrepreneurs et terrassiers," *L'aurore*, 22 August 1908; "Pas de Lock-Out!" *Le radical*, 22 August 1908; "Chez les terrassiers," *L'aurore*, 23 August 1908 (all 1908 newspapers in AP V.ONC 131). On health and safety, see "Au métropolitain," *L'humanité* 16 August 1907; "Questions Parisiennes: l'hygiène et le métropolitain," *La patrie*, 16 August 1908; and "L'hygiène au métropolitain," *L'humanité*, 3 September 1910.

28. Lafargue, "La communisme et les services publics," *L'égalité*, 25 June and 2 July 1882.

29. Nord, *Republican Moment*; and Christianson, *Paris Babylon*.

30. "Au métropolitain," *L'humanité*, 16 August 1907; and "L'hygiène au métropolitain," *L'humanité*, 3 September 1910.

31. Blau, "City as Protagonist," 16; and Jordan, "Haussmann and Haussmannisation," 112.

32. "Le métropolitain," *Le temps*, 26 March 1896, 3; Romains quoted in Pike, *Subterranean Cities*, 51 and Williams, *Notes on the Underground*, 79.

33. AP V.O8 15 and V.ONC 78; and "Les embarras de Paris," *Le radical*, 29 August 1908.

34. "Chronique de Paris," *Le temps*, 16 February 1872; "Chronique de Paris," *Le temps*, 8 March 1872; prodigies, in "Flâneries et curiosités," 212–14; Claretie, *La vie à Paris* (1897), 110; and Harrison, "Paris in 1851."

35. "Les embarras de Paris," *L'aurore*, 23 August 1909; "Les travaux de Paris: on n'en finira jamais!" *La libre parole*, 22 August 1909; and "Les propos du lanternier," *La lanterne*, August 18, 1909 (AP V.ONC 129). See also "Paris impracticable," *Le XIXe siècle*, 14 April 1909, 1; and "Un musée, une bibliothèque et un lycée menacés d'un chantier," *Le matin*, 25 August 1911 (AP V.ONC 131).

36. Series of tramway requests to modify itineraries, 1888–1897 (AP V.ONC 79); more tramway requests for new routing, 1899–1901 (AP V.ONC 78); tramway "restructuring" of 1902–1910 (AN F[14], cartons 15024–15028); see also Conseil municipal de Paris, *Rapport* no. 60 (25 July 1907), 12, where Duval-Arnould mentions CGO lines rendered "inoperable" (*inexploitable*) by Métro works.

37. "Les embarras de Paris," *Le radical*, 29 August 1908, 1–2.

38. Métro-related complaints (*contentieux*), AP V.ONC 78 and V.O8 15.

39. On 1899, see BA Métro Photo no. 182. The credited photographer, Daniel Lieferman, is not a known historical figure. To document similar public works, especially roadworks and building demolitions, the Office of Works hired the Union photographique française (UPF), a worker-run photographic cooperative, to which Liferman likely belonged. More UPF Métro photos are held in AP D10.S9, cartons 6–19, and published in Tricoire, *Le métro de Paris*. See also J-P. Dumas, *Paris la rue*. On 1915, see Agence Rol photograph, https://commons.wikimedia.org/wiki/File:Fontis_Place_de_l%27Alma_8_novembre_1915.jpg. See also Bibliothèque nationale de France, "Métros—images sur ce thème," https://data.bnf.fr/fr/documents-by-rdt/11950823/i/page1.

40. "L'hygiène et le métropolitain," *La patrie*, 16 August 1908; "L'hygiène au métropolitain," *L'humanité*, 3 September 1910; C. Vibert, *La catastrophe du métropolitain*; and Thierry, *Étude sur le métropolitain*.

41. Letter, Comité des Intérêts du VIIe Arrondissement to the municipal council, ca. 3–4 July 1900 (AP V.ONC 78).

42. Darcy, "Le 'ratarium' de la ville de Paris," *L'événement*, 22 August 1909.

43. "Affaires municipals: la toilette de Paris," *Les nouvelles*, 9 August 1911.

44. "Le paradis de capucines," *L'événement*, 19 June 1909; "La capitale sabotée," *L'intransigeant*, 5 August 1911; "Affaires municipales," *Les nouvelles*, 9 August 1911; and "Les bûcheronnes de Paris," *L'intransigeant*, 30 September 1911 (all sources AP V.ONC 131).

45. Nansouty, "Causerie scientifique"; Bonnin, "Les travaux du métropolitain"; Biette, "Chemins de fer"; Pollet, "La nouvelle ligne métropolitaine"; Loyselles, "Le métropolitain"; and Hervieu, *Le chemin de fer*, 2:308, 310, 317, 348–52. For photographs, see Union photographique française, *Chemin de fer métropolitain municipal*, vol. 3. Foreign sources: Bonnin, "Work of Superposing Three Lines" (US); "Paris Metropolitan Railway" (UK); and "Busiest Underground Corner" (US).

46. On meanings of "trou," see Larchey, *Dictionnaire historique d'argot*, 353; and Barrère, *Argot and Slang*, 462. For "trou de l'Opéra," see "La gare de l'Opéra"; Bailby, "Le 'trou' de l'Opéra," *La presse*, 3 October 1904, 1 (see this source for "mouth and orifice"); and Puck, "Le trou de l'Opéra," *Gil Blas*, 17 July 1910, 1.

47. "Le trou de l'Opéra," *L'humanité*, 19 October 1904, 1.

48. Nansouty, "Causerie scientifique"; "Paris éventré," *La liberté*, 26 August 1909; and "Petits paradoxes de Paris," *Le matin*, 18 August 1909 (both 1909 sources are in AP V.ONC 129); "La place de l'Opéra n'est plus qu'un trou immense!" *L'intransigeant*, 25 August 1910; and "Partout des trous," *L'intransigeant*, 2 September 1910 (both 1910 AP V.ONC 131).

49. "Opéra-chantiers," *L'écho de Paris*, 25 August 1909 (AP V.ONC 129); Tricoire, *Le Métro de Paris*, 19, 23, 35, 74, 90, 130–33, 190, 198; Hovey, *Paris Underground*, 20: illustration of the Place de la Concorde 1899; figure 5.5 in this chapter; and https://en.m.wikipedia.org/wiki/File:Paris_Metro_construction_03300288-5.jpg.

50. "Paris en démolition," *Le temps*, 19 August 1909; "Paris-Chantiers," *L'éclair*, 6 August 1911; and "Les travaux de Paris: on n'en finira jamais!" *La libre parole*, 22 August 1909 (all sources AP V.ONC 129).

51. "Les arbres de l'avenue de Clichy," *La patrie*, 16 August 1908, complains that construction of the Nord-Sud subway threatened the avenue's planted trees. Préfecture de la Seine, Direction administrative des travaux de Paris, Service technique de la voie publique et de l'éclairage, *Rapport de l'inspecteur général*, 26 November 1908 (AP V.ONC 129); and Howe et al., "Paradoxical Infrastructures."

52. Pinning a newspaper clipping to administrative correspondence, often to justify claims about public opinion, was common administrative practice. Secretaries in public works administration regularly kept press reviews and files of news clippings about controversial topics. These press reviews are the governor watching the governed, which makes the press the arbiter of public opinion, with governors relying on the press to provide accurate information about the public. Press reviews *do* provide historians with access to public opinion, but it is public opinion as paraphrased by the press and then sorted by officials.

53. "Petits paradoxes de Paris," *Le matin*, 18 August 1909.

54. "Nos échos," *Le journal* 28 June 1898, 1; and "Le métropolitain," *Le petit Parisien* 18 February 1903, 1.

55. "Notes de la semaine: les embarras de Paris," *Les annales politiques et littéraires* 49 (22 December 1907): 589–90; "Les travaux de Paris," *La petite république* 7 September 1909; "Les travaux de Paris," *La liberté*, 9 September 1909; "Paris-Fondrière," *L'autorité*, 9 September 1909; "L'encombrement de la voie publique," *Le petit Parisien*, 10 September 1909; and "Les embarras de Paris," *Le moniteur de Paris*, 19 September 1909 (all 1909 sources are in AP V.ONC 129).

56. "Petits paradoxes de Paris," *Le matin*, 18 August 1909; "Les embarras de Paris," *L'aurore*, 23 August 1909; "La place de l'Opéra n'est plus qu'un trou immense!" *L'intransigeant*, 25 August 1910; and "Sabotage stupide," *L'éclair*, 1 August 1911.

57. "Nouvelles diverses," *Le Figaro*, 28 October 1908, 4; "Les embarras de Paris," *Le mois littéraire et pittoresque* 14 (1905), 347; "La réglementation des automobiles," *Gil Blas*, 15 October 1905, 1; "À la brigade des voitures," *Le Gaulois*, 3 August 1896, 1; "Gazette des tribunaux," *Le Figaro* 5 January 1884, 3; and "Les on-dit," *Le rappel* 10 May 1879, 2.

58. Lefevbre, *Rhythmanalysis*; Jalas et al., "Rhythms of Infrastructure"; and J. Engels, "Rhythm Analysis."

59. "Sabotage stupide," *L'éclair*, 1 August 1911; "La capitale sabotée: ce que l'administration à fait du splendide boulevard des Capucines," *L'intransigeant*, 5 August 1911; and "Partout des trous," *L'intransigeant*, 2 September 1910.

Chapter 6. Living with Risk

1. "Les méfaits de l'électricité," *Le matin*, 27 July 1904, 4; "Les méfaits de l'électricité," *L'aurore*, 28 July 1904, 3; "Les méfaits de l'électricité," *L'univers*, 28 July 1904, 4; and "Les méfaits de l'électricité," *Le radical*, 29 July 1904, 3.

2. "Partie financiere," 211.

3. "Sur la traction mécanique à Paris," 460; "La traction mécanique dans Paris; discussion," 20; D., "Le trolley mortel," *L'électricien* (1897): 30.

4. Blondin, "La traction électrique," 126; and "Renseignements divers."

5. Floriane, "La vie mondaine."

6. Green, *Risk and Misfortune*, 2.

7. Blanchot, *Writing of the Disaster*; and Niget and Petitclerc, "Introduction—Le risque comme culture de la temporalité."

8. Datta, *Heroes and Legends*; Garner and Abel, "Regulating a Risky Business"; Cubitt, "*Martyrs of Charity*"; and Gosser, "The Bazar de la Charité Fire."

9. Beck, *Risk Society*; Fressoz, "Beck Back in the Nineteenth Century"; Fressoz, *L'apocalypse joyeuse*; and Esbester and Wetmore, "(Auto)mobility."

10. Perrow, *Norman Accidents*; Schivelbusch, *Railway Journey*, 129–49; Rieger, *Technology*, 51–85; Killen, *Berlin Electropolis*; L. Simon, *Dark Light*; Cartwright, *Screening the Body*; and D. Nye, *American Technological Sublime* and *Electrifying America*.

11. *Rapport au nom de la Commission du métropolitain sur l'accident*, 32, 37; Perrow, *Normal Accidents*; and Soppelsa, "Paris's 1900 Universal Exposition." Launched plates in "Les méfaits de l'électricité," *Le XIXe siècle*, 18 June 1892, 3; and "Explosion sous un trottoir," *Le petit Parisien*, 25 November 1905, 2.

12. "Tour du monde: Électricité," *Cosmos*, no. 522 (26 January 1895): 255–57.

13. "Les méfaits de l'électricité," *Le XIXe siècle*, 17 December 1889, 2; "À la Porte-Saint-Martin," *Le matin*, 30 December 1889, 2; "Les méfaits de l'électricité," *Le XIXe siècle*, 23 January 1890, 1; "À l'Hôtel de ville: Les méfaits de l'électricité," *Le XIXe siècle*, 10 July 1890, 2; "Les méfaits de l'électricité," *Le XIXe siècle*, 20 July 1890, 2; "Les méfaits de l'électricité," *Le matin*, 27 September 1890, 3; "Les méfaits de l'électricité," *Le XIXe siècle*, 25 November 1890, 2; "Les méfaits de l'électricité," *Le XIXe siècle*, 20 November 1891, 2; "Les méfaits de l'électricité," *Le XIXe siècle*, 28 November 1891, 1; "Les méfaits de l'électricité," *Le XIXe siècle*, 16 December 1891, 2; "Les méfaits de l'électricité," *Le XIXe siècle*, 6 March 1892, 2; "Les méfaits de l'électricité," *Le XIXe siècle*, 26 May 1892, 3; "Les méfaits de l'électricité," *Le XIXe siècle*, 18 June 1892, 3; "Les

méfaits de l'électricité," *Le XIXe siècle*, 9 July 1894, 3; "Les méfaits de l'électricité," *Le XIXe siècle*, 20 September 1894, 3; "Les méfaits de l'électricité," *La presse*, 6 December 1895, 1; "Les méfaits de l'électricité," *Le matin*, 18 April 1896, 3; "Les méfaits de l'électricité," *Le matin*, 27 April 1897, 3; "Les méfaits de l'électricité," *La lanterne*, 11 July 1898, 1; "Les méfaits de l'électricité" *Le petit Parisien*, 2 June 1900, 3; "Les méfaits de l'électricité" *Le petit Parisien*, 4 July 1900, 3; "Les méfaits de l'électricité," *Le matin*, 4 July 1900, 3; "Les méfaits de l'électricité," *Le matin*, 27 August 1900, 5; "Les méfaits de l'électricité," *La lanterne*, 3 Feburary 1901, 2; "L'Incendie du Faubourg St-Antoine," *Le matin*, 17 December 1901, 3; "Les méfaits de l'électricité," *Le matin*, 28 December 1901, 4; "Ivry—Les méfaits de l'électricité," *Le petit Parisien*, 6 April 1902, 4; "Les méfaits de l'électricité," *Le XIXe siècle*, 19 April 1902, 3; "Les méfaits de l'électricité," *La lanterne*, 20 May 1902, 3; "Les méfaits de l'électricité," *Le XIXe siècle*, 4 June 1902, 3; "Les méfaits de l'électricité," *Le XIXe siècle*, 16 November 1902, 3; "Les méfaits de l'électricité," *Le XIXe siècle*, 15 December 1902, 3; "Les méfaits de l'électricité," *Le petit Parisien*, 10 January 1903, 4–5; "Les méfaits de l'électricité," *La lanterne*, 6 June 1903, 2; "Les méfaits de l'électricité," *Le petit Parisien*, 27 October 1903, 4; "Tuujours [*sic*] le métro: Les accidents quotidiens; Les méfaits de l'électricité," *La presse*, 1 November 1903, 3; "Les méfaits de l'électricité," *La lanterne*, 12 December 1903, 3; "Les méfaits de l'électricité," *La lanterne*, 12 June 1904, 3; "Les méfaits de l'électricité," *Le matin*, 27 July 1904, 4; "Les méfaits de l'électricité," *L'aurore*, 28 July 1904, 3; "Explosion sous un trottoir," *Le petit Parisien*, 25 November 1905, 2; "Les méfaits de l'électricité," *Le petit Parisien*, 5 February 1913, 4; "L'électricité meutrière," *Le rappel*, 22 October 1913, 1–2; "Les méfaits de l'électricité" *Le Gaulois*, 11 August 1924, 3; and "Les méfaits de l'électricité," *Le matin*, 12 May 1930, 1.

14. Figuier, *L'année scientifique* (1890), 135–36, 139–41; and Francq, *Chemin de fer métropolitain*, "dream" on 34.

15. Beltran and Carré, *La fée et la servante*; Rabinbach, *Human Motor*; Prochasson, *Les années électriques*; Cordulack, "Franco-American Battle"; Robida, *Le vingtième siècle*; Weber, *France, Fin de Siècle*, 67–77; and 1878 lights, in Beltran, "La difficile conquête d'une capitale."

16. "La grève des omnibus," *Le Figaro*, 24 April 1895, 1–2; and "Le métropolitain urbain," *Le temps*, 16 April 1896, 3.

17. Monnory, "La locomotion électrique dans le monde," 93.

18. Petitjean, *Les grands travaux*, 19; P. Vibert, *La concurrence* étrangère, 2:107; d'Avenel, *Le mécanisme de la vie moderne* 5:182; and Verne, *Robur-le-Conquérant*, 108.

19. Beltran and Carré, *La fée et la servante*, 188–89; Lopez, "Belgian Investment"; Lanthier, "Relationship Between State and Private Electric Industry"; and Schatzberg, "Culture and Technology."

20. Dupuy, *La traction* électrique; H. Martin, *Production et distribution*; Barbillion and Griffisch, *Traité pratique*; Rider, *Electric Traction*; and Davies, *Electric Power and Traction*.

21. Quote from d'Avenel, *Le mécanisme de la vie moderne*, 5:182; Vignes, *La traction méchanique*, 8; P. Vibert, *La concurrence* étrangère, 2:134–37; and McKay, *Tramways and Trolleys*, 96–97. On risk, see "Questions scientifiques," 458.

22. Compagnie des Tramways de Paris et du Département de la Seine. *Assemblées générales extraordinaire et ordinaire du 15 Juin 1894* (Paris: Chaix, 1894), quotes on 15 and 5, respectively; false generalization on 6–11 (AN F^14 8587).

23. P. Vibert, *La concurrence* étrangère 2:73–75; Tramways 1896 *enquête d'utilité publique* (AN F^14 14999); quotes from McKay, *Tramways and Trolleys*, 86.

24. D., "Le trolley mortel."

25. Mouchelet, "Doléances."

26. Figuier, *L'année scientifique* (1890), 136–37; Dupuy, *La traction électrique*, 150; McKay, *Tramways and Trolleys*, 72, 86–87; and Schatzberg, "Culture and Technology."

27. Paris used the Diatto, Dolter, Claret-Vuilleumier, and Védovélli systems; see Maréchal, *Les tramways électriques*, 117–45; Barbillion and Griffisch, *Traité pratique*, 451–513; and R. Smith, *Electric Traction*, 167–206.

28. "Tour du Monde," *Cosmos* (1895), 255; "Les méfaits de l'électricité," *L'éclairage électrique*, 2, no. 12 (23 March 1895): 574–75; "Les méfaits de l'électricité," *Le matin*, 27 September 1890, 3; and "Les méfaits de l'électricité," *Le XIXe siècle*, 28 November 1891, 1.

29. "Les méfaits de l'électricité," *Le matin*, 27 April 1897, 3; "Les méfaits de l'électricité," *La lanterne*, 20 May 1902, 3; and "Les méfaits de l'électricité," *Le petit Parisien*, 27 October 1903, 4.

30. "Le métropolitain," *Le petit Parisien*, 23 July 1900, 2; *Le temps*, 21 July 1900; and "Smaller Crowd at Paris Fair," *Chicago Daily*, 21 July 1900, 6. *Le rappel*, quoted in Hausser, *Paris*, 37.

31. Schroeder-Gudehus and Rasmussen, *Les fastes du progrès*; and quotes from Petitjean, *Les grands travaux*, 7.

32. "Le métropolitain," *Le petit Parisien*, 23 July 1900, 2; and toy/tool quote, in Baumier, "Le métropolitain: Les revendications du public," *La presse*, 27 July 1900, 1.

33. Tricoire, "L'exploitation du métropolitain." See also *Le petit Parisien*, 13 July 1900; and Métro accident reports, 1899–1905 (AP V.ONC 129).

34. Quote from "Les tramways meurtières," *Le petit Parisien*, 22 August 1900, 1; see also: "Les tramways meurtriers," *Le petit Parisien*, 5 October 1900, 1; and "Rencontre de Tramways," *Le petit Parisien*, 24 July 1900, 3.

35. "Collision de tramways," *Le petit Parisien*, 3 July 1900, 3; "Les méfaits de l'electricité," *Le petit Parisien*, 4 July 1900, 3; "Un Cheval dans un Omnibus," *Le petit Parisien*, 7 July 1900, 3; "Collision de Tramways," *Le petit Parisien*, 9 July 1900, 1; "Enfant écrasé," *Le petit Parisien*, 11 July 1900, 3; "L'accident de la place de Clichy," *Le petit Parisien*, 16 July 1900, 3; "L'accident du quai de la conférence," *Le petit Parisien*, 22 July 1900, 3; "Rencontre de tramways," *Le petit Parisien*, 23 July 1900, 3; "Rencontre de tramways," *Le petit Parisien*, 1 August 1900, 3; "L'accident du boulevard Ornano," *Le petit Parisien*, 5 August 1900, 3; "Ecrasée par un tramway," *Le petit Parisien*, 8 August 1900, 3; "Une collision avenue de Versailles," *Le petit Parisien*, 13 August 1900, 3; "Grave collision de tramways," *Le petit Parisien*, 16 August 1900, 1; "Les tramways meurtriers," *Le petit Parisien*, 22 August 1900, 1; Jean de Paris, "Cycliste renversé par un tramway," *Le Figaro*, 11 August 1900, 4; and "Tamponnements de tramways," *Le Figaro*, 24 July 1900, 4.

36. Wolff quote, in Hausser, *Paris*, 39; "Les accidents de tramways," *L'aurore*, 8 October 1900, 1.

37. First quote from Councilor Duval-Arnould, 15 December 1900; *Bulletin municipal officiel de la ville de Paris*, 16 December 1900, 4081; "La marche des tramways," *Le petit Parisien*, 8 September 1900, 3; and final quote from "Les tramways meurtriers," *Le petit Parisien*, 5 October 1900, 1.

38. "Un tramway pris d'assaut" and "Le feu au métropolitain," *Le petit Parisien*, 12 October 1900, 1 and 3, respectively; quote from "Mésaventures électriques," *Le temps*, 19 October 1900, 3; and "Le métropolitain et les tramways" *Le petit Parisien*, 25 October 1900, 2.

39. Rieger, *Technology*; Adams, "The Dynamo and the Virgin"; Geddes, "Man and the Environment"; "Collision de tramways," *Le petit Parisien*, 4 November 1900, 1; and "Accidents sur le métropolitain," *Le petit Parisien*, 11 November 1900, 3.

40. "Un tramway pris d'assaut"; and "La police en tramway," *Le temps*, 13 December 1900, 4.

41. "Les tramways électriques à contacts superficiels à Paris," *La nature*, no. 1464 (15 June 1901): 33–35.

42. Quote from "Le trolley des rues Réaumur et du Quatre-Septembre," *Le journal des transports*, (27 August 1904), 433. See also "Les plots meurtriers," *La croix*, 24 October 1900, 1; Le Masque de Fer, "Nouvelles à la main," *Le Figaro*, 24 December 1900, 1; "Conseil municipal," *Le rappel*, 27 March 1901, 2; "Conseil municipal," *Le XIXe siècle*, 27 March 1901, 2; "Les plots meurtriers," *La justice*, 6 October 1901, 3; "Les plots meurtriers," *La justice*, 9 December 1901, 3; "Les plots meurtriers," *L'aurore*, 27 June 1902, 3; "Les surprises de l'électricité," *Journal des débats politiques et littéraires*, 14 August 1902, 3; "Faits divers: plots meurtriers," *Le XIXe siècle*, 13 September 1902, 3; "Faits divers: les plots meurtriers," *Le XIXe siècle*, 19 July 1903, 3; "A travers Paris," *Le Figaro*, 6 November 1902, 1; "Les plots meurtriers," *L'aurore*, 11 November 1905, 3; "Petits faits-divers," *Le matin*, 6 December 1908, 5; "Les plots meurtriers," *La justice*, 6 December 1908, 3; "Les plots meurtriers," *Le Figaro*, 15 June 1910, 5; "Paris," *La croix*, 16 June 1910, 6.

43. "Mésaventures électriques," *Le temps*, 19 October 1900, 2–3; "Les accidents des tramways," *L'aurore*, 8 October 1900, 1–2; "Hôtel de ville: la question des 'plots,'" *Le Gaulois*, 17 December 1900, 4; "Les tramways meurtriers," *Le petit Parisien*, 5 October 1900, 1; "Les méfaits de l'électricité," *Les annales politiques et litteraires* 28 (1897), 46; and "Les méfaits de l'électricité," *Le petit Parisien*, 4 July 1900, 3. On stock prices, see "Valeurs industrielles," *La lanterne*, 16 October 1900, 4; "La finance," *La lanterne*, 8 December 1900, 3; "Valeurs industrielles," *La lanterne*, 18 December 1900, 4; "Valeurs industrielles," *La Lanterne*, 30 April 1901, 4; *Banque industrielle & coloniale* 3, no. 6 (3 April 1901); and *Banque industrielle & coloniale* 3, no. 9 (15 May 1901).

44. *L'assiette au beurre*, no. 35, 30 November 1901, 553; quote in "Les méfaits de l'électricité," *Le matin*, 4 July 1900, 3; "Les dangers de l'électricité," *L'année scientifique et industrielle* 34 (1890): 155–59; Conseil municipal de Paris, *Note*, no. 35 (10 June 1891), 7; on 1889, see Beltran, "Les débuts du réseau électrique," 82.

45. *Bulletin municipal officiel de la ville de Paris*, 16 December 1900, 4081–96; *Rapport de l'Ingénieur ordinaire*, Paris le 25 Décembre 1900; Letter from the Prefect of Police to the Minister of Public Works, 4 February 1901 (all previous sources AN F^{14} 14999); "Mésaventures électriques," *Le temps*, 19 October 1900, 2–3; and Reiger, *Technology*.

46. *Bulletin municipal officiel de la ville de Paris*, 26 March 1901, 1095–99, "rainy periods" on 1096; Chauveau, "Les tramways," quote 128; "Après la catastrophe: la recherche des responsabilités," *Le matin*, 13 August 1903, quote 1. More rain concerns, in "Les méfaits de l'électricité," *Le matin*, 27 April 1897, 3; and "Les méfaits de l'électricité," *Le petit Parisien*, 27 October 1903, 4.

47. Staunch criticism, in "Hôtel de ville: la question des 'plots'" *Le Gaulois*, 17 December 1900, 4; *L'assiette au beurre*, no. 35, 30 November 1901, 553.

48. *Rapport au nom de la Commission du métropolitain*, Annex no. 4, 410–14. Illustrations: *L'actualité*, no. 187 (16 August 1903); *Le petit journal* supplément illustré, 23 August 1903, cover; and *L'assiette au beurre*, no. 125 (22 August 1903), back cover.

49. Tricoire, "L'exploitation du métropolitain"; and *Rapport au nom de la Commission du métropolitain sur l'accident*.

50. "The Fire on the Paris Underground," *Times*, 13 August 1900, 3; and Ghandi, *Collected Works* 3:414–15.

51. "La catastrophe du métropolitain," *Le temps*, 12 August 1903, 2; "La catastrophe du métropolitain," *Le temps*, 13 August 1903, 1–2; "La catastrophe," *Le temps*, 14 August 1903, 1–2; "Cri de douleur" and "Le Métro en feu," *La croix*, 12 August 1903, 1; "Lendemain de

catastrophe," *La croix*, 13 August 1903, 3; and "Catastrophe: L'incendie du métro," *Le matin*, 12 August 1903, 1.

52. *L'assiette au beurre*, no. 125 (22 August 1903), 2017, 2117, and back cover.

53. "Responsabilités," *La croix*, 1 May 1900, 1; "Après la catastrophe: la récherche des résponsabilités," *Le matin*, 13 August 1903, 1; and "Responsabilités," *L'assiette au beurre* 125 (22 August 1903), 2114; see also Henry De Larègle, "Les résponsabilités," *Le soleil*, 11 February 1910, 1.

54. Bizarre, in "Les méfaits de l'électricité," *La lanterne*, 20 May 1902, 3; and "Les méfaits de l'électricité," *Le petit Parisien*, 27 October 1903, 4. Curious, in "Les méfaits de l'électricité," *Le XIXe siècle*, 20 September 1894, 3; "Accident causé par l'électricité," *L'éclairage électrique*, year 2, no. 25 (2 February 1895): 239–40; "Les méfaits de l'électricité," *Le matin*, 18 April 1896, 3; and "Les méfaits de l'électricité," *La lanterne*, 12 June 1904, 3. Extraordinary, in "Les méfaits de l'électricité," *L'aurore*, 28 July 1904, 3 (*peu ordinaire*); and "Explosion sous un trottoir," *Le petit Parisien*, 25 November 1905, 2 (*extraordinaire*). Singular, in "Les méfaits de l'électricité," *La lanterne*, 11 July 1898, 1; "Les méfaits de l'électricité," *Le matin*, 27 August 1900, 5; and "Les méfaits de l'électricité," *Le matin*, 28 December 1901, 4. Uncommon, in "Les méfaits de l'électricité," *Le matin*, 27 July 1904, 4.

55. Soppelsa, "Paris's 1900 Exposition"; Soppelsa, "Finding Fragility in Paris."

56. "Les trous de Paris," *Le XIXe siècle*, 30 August 1910; Judet, "Les cinq pouvoirs," *L'éclair*, 29 January 1910; "L'orage d'hier, par ses consequences, a pris les proportions d'une catastrophe," *Le petit Parisien*, 16 June 1914, 1–2; and "L'électricité meurtrière," *Le rappel*, 22 October 1913, 1–2.

57. "Les méfaits de l'électricité," *Le XIXe siècle*, 20 November 1891, 2.

58. "Les méfaits de l'électricité," *Le matin*, 4 July 1900, 3.

59. Everyday risk, in "Toujours le métro," *La presse*, 17 August 1903; "Chronique de Paris," *Le temps*, 8 March 1872; and Hausser, *Paris*, 39. Frequent accidents, in "Les méfaits de l'électricité," *Le XIXe siècle*, 17 December 1889, 2; and "À l'Hôtel de ville: les méfaits de l'électricité," *Le XIXe siècle*, 10 July 1890, 2.

60. Hussey, *Paris*, 243; Chevalier, *Laboring Classes*, 11–23; and Harvey, *Paris, Capital of Modernity*, ch. 1. On the Great War, see Zahra, "Behind the Storm."

61. Soppelsa, "Paris's 1900 Exposition."

62. Rambaud, in Sizeranne, "Les prisons de l'art," 135.

Chapter 7. Making a Stink

1. Halliday, *Great Stink of London*; Ashton, *One Hot Summer*; D. Barnes, *Great Stink of Paris*; and Soppelsa, "Paris's 1900 Universal Exposition."

2. Nixon, *Slow Violence*; and Knowles, "Learning from Disaster?"

3. Pritchard, "Envirotechnical Disaster." Cascading failures, in Little, "Managing the Risk."

4. "L'hygiène publique." For uses of "crisis," see Frollo, "Après la crise . . ." *Le petit Parisien*, 5 October 1895, 1; Bechmann, "Rapport de l'ingénieur," 13; and Commission scientifique de perfectionnement de l'observatoire municipal de Montsouris, *Travaux de l'année 1902 sur les eaux d'alimentation et les eaux d'égouts de la ville de Paris*, 21. See also Frollo, "L'eau a Paris," *Le petit Parisien*, 19 August 1888, 3; Frollo, "Les eaux potables," *Le petit Parisien*, 31 May 1892, 1; Frollo, "L'eau bouillie," *Le petit Parisien*, 12 March 1894, 1; Frollo, "Ce qu'on boit," *Le petit Parisien*, 8 July 1898, 1; Petitjean, "La disette d'eau," *Le Figaro*, 23 July 1900, 2; L. R., "Une visite à Paris," 332; "Chronique Parisienne," *Bibliothèque universelle* (1900), 611; "La chaleur," *Le petit Parisien*, 20 July 1904, 4; "À travers la science," *Le petit Parisien*, 27 August 1906, 5; R. Léon, "Paris et son alimentation"; and Hausser, *Paris*, 37–40, 165, 200, 418.

5. Montorgeuil, *Les eaux*, 7–9; P. Léon, *Eaux et fontaines*, 11; Boucard, *La vie de Paris*, 238; and P. Strauss, "Les eaux de Paris" (1893), 430.

6. Chatzis, "Brève histoire des compteurs," 163; and Ls., "Hygiène."

7. Pritchard, *Confluence*, 19–21; Pritchard, "Envirotechnical Disaster"; and D. Nye, *When the Lights Went Out*.

8. Couche, *Le service des eaux*, 1.

9. P. Strauss, "La disette d'eau." See also P. Strauss, "La vie municipal"; Strauss, "Les eaux de Paris."

10. Humblot, quoted in "L'eau de Source: Paris qui boit et Paris qui gaspille," *Le matin*, 12 September 12, 1895, 1.

11. "Industrial product," in Goubert, *Conquest of Water*; Bocquet et al., "From Free Good to Commodity"; and Le Roy Ladurie, *Le réchauffement*, 126–28.

12. P. Léon, *Eaux et fontaines*, 13; and du Camp, *Paris*, 5:348.

13. Lorang, "Les fontaines Wallace"; Kaufmann, *Paris of To-day*, 166–72; Hussey, *Paris*, 277; and https://wallacefountains.org/about-sir-wallace-and-his-fountains/. The 1904 figure is from L. Martin, *Encyclopédie municipale*, 1027.

14. Du Camp, *Paris*, 5:340; and Direction des eaux et des égouts, *Reneignements généraux sur les eaux et les* égouts *de la ville de Paris*, 11 (AP V.O3 220).

15. Worms, *Rapport adressé a M. Le Préfet*; "Chronique: le choléra"; "Hygiene publique: choléra"; "Société médicale des hôpitaux"; *Gazette hebdomadaire de médecine et de chirurgie*, série 2, tome 11, no. 5 (30 January 1874): 74–75; Aisenberg, *Contagion*, 118–29, 164–76; Bouchardat, in Latour, *Pasteurization of France*, 20; typhoid, in Perrot, *History of Private Life* 397.

16. Zola, *"Assommoir."*

17. *Annuaire de l'observatoire de Montsouris*, Introduction, 3–5; and Chatzis, "Brève histoire des compteurs," 160–61.

18. Chatzis, "Brève histoire des compteurs," 164.

19. Chatzis, "Brève histoire des compteurs," 165–66.

20. Direction des eaux et des égouts, *Renseignements généraux sur les eaux et les* égouts *de la ville de Paris*, 4; and Couche, *Le service des eaux* (both AP V.O3 220).

21. Geffroy, "Chronique: pas d'eau," *La justice*, 1 July 1890, 1.

22. Bechmann, *Salubrité urbaine*, 430–33, 436–38; Couche, *Les eaux de Paris*, 90; "Robinet intermittent"; André, "Hydraulique"; Tresca, "Arts mécaniques"; and Chatzis, "Brève histoire des compteurs," 170–71.

23. Mortality figures, in Perrot, *History of Private Life*, 397.

24. "L'eau à Paris," *Le petit Parisien*, 21 July 1881, 2. More critique, in "Ce que nous buvons," *Le petit Parisien*, 10 October 1884, 3.

25. *Bulletin municipal official de la ville de Paris*, 25 July 1881, 233.

26. Sarcey, "Compteurs de la Compagnie des eaux," *Le XIXe siècle*, 19 May 1884, 1; and "La disette d'eau," *Le Gaulois*, 24 May 1889.

27. Poubelle, "Mémoire au Conseil municipal," 3, 6–7; and *Notice sur le service des eaux*, 60–62.

28. Perhaps as early as 1884, see Sarcey, "La question des eaux," *Le XIXe siècle*, 24 July 1884; and Goubert, *Conquest of Water*, 49–50.

29. Figuier, "La fièvre typhoïde," 307; and "L'eau à Paris," *Le petit Parisien*, 19 August 1888, 3.

30. "Disette d'eau de source," *Bulletin médical* 3 (1889): 685; and Conseil municipal de Paris, *Procès-verbaux* (1889): 814–23.

31. Weiss, "Making Engineering Visible."

32. Perrot, *History of Private Life*, 397.

33. Duprey, *L'Oeuvre sociale*, 49–60, esp. 51; Strauss, "Les eaux de Paris," 434; and Goubert, *Conquest of Water*, 67.

34. Conseil municipal de Paris, *Rapport*, no. 99 (1893), 3–11.

35. Water system, in Boucard, *La vie à Paris*, 215; and typhoid, in *Annuaire statistique de la ville de Paris*, year 15 (1894): 146–49, 156–57, 164.

36. "Les eaux contaminées," *Le matin*, 11 March 1894, 1; "Oh! La statistique!" *Le matin*, 18 June 1894, 2; Frollo, "L'eau bouillie," *Le petit Parisien*, 12 March 1894, 1; and Collin, *L'eau potable*, 11.

37. See http://www.meteofrance.fr/actualites/75667423-paris-plus-d-un-mois-sans-pluie.

38. "Les odeurs de la Seine," *Le petit journal*, 8 September 1895; and "La baisse de la Seine," *La petite république*, 9 September 1895.

39. "La chaleur," *La petite république*, 10 September 1895; quote from Charnay, "De l'eau!" *La petite république*, 13 September 1895; see also Charnay, "Le retour de l'eau de source," *La petite république*, 18 September 1895, 1.

40. Quotes from "Le régime de l'eau de Seine," *Le petit journal*, 10 September 1895; "En voulez-vous des microbes?" *Le petit journal*, 11 September 1895; and "L'eau de source: Paris qui boit et Paris qui gaspille," *Le matin*, 12 September 1895, 1.

41. Frollo, "Après la crise . . ." *Le petit Parisien*, 5 October 1895, 1.

42. Galland, *Vingt mois au pays de la soif*; Gondrecourt, *Le pays de la soif*; Noir, *L'Amour au pays de la soif*; and Toulhouse, *Aux pays de la soif*. See also Schneider, *Empire for the Masses*, 3; Ngalamulume, "Coping with Disease"; and Pritchard, "From Hydroimperialism to Hydrocapitalism."

43. For "chaleur sénégalienne," see A. Dumas, "Le Speronare"; for variant "une temperature sénégambienne," see Laumann, "À Paris & ailleurs: de l'eau! . . ." *La justice*, 18 June 1885, 2; *Almanach du Rire*, 14; "Nîmoiserie"; "Vision Parisienne," *Le Gaulois*, 15 July 1911, 1; "Sierre," *Le Confédéré*, year 51, no. 58 (22 July 1911), 2; "La Brouette," *L'indicateur de la Savoie*, year 34, no. 1673 (26 August 1911), 1; "Les prognostics de juillet"; "Courrier des Théatres," *Le Figaro*, 19 June 1917, 3; "Et le 22 Juillet, il fera encore plus chaud!" *La presse*, 11 July 1921, 1; and "Pour lire avant la sièste," 1. The phrase "température saharienne et sénégalienne" is used in "Aux Antilles françaises," *Le petit Parisien*, 18 September 1923, 2; and Haroudin, *La vie des abeilles solitaires*, 196. These are only a few of hundreds of similar usages.

44. Gautier, "Paris qui boit," *Le matin*, 16 September 1895; and "Actualités," *L'éclair*, 13 September 1895 (colonization in both sources).

45. Charnay, "De l'eau!"; and Charnay, "Le retour de l'eau de source."

46. "L'eau de source," *Le matin*, 12 September 1895, 1.

47. Quotes from "L'eau de Seine," *Le petit journal*, 13 September 1895.

48. Judet, "L'eau et la politique," *Le petit journal*, 14 September 1895; Frollo, "Ce qu'on boit," *Le petit Parisien*, 8 July 1898; see also Charnay, "De l'eau!" and Charnay, "Le retour de l'eau de source."

49. Curtin, *Disease and Empire*, ch. 7, 175–94; and Rabinow, *French Modern*, 154.

50. "Retour de l'eau de source," *La petite république*, 18 September 1895; and "L'eau de source: Plus d'eau de Seine," *L'intransigeant*, 18 September 1895.

51. Strauss, "La disette d'eau."

52. "Paris sans eau!" *Le matin*, 16 July 1896, 1; and Frollo, "À Boire!" *Le petit Parisien*, 22 July 1896, 1.

53. Poubelle, "Mémoire au Conseil municipal" (8 December 1884); and Humblot, Étude &

programme pour le complément de l'alimentation de Paris en eaux de source et de rivière. Rapport de l'inspecteur général (4 November 1896),12–14.

54. Humblot, Étude & *programme*, 4–5; Bienvenüe, in *Aqueduc de dérivation des sources du Loing et du Lunain. Projet d'exécution. Bordereau des pièces*, 22 November 1895, 3.

55. "La question de l'eau," *Le matin*, 16 August 1897, 1.

56. Frollo, "Ce qu'on boit," *Le petit Parisien*, 8 July 1898; "Fermez vos robinets," *Le matin*, 22 August 1898, 1; and "Les odeurs de Paris," *L'intransigeant*, 31 August 1898.

57. Aucoc, *La dérivation des sources*, 1; and "L'eau de source," *La justice*, 22 February 1890, 1–2.

58. "L'eau à Paris," *Le temps*, 24 August 1898; and *Commission technique d'eau potable* (AP V.O3 1221), 1897 to 1900.

59. M. Félix Marboutin, *Note sur le captage des eaux de Val d'Orléans* (24 May 1903), AP V.ONC 1217.

60. Commission technique chargée d'étudier les diverse questions se rattachant à l'Alimentation de Paris et de la Banlieue en Eaux potables, *Compte-Rendu de la séance du 30 Novembre 1900* (AP V.O3 1221); and Thoinot, "La fièvre typhoïde."

61. Frollo, "L'eau à Paris," *Le petit Parisien*, 15 January 1899, 1; Saint-Réal, "Hôtel de ville," *Le Gaulois*, 17 March 1899, 3; and "L'eau à Paris—l'opinion du conseil d'hygiène," *Le temps*, 19 August 1899, 2.

62. Humblot, *Étude & programme*. On the heat, see Rousseau, "Les temperatures mensuelles," 53. For the water system in 1900, see *Notice sur le service des eaux*; Commission technique chargée d'étudier les diverse questions se rattachant à l'Alimentation de Paris et de la Banlieue en Eaux potables, *Compte-Rendu de la séance du 30 Novembre 1900*, AP V.O3 1221.

63. "Sans eau!" *Le Figaro*, 19 July 1900.

64. Petitjean, "De l'eau! De l'eau!" *Le Figaro*, 22 July 1900, 2–3. On pressure, see Couche, *Les eaux de Paris*, 51–55 (AP V.O3 220); Strauss, "La disette d'eau"; and Nansouty, "La question de l'eau," 105.

65. "Victimes de la Chaleur," *La croix* (supp.), 19 July 1900, 4; Petitjean, "De l'eau! De l'eau!" *Le Figaro*, 22 July 1900, 2–3; "L'eau à Paris," *La croix* (supp.), 23 July 1900, 4; Larbaud, "Les rigueurs du thermomètre: La question de l'eau," *La presse*, 24 July 1900; Larbaud, "La chaleur qui tue," *La presse*, 28 July 1900; and "De l'eau! De l'eau!" *La presse*, 30 July 1900.

66. Hausser, *Paris*, 37; "Le désespoir de M. Picard: coquin de soleil!" *La croix*, 25 July 1900, 4; and "L'eau à Paris," *Le petit Parisien*, 22 July 1900.

67. "Paris sans eau," *La presse*, 22 July 1900; and "Chronique," *Revue générale de chimie pure et appliquée* 2 (1900), 27.

68. "À travers Paris," *Le Figaro*, 29 July 1900, 1; and Cornély, "Échos: la question de l'eau," *Le Figaro*, 29 July 1900, 1. On Lake Léman, see also "L'eau à Paris," *Le temps*, 24 August 1898.

69. "L'eau à Paris: La fermeture des conduits pendant la nuit. Les précautions à prendre," *La presse*, 22 July 1900, 1; and "La question de l'eau," *Le Figaro*, 6 August 1900, 3.

70. Poubelle, "Mémoire au Conseil municipal"; Petitjean, "La question de l'eau," *Le Figaro*, 24 July 1900, 3–4; "Chronique Parisienne," *Bibliothèque universelle* (1900); and Jacquemet, "Urbanisme Parisien," 518.

71. Le Masque de Fer, "Nouvelles à la main: les gaspilleurs d'eau," *Le Figaro*, 23 July 1900. See also Larbaud, "Les rigueurs du thermomètre."

72. "Victime du devoir," *La presse*, 29 July 1900, 3.

73. Cornély, "Échos: la question de l'eau," *Le Figaro*, 29 July 1900, 1.

74. Petitjean, "La disette d'eau," *Le Figaro*, 23 July 1900, 2.

75. Petitjean, "La question de l'eau," *Le Figaro*, 24 July 1900, 3–4; "L'eau et les pompiers,"

La croix (supp.), 24 July 1900, 4; "Les incendies et le manque d'eau," *Le temps*, 25 July 1900; and "La disette de glace," *Le temps*, 26 July 1900.

76. Petitjean, "La disette d'eau," *Le Figaro*, 23 July 1900, 2; L. R., "Une visite à Paris"; and "Chronique Parisienne," *Bibliothèque universelle* (1900), 611.

77. J. B., "À propos des eaux," *La croix* (supp.), 24 July 1900, 1.

78. Henriot, "La question des eaux! . . ."

79. Quotes from "Mauvaise odeur," *La croix*, 27 July 1900, 4. See also from *Le petit Parisien*: "L'eau à Paris," 22 July 1900; "L'empoisonnement de la Seine," 26 July 1900; "L'eau de source," 27 July 1900; and "L'eau à Paris," 28 July 1900; and from *Le temps*: "L'eau à Paris," 22 July 1900; "L'eau à Paris," 24 July 1900; "La secheresse et la Seine," 24 July 1900; "L'infection de la Seine," 26 July 1900; "L'infection de la Seine," 27 July 1900; "L'eau à Paris," 28 July 1900; and "L'eau à Paris," 30 July 1900.

80. "Toujours l'eau," *La croix* (supp.), 26 July 1900, 4; and Hausser, *Paris*, 38.

81. "Au jour le jour: l'assainissement de la Seine," *Le temps*, 9 July 1899, 3; "L'épuration de la Seine," *Le petit Parisien*, 9 July 1899, 1; and Reid, *Paris Sewers*, 82.

82. Basset, "La Seine empoisonnée," *Le Figaro*, 25 and 26 July 1900, both p. 2.

83. Quotes from Frollo, "L'empoisonnement de la Seine," *Le petit Parisien*, 27 July 1900, and "L'infection de la Seine," *La croix* (supp.), 26 July 1900, 4, respectively.

84. Basset, "Pour l'assainissement," *Le Figaro*, 27 July 1900, 2; "La chaleur," *La presse*, 27 July 1900; and "L'empoisonnement de la Seine," *Le petit Parisien*, 27 July 1900.

85. On decentralism, see Mandell, *Paris 1900*, 25–51; and Geppert, *Fleeting Cities*, 71–74. On Orléans, see Félix Marboutin, *Note sur le captage des eaux de Val d'Orléans*, 24 May 1903 (AP V.ONC 1217); Commission technique chargée d'étudier les diverse questions se rattachant à l'Alimentation de Paris et de la Banlieue en Eaux potables, *Compte-Rendu de la séance du 30 Novembre 1900* (AP V.O3 1221).

86. "La fièvre typhoïde à Paris" *Le matin*, 27 August 1900.

87. "La fièvre typhoïde" *Le matin*, 29 August 1900; Thoinot, "La fièvre typhoïde"; *Annuaire statistique de la ville de Paris*, year 21 (1900), 187, 189; and Goubert, *Conquest of Water*, 24, 49.

88. Officials counted 75 heatstroke deaths, and 126 deaths of people unidentified at the morgue, who died of (undifferentiated) temperature-related causes including hypo- and hyperthermia; *Annuaire statistique de la ville Paris*, year 21 (1900), 152–53, 164–65, 379.

89. Larbaud, "La chaleur qui tue."

90. Bechmann, "Rapport de l'ingénieur" (8 October 1902).

91. Pritchard, *Confluence*, 23–24; and Mukerji, *Impossible Engineering*.

92. "La question des eaux," *Le matin*, 9 April 1901, 4; "Ça va recommencer," *Le matin*, 13 April 1901, 1–2; "Le service de l'assainissment," *L'aurore*, 7 July 1901, 1–2; "À l'Hôtel de ville," *La lanterne*, 14 July 1901, 2; Bailby, "Tout Arrive!" *La presse*, 30 October 1901, 1; "À l'Hôtel de ville," *Le matin*, 16 November 1901, 2; Conseil municipal de Paris, *Procès-verbaux* (1901): 889–911, bathtub, on 892; address in *Le livre d'or*, 89.

93. Cadoux, "Les eaux à Londres," 602; and Commission scientifique de perfectionnement, *Travaux de l'année 1902*, 21.

94. "De l'eau pure, S.V.P.," *Le matin*, 18 March 1903, 1.

95. "La fièvre typhoïde," *Le matin*, 13 March 1904, 1; and "La fin du mal," *Le matin*, 9 April 1904, "châtelain" on 2.

96. "La chaleur," *Le petit Parisien*, 17 July 1904, 4; "La chaleur," *Le petit Parisien*, 20 July 1904; "L'eau à Paris," *Le petit Parisien*, 17 August 1904; and Rousseau, *Les temperatures mensuelles*, 53.

97. "L'eau," *Le petit Parisien*, 5 August 1905; and "L'eau à Paris," *Le temps*, 11 August 1905. On Bechmann, see "La fin de Bechmann," *Le matin*, 13 June 1904, 1 (published before that year's shortage); "La dernière de Bechmann," *Le matin*, 23 June 1904, 1; "Le service des eaux," *Le matin*, 17 June 1905, 4; "Correspondence," *Le matin*, 29 November 1905, 4 (letter from Bechmann).

98. "À traverse la science," *Le petit Parisien*, 27 August 1906.

99. "Un Sahara à Pantin," *Le petit Parisien*, 13 August 1906.

100. Cannon, *Paris Zone*; Sante, *Other Paris*; Wakeman, "Nostalgic Modernism"; Evenson, *Paris*, 204–7; Merriman, *Margins of City Life*, 6, 34, 59, 65, 81, 121; and Horne, *Social Laboratory*, 250–52.

101. Knowles, "Learning from Disaster?"; and Nixon, *Slow Violence*.

102. Paul Bersonnet, "Parisien, bois ton eau sans crainte!" *L'intransigeant*, 12 August 1907; "La toilette de Paris," *Le Figaro*, 31 August 1908; and "Aux service des eaux: un scandale administratif; une epidémie de typhoide," *L'autorité*, 23 August 1909.

103. *Annuaire statistique de la ville de Paris*, year 32 (1911): 44–45; Flammarion, "L'été de 1911"; Rollet, "La canicule de 1911"; and Le Roy Ladurie, *Le réchauffement*, 3:99–101.

104. Shutoffs announced, in "Ce soir, à minuit, plus d'eau," *Le petit Parisien*, 9 June 1911, 1; "Bloc-notes Parisiens: Paris sans eau," *Le Gaulois*, 10 June 1911, 1; and "À l'Hôtel de ville," *Le rappel*, 11 June 1911, 2. Water restored, in "La vie municipal," *Le petit Parisien*, 26 July 1911, 2; and drought announced in "36° à l'ombre," *Le petit journal*, 31 July 1911. In the press reviews in AP V.ONC 131, 1911 newspaper articles often lack page numbers.

105. Waste warning: "Eau," *Le confedéré*, year 51, no. 58 (22 July 1911), 3; boil warning: Lefebvre, "Nous buvons de l'eau de Seine," *Gil Blas*, 27 July 1911, 1. "Temperature Senegalienne," in "Nîmoiserie," 3; "Vision Parisienne," *Le Gaulois*, 15 July 1911, 1; "Sierre," 2; and "La Brouette," 1. Firefighters, in "Pas d'eau!" *Le petit journal illustré*, 30 July 1911.

106. Boudouard, "Recherches sur les odeurs," 614; and T., "Les odeurs de Paris en 1911."

107. "Les résignés," *L'éclair*, 1 August 1911.

108. Latapie, "Canicula," *La liberté*, 31 July 1911.

109. Latapie, "Canicula," *La liberté*, 31 July 1911.

110. Judet, "À l'eau! À l'eau!" *L'éclair*, 31 July 1911.

111. "36° à l'ombre," *Le petit journal*, 31 July 1911; and "Paris manque d'eau," *Le petit Parisien*, 31 July 1911.

112. "La vie municipale: Paris aura-t-il assez d'eau?" *Le radical*, 1 August 1911.

113. "La question de l'eau," *La petite république*, 2 August 1911; and "La Seine empoisonnée," *Le petit Parisien*, 23 September 1911.

114. "Les bassins filtrants qui ne filtrent pas," *L'intransigeant*, 11 August 1911.

115. Duchemin, "Grace à M. Delauney, les Parisians auront de l'eau," *Le rappel*, 5 August 1911, 2; "À l'Hôtel de ville: la question des eaux," *Le rappel*, 23 December 1911, 3.

116. Surville, "Pour l'eau à Paris: conclusion de notre campagne," *L'éclair*, 12 August 1911; Surville, "Eau du Rhône ou du Lac de Genève?" *L'éclair*, 16 August 1911; Denais, "Le Lac de Genève à Paris?" *La libre parole*, 26 August 1911; "L'eau du Rhône à Paris," *Les nouvelles*, 26 August 1911; and "Les eaux du Lac Léman à Paris," *Le Gaulois*, 27 August 1911, 2.

117. "Le chaleur," *Le temps*, 11 September 1911, 4; and "La Seine empoisonnée," *Le petit Parisien*, 23 September 1911.

118. "À l'Union du commerce et d'industrie," 547.

119. R. Léon, "Paris et son alimentation," 857; and Conseil municipal de Paris, *Note* no. 80 (30 August 1911): 44–45.

120. Conseil municipal de Paris, *Note* no. 80 (30 August 1911): 54; and France, Ministère des travaux publics, *Le développement des travaux publics en France sous la IIIe République*, 154–55.

Chapter 8. Resources and Risks of Rivers

1. Disco, "Taming the Rhine"; Schott et al., *Resources of the City*; Juuti and Katko, *Water, Time and European Cities*; Goodman, *European Cities*; and Worster, *Rivers of Empire.*

2. White, *Organic Machine*, ix.

3. Pritchard, *Confluence*; Mukerji, *Impossible Engineering*; Soppelsa, "Water and Power"; "Water," *Technology and Culture*, special issue; *Cultural Anthropology*, special section "water"; Barnes and Alatout, "Water Worlds"; and Kinkela et al., "Water: History, Power, Crisis."

4. Cardoso da Silva and Wheeler, "Ecosystems as Infrastructure"; Carse, "Nature as Infrastructure"; and Bélanger, "Landscape as Infrastructure."

5. Hughes, *Human-Built World*, 7, 48–52; Schott et al., "Resources of the City," 3–4; Knoll et al., *Rivers Lost*; Castonguay and Evenden, *Urban Rivers*; and Evenden, "Beyond the Organic Machine?" See also Amin and Thrift, *Cities*; Whatmore, *Hybrid Geographies*; Graham and Marvin, *Splintering Urbanism*; Graham, *Disrupted Cities*; Cronon, *Nature's Metropolis*; Kaika, *City of Flows*; and Gandy, *Fabric of Space.*

6. Du Camp, *Paris*, quotes 1:283, 313–18; and 5:283; Backouche, *La trace du fleuve*; Heynen et al., *In the Nature of Cities*; and Barles, "Urban Metabolism."

7. Ford, *Natural Interests*, 90–91; and Fressoz, *L'apocalypse jouyeuse.*

8. M. Cohen, "Waterways of Modernity," 22–23, and M. Cohen, "Modernity on the Waterfront," 64; and Hopkins, *Planning the Greenspaces.*

9. Muntz and Girard, *Les engrais*, 351; Gandy, "Paris Sewers," 32; Barles, *L'invention des déchets*; and Barles, "Urban Metabolism."

10. *Annuaire stastique de la ville de Paris*, year 1 (1880), 44; Jerrold, *Imperial Paris*, 176–92; Johnson, *Ghost Map*, 1–24; Reid, *Paris Sewers*, 10; Jacquemet, "Urbanisme Parisien"; Cronon, *Nature's Metropolis*; and tobacco, from Walton, *Paris Known and Unknown*, 107–8.

11. Reid, *Paris Sewers*, pipe length on 30, Haussmann on 80; Gandy, "Paris Sewers," 32–33; Douglas, *Purity and Danger*; and J. Strauss, *Human Remains.*

12. "Hygiène publique."

13. *Rapport de l'ingénieur ordinaire: Assainissement de la Seine en aval de Paris*, 31 August 1892, 3 (AP D1.S8.6).

14. Veron, "Humble requête d'un barbillon," 1–2.

15. *Collection officielle des ordonnances de police*, 391; Conseil municipal de Paris, *Rapport* no. 5 (1890): 117; Carré and Lestel, *Les rivières urbaines*, 71–74; and Mouchel et al., "Bathing Activities," 329.

16. Conseil municipal de Paris, *Rapport* no. 5 (1890): 54, 55, 113, 118; and Hopkins, *Planning the Greenspaces*, 143–46.

17. Meyer, *Dividing Paris*, 152–54.

18. *Rapport de l'ingénieur ordinaire*, 21 July 1874, and 4 November 1874 (both sources AP D1.S8.6).

19. Commission de l'assainissement de Paris, Séance du 15 janvier 1885, 5 (AP VI.5.1); Prefecture de la Seine, imprint No. 5. Direction des Travaux de Paris. Commission technique de l'assainissement de Paris, 4ème Sous-Commission. Séance du 9 décembre 1882. Rapport présenté par M. Mille, 5–8 (AP D1.S8.6).

20. Dickens, *Dickens's Dictionary of Paris*, 7–9, 95–96, 247–48.

21. "Au jour le jour: l'assainissement de la Seine," *Le temps*, 9 July 1899, 3.

22. For sewer design and construction, see Reid, *Paris Sewers*; Jacquemet, "Urbanisme Parisien"; Guerrand, "La bataille du tout à l'égout"; and Gandy, *Fabric of Space.*

23. See "Tout à l'égout, rien à la Seine"; Ponts et Chaussées—Direction des Travaux de Paris—Service de l'Assainissement de la Seine. *Rapport de l'ingénieur ordinaire, Assainissement de la Seine en aval de Paris* (31 August 1892), AP D1.S8.6; and Mitchell, *Rêves Parisiens*, 48–64.

24. Préfecture de la Seine, *Assainissement de Paris. Réforme du mode actuel de la vidange. Projet de règlement relatif aux cabinets d'aisance, aux tuyaux d'évacuation, etc. Projet de loi autorisant une taxe par tuyau de chute*, 1883. Strauss, quoted in Jacquemet, "Urbanisme Parisien," 518.

25. Nansouty, "Tout à l'égout?"; and "A propos du 'tout à l'égout.'"

26. Pike, *Subterranean Cities*, 201–3, 245–51, 289–90; Reid, *Paris Sewers*, 37–52; and J. Strauss, *Human Remains*, 105–7, 249–51.

27. Bruant and de Bercy, *L'argot au XXe siècle*, 67; and Virmaitre, *Dictionnaire d'argot*, 122.

28. Jouy, "Tout à l'égout!"; *Almanach du père peinard* (1894), 17; *Le père peinard*, no. 63 (2 January 1898), back cover; "Le veritable tout à l'égout," *La libre parole* (illustrée), 17 February 1894, 1.

29. "L'épuration de la Seine," *Le petit Parisien*, 9 July 1899, 1; F. Launay, *Les champs d'épandage*, 5; Conseil général de la Seine, *Rapport* no. 29, *Rapport au nom de la Commission départementale des eaux sur l'assainissement du département de la Seine*, presented by MM. Paris and Carmignac, 1909, 3; see also Moses Baker, "Sewage Farms of Paris"; and Jacquemet, "Urbanisme Parisien," 509.

30. Jacquemet, "Urbanisme Parisien," 509–10; Durand-Claye et al., "Avant-project d'un canal d'irrigation à l'aide des eaux d'égout entre Clichy et la forêt de Saint-Germain," in *Assainissement de la Seine* (1876), 8–11; Salet, *Mémoire sur l'avant-projet.*

31. *La construction moderne*, special issue, "Tout à l'égout," 3 March 1894.

32. Ramsay, "L'épuration des eaux."

33. The figure 919,800 m^3 is cited in *Observations des ingénieurs du service municipal*, 5–6; the figure 1,126,706 m^3 is cited in *Annuaire statistique de la ville de Paris*, year 2 (1881), 134–36.

34. Jacquemet, "Urbanisme Parisien," 543; *Annuaire statistique de la ville de Paris*, year 23 (1902), 94.

35. Poubelle, *Mémoire au conseil municipal*, 19 December 1892 (AP D1.S8.6).

36. Brelay, "Affaires municipales," 300.

37. Judet, "L'eau et la politique," *Le petit journal*, 14 September 1895, 1.

38. The first two quotes are from Benjamin, *Arcades Project*, 797 and 98, respectively; and for "cloaca maxima," see Pike, *Subterranean Cities*, 247.

39. Reid, *Paris Sewers*, 82; "Au jour le jour: l'assainissement de la Seine," *Le temps*, 9 July 1899, 3; and "L'épuration de la Seine," *Le petit Parisien*, 9 July 1899, 1.

40. *Rapport de l'ingénieur ordinaire* and *Rapport de l'ingénieur en chef*, 15 September 1899; *Etat d'infection de la Seine en aval de Paris: Réponse au Service municipal d'assainissement: Avis de l'ingénieur en chef* (Résal), 3 October 1900 (both sources AP D1.S8.6).

41. Lavigne, "La brigade fluviale"; Conseil municipal de Paris, *Rapport* no. 129 (21 December1901), 38; "Les chiens plongeurs"; and "La brigade fluviale de Paris."

42. Conseil municipal de Paris, *Procès-verbaux* (1906), 113.

43. Conseil général de la Seine, *Compte rendu de la séance du samedi 24 Décembre 1904* (AP V.ONC 1217); Conseil général de la Seine, *Rapport* no. 4 (29 June 1905); Conseil général de la Seine, *Rapport au nom de la Commission des eaux et de l'assainissement, sur l'établissement d'un programme avec épuration biologique des eaux d'égout pour l'assainissement de la Seine, de la Marne et de la Bièvre*, no. 13 (1907) (both sources AP D1.S8.6).

44. Conseil général de la Seine, *Rapport* no. 29 (3 December 1909); Conseil général de la Seine, *Rapport* no. 7 (17 March 1911); and Conseil général de la Seine, *Rapport* no. 3 (8 June 1914) (latter two AP D1.S8.7). On urbanism and scale, see Charvet, *Les fortifications de Paris*; Papayanis, "César Daly"; and Wolf, *Eugène Hénard.*

45. Withington, *Flood*, 117.

46. Lechalas, *Hydraulique fluviale*, 287; Vernon-Harcourt, *River Seine*, 8–17; and Paris Préfecture de Police, *Plan de Secours*, 25–28.

47. "Faits divers," *Le temps*, 18 March 1876, 2; and "Moyens de sécher."

48. Zola, *Flood*; and E. F., "Les inondations."

49. Withington, *Flood*, 100–103.

50. "Notes sur les hauteurs de la Seine "; and Chateaublanc, "Le crue de la Seine," 400.

51. G. Lemoine and Preaudeau, "Etude sur les crues." On Belgrand, see Lechalas, *Hydraulique fluviale*, 287; Preaudeau, *Manuel hydrologique*, 56; Vernon-Harcourt, *Rivers and Canals*, 151; and McClelland, *Floods and Flood Protection*, 39.

52. Commission des inondations, *Rapport général de M. Alfred Picard.*

53. "Les inondations: Le désastre s'étend à la region Parisienne," *Le petit Parisien*, 22 January 1910.

54. Flammarion, "Le fleau qui rampe," quotes 177 and 180; and handbill, in Jackson, *Paris Under Water*, 99.

55. Gibbons, *Paris Vistas*, 161.

56. "Les inondations: le désastre s'étend à la région Parisienne," 1; Goubert, *Conquest of Water*, 61–67; Graham and Marvin, *Splintering Urbanism*, 53–55.

57. "Les inondations: le désastre s'étend à la région Parisienne," 1; "Les inondations: la Seine continue à monter, le désastre s'accentue partout," *Le petit Parisien*, January 23, 1910, 1.

58. Commission des inondations, *Le chemin de fer*, 241, 245, quote 235.

59. Commission des inondations, *L'évacuation* , 163; Commission des inondations, *Les communes suburbaines*, 422–500; *Crue de la Seine*; Jackson, *Paris Under Water*, 64–5; for boil notice, see *Le petit Parisien*, 23 January 1910, 2.

60. On Venice, see *Je sais tout* 62, year 6 (15 March 1910), 160; *Le petit Parisien*, 22 and 23 January 1910; and Gibbons, *Paris Vistas*, photographers 155, Venice 159, and Notre Dame 163. Photo albums: *Crue de la Seine*; Michel, *Paris dans l'eau*; and *Paris inconnu*, which the Red Cross sold to raise relief money.

61. "Les inondations: le désastre s'étend à la région Parisienne," 1.

62. Henry De Larègle, "Les résponsabilités," *Le soleil*, 11 February 1910, 1; and Gibbons, *Paris Vistas*, 158.

63. Quote from Judet, "Les cinq pouvoirs, *L'éclair*, 29 January 1910." See also "L'hygiène au Métropolitain" *L'humanité*, 3 September 1910; and "Un musée, une bibliothèque et un lycée menacés d'un chantier," *Le matin*, 25 August 1911.

64. Jouvenal, "Maintenant, l'avenir," *Le matin*, 31 January 1910.

65. Klinenberg, *Heat Wave*; Steinberg, *Acts of God*; Pritchard, "Envirotechnical Disaster"; and Knowles, "Learning from Disaster?"

66. Poggiale in *Bulletin de l'Académie nationale de médecine* (session of 30 November 1875), 1370.

67. The same idiom of life and death was used to describe diversions of the Rhône, see Pritchard, *Confluence*, 93–96, 99, 103, 110. See also J. Strauss, *Human Remains.*

68. Cronon, *Nature's Metropolis*; Heynen et al., *In the Nature of Cities*; Brantz and Dumpelmann, *Greening the City*; and Hopkins, *Planning the Greenspaces.*

69. Colin Jones, *Paris*, 108–13, 203–4; and Le Roux, "Une rivière industrielle."

70. Parent-Duchatelet, *Hygiène publique* (1836), 116; Delvau, *Au bord de la Bièvre*, 75; Bonnet, "La carte des pauvres," 386, 397, 399; and d'Haussonville, "La misère à Paris," 836–37. On public health, see La Berge, *Mission and Method*, 202–7; and D. Barnes, *Great Stink of Paris*, 79–80.

71. Shukin, *Animal Capital*; Popelin, *Le jour où l'eau reviendra*; Conseil général de la Seine, Séance du 15 novembre 1847, *Amélioration du cours de la Bièvre: Rapport de la commission.*

72. Du Camp, *Paris*, 6th ed., 5:332.

73. *Bulletin de l'Académie nationale de médecine* (30 November 1875), 1372–73 and 1368, respectively.

74. Conseil municipal de Paris, 6ᵉ Commission, *Rapport* no. 37 (1907), 3. For literary soures, see Delvau, *Au bord de la Bièvre*, 19; Cain, *Nouvelles promenades*, 68; Callet, *L'agonie du vieux Paris*, 104; "Courrier de Paris," *Le monde illustré*, 27 July 1895, 50; and "Courrier de Paris," *Le monde illustré*, 13 February 1897, 98. On mosquitoes, see *Bulletin de l'Académie nationale de médecine* (session of 16 April 1901), 491–92.

75. Belgrand, *Les travaux souterrain*, 5:178n1; Colin, *Paris*, 9; Dupain, *La Bièvre*, vii; *Times* (London), 30 June 1885, 5; "Courrier de Paris," *Le monde illustré*, 13 February 1897, 98; "Tempêtes et inondations," *Le monde illustré*, 7 January 1899, 9; Béry, *La Bièvre autrefois*, 187–88; Callet, *L'agonie du vieux Paris*, 111–12; Jackson, *Paris Under Water*, 45–46, 175; AP D3.S3, carton 3; and V.O3, cartons 73, 340, 342–45.

76. *Plan indicative des adherents et des opposants a la couverture de la Bièvre* (15 June 1874); *Petition des industriels riverains de la Bièvre . . .* (10 June 1882); *Petition tendant a faire couvrir la Bièvre . . .* (19 April 1883); all previous sources AP V.O3 806. See also *Bulletin municipal officiel de la ville de Paris*, year 2, no. 82 (23 March 1883), 472; year 2, no. 357 (24 December 1883), 2058; and year 3, no. 43 (12 February 1884), 236.

77. *Bulletin municipal officiel de la ville de Paris*, year 13, no. 141 (29 May 1894), 1188; year 14, no. 314 (21 November 1895), 2895.

78. "Tribunal Civil de la Seine," 71–72; Bery, *La Bièvre autrefois*, 207; and Conseil municipal de Paris, 6ᵉ Commission, *Rapport* no. 37 (1907), 8–15.

79. *Bulletin municipal officiel de la ville de Paris*, 25 March 1909, 1401; see also notes and clippings in AP V.O3 807.

80. Culbert, "Nowhere in Paris"; and Koos, "River Runs Through It."

81. For "pauvre Bièvre," see "Courrier de Paris," *Le monde illustré*, 27 July 1895, 50; Cain, *Nouvelles promenades*, 75. Henriet, "Au fil de la Bièvre," 565: "la pauvrette." Mithouard, *La perdition de la Bièvre*, "la pauvre petite" on 14, 22, among many others; Callet, *L'agonie du vieux Paris*, 107; and Capitan, *Notules sur la Bièvre*, 12.

82. Huysmans, *La Bièvre*, quotes on 8 and 11; and Maisonneufve, "Les tanneries de la Bièvre."

83. Huysmans, *La Bièvre*, 13–15; "Courrier de Paris," *Le monde illustré*, 13 February 1897, 98; Culbert, "Nowhere in Paris"; and Koos, "River Runs Through It."

84. Lenotre, "La Bièvre"; and Mithouard, *La perdition de la Bièvre*, 25–32.

85. Baedeker, *Paris and Environs* (1907), 334; Henriet,"Au fil de la Bièvre," 565; "Art et hygiène"; and Lucas, *Wanderer in Paris*, 187.

86. Lehaguez, *Le nouveau Paris*, 30; Lansdale, *Paris*, 522; Cain, *Nouvelles promenades dans Paris*, 67, 77; Rochegude, *Guide pratique*, 366–68; Lucas, *Wanderer in Paris*, 186–87; Bicknell, *Paris and Her Treasures*, 110; and Taylor, *Paris*, 8.

87. W. Cohen and Johnson, *Filth*; D. Barnes, *Great Stink of Paris*; J. Strauss, *Human Remains*; Douglas, *Purity and Danger*; Elias, *Civilizing Process*; Goubert, *Conquest of Water*; Reid, *Paris Sewers*; and Barles, *L'invention des déchets.*

88. Gibbons, *Paris Vistas*, 154.

Conclusion: Inverting Haussmannization

1. Harvey et al., "Introduction," 4.

2. Merriman, "French Revolutions," 206; and Terdiman, "Mnemonics of Dispossession," 176.

3. Kafka, *Diaries of Franz Kafka*, 281.

4. Cottereau, "Les batailles pour la création du Métro"; and Pike, *Subterranean Cities*, 57.

5. Evenson, *Paris*, 208.

6. Edwards, "Infrastructure and Modernity."

7. Canguilhem, *Knowledge of Life*.

8. Palus, "Networked Infrastructure"; Mehta, "Liberal Strategies of Exclusion"; Joyce, *Rule of Freedom*; Nord, *Republican Moment*; and Wileman, "Not the Radical Republic."

Bibliography

Archival Sources

Archives Nationales (AN)

Series F14—Ministry of Public Works

F14 9153: Métro plans 1872 and 1876–1877

F14 9154: Métro plans 1880s–1890s; Enquêtes of 1883, 1890

F14 9189: Tramways 1846–1885

F14 8587: Compagnie des tramways de Paris et du Département de la Seine, 1893–1895

F14 8588: Local interest railways—statistics, 1894–1895

F14 14999 and 15000: Affaires générales, 1855–1915

F14 15024 to 15030: Reorganization of tramways in Paris and the department of the Seine, 1899–1920

F14 16584: protection of Paris against future floods, 1910–1914

Series F8—Police sanitaire

F8 209: Water and sanitation

Archives de Paris (AP)

Series D.M5 Travail hygienique et Santé Publique

D.M5.11—Disease reporting 1893, Voirie de Bondy 1881

D.M5.13—Laws of 15 February 1902 and 7 April 1903, plus related 1902–1909

D.M5.15—Cholera, 1873–1884

Series D3S3 Service Hydraulique

D3S3.3—Bièvre

Series D.S8 Assaissement Général, 1808–1934

D1.S8.1—Affaires générales

D1.S8.3—human waste disposal (vidanges), 1892–1921

D1.S8.6, 7, 8—Sanitation of the Seine (assainissement de la Seine)

Series D.S9—Prefecture of the Seine, railways, 1827–1922

D10.S9 6–19: photographs of Métro construction

Series VI.5 Public Health

VI.5.1 direct to sewer drainage (*tout à l'égout*), 1885

VI.5.3 Ambulances and varia

VI.5.4 1873 Cholera

VI5.5 Hygiene and disease, various

Series V.ONC Prefecture of the Seine—Municipal service for public works, 1810–1950

V.ONC 16: public urinals and toilets

V.ONC 67: working conditions, salaries

V.ONC 69: brochures—streets, 1888–1900

V.ONC 74: brochures—streets, 1875–1916

V.ONC 78: complaints about, and demands for indemnities for damages caused by, tramway and Métro construction, 1894–1912

V.ONC 79: omnibus and tramways, 1888–1900

V.ONC 84: Métro Technical Service (Service technique du métropolitain), 1897–1912
V.ONC 110: strikes, 1905–1908
V.ONC 112: strikes, 1905–1908
V.ONC 129: work accidents during Métro construction, 1899–1905, press review 1909
V.ONC 131: press reviews, 1905–1911
V.ONC 221: daily accounts of the 1898 *terrassier* strike
V.ONC 227: traffic accidents—notes, reports and complaints, 1883–1904
V.ONC 239: brochures—streets, 1888 and 1896
V.ONC 826: various floods—requests for information about water height from the Travaux de Paris, 1872–1920 (excluding 1910)
V.ONC 834: 1910 flood
V.ONC 1201: water supply, 1906–1907
V.ONC 1206: Commission for the Study and Surveillance of Freshwater and Wastewater (Commission d'étude et de surveillance des eaux d'alimentation et d'assainissement).
V.ONC 1211: water quality control and sterilization, 1900–1913
V.ONC 1217: press reviews—water and sewers
V.ONC 1249: surveillance of freshwater: correspondence, 1900–1916
V.ONC 1342: sanitation of the îlots insalubres (1st and 4th arrondissements), 1909–1915
V.ONC 1350: washing and disinfection of tramway and omnibus stops, 1894–1895
Series V.O3: Water, canals and sewers
V.O3 51, 57: direct to sewer drainage (*tout à l'égout*), 1886–1895
V.O3 52: sanitation, main sewers, 1893–1899
V.O3 59, 60: water service—brochures, 1909–1914
V.O3 62: human waste disposal (*vidanges*), 1871–1874; household garbage disposal (*ordures ménagères*), 1887–1914
V.O3 63: printed material on unclean dwellings and the Seine
V.O3 73: Bièvre
V.O3 126: Potable water
V.O3 133: maps and atlas 1883
V.O3 167: High Commission on Sanitation (Commission superieure de l'assainissement), 1882–1883, 1885.
V.O3 169bis: 1884 sanitation enquête by district
V.O3 220: water supply, 1897–1898
V.O3 340: Bièvre—storms, damages
V.O3 342: Bièvre—canalization, complaints
V.O3 343: Bièvre—notes
V.O3 344: Bièvre—construction, sanitation, floods
V.O3 345: Bièvre—authorizations, contraventions, various, 1871–1886
V.O3 346: Bièvre—storms, damages
V.O3 347: Bièvre, curage 1870–1880
V.O3 569: flooding damages
V.O3 610: water supply, Navarre report, 1902–1903
V.O3 620: Deliberations, Service des eaux, des égouts et de l'assainissement, 1894–1900s
V.O3 802: Bièvre, affaires diverses
V.O3 804: Bièvre, affaires diverses
V.O3 805: Bièvre, affaires diverses, plans

V.O3 806: Bièvre, couverture
V.O3 807: Bièvre, couverture
V.O3 1221: Technical Commission on Paris water supply, 1897–1900
Series V.O8: Métropolitain, 1897–1914
V.O8 13—brochures and various information
V.O8 15—accidents, complains, damages, indemnities, 1899–1913
Series 25W: Direction of Departmental Affairs, Office of Public Works, Direction of Paris Works, Service of Public Streets
25W 100—electrification of tramways, press review 1899–1903
Series Perotin 10653: Prefecture of the Seine. Architecture, promenades and plantation services, 1860–1930
Perotin 10653/347: ventilation, access, and elevators in Métro line 3, 1909–1913
Perotin 10653/348: ventilation, access, and elevators in Métro lines 4 and 8, 1908–1914
Perotin 10653/150: îlot insalubre no. 9—construction de HBM (1909, 1921–7)
Perotin 10653/156, 300: îlots insalubres, plans, no date

Published Sources

"A Paris impromptu." *Punch*, 29 May 1897, 262.

"À l'Union du commerce et d'industrie." *Le monde économique* 22, no. 18 (14 May 1912): 545–49.

"À propos du 'tout à l'égout.'" *Cosmos*, no. 411 (1893): 41–44.

Acheray, Paul Edouard. *La lutte contre la tuberculose dans les milieux populaires: dispensaires antituberculeux.* Thèse no. 39. Lille Faculty of Medicine, 1901.

"Actualités." *La construction moderne* (29 June 1907): 458.

Adams, Annemarie, Kevin Schwartzman, and David Theodore, "Collapse and Expand: Architecture and Tuberculosis Therapy in Montreal, 1909, 1933, 1954." *Technology and Culture* 49, no. 4 (October 2008): 908–42.

Adams, Henry. "The Dynamo and the Virgin." In *The Education of Henry Adams: An Autobiography*, 379–90. Houghton Mifflin, 1918.

Adas, Michael. *Machines as the Measure of Men: Science, Technology and Ideologies of European Dominance.* Cornell University Press, 1989.

Ahmed, Sara. *The Cultural Politics of Emotion.* Edinburgh University Press, 2014.

Aisenberg, Andrew. *Contagion: Disease, Government, and the "Social Question" in Nineteenth-Century France.* Stanford University Press, 1999.

Akrich, Madeleine. "The De-Scription of Technical Objects." In *Shaping Technology/Building Society: Studies in Sociotechnical Change*, edited by Wiebe Bijker and John Law, 205–44. MIT Press, 2000.

Almanach du Rire & de Fantasio. Bureau de Rire, 1910.

Amin, Ash. "Lively Infrastructure." *Theory, Culture & Society* 31, no. 7/8 (2014): 137–61.

Amin, Ash, and Nigel Thrift. *Cities: Reimagining the Urban.* Polity Press, 2002.

Anand, Nikhil, Akhil Gupta, and Hannah Appel, "Introduction: Temporality, Politics, and the Promise of Infrastructure." In *Promise of Infrastructure*, 1–40.

Anand, Nikhil, Akhil Gupta, and Hannah Appel, eds. *The Promise of Infrastructure.* Duke University Press, 2018.

Anderson, Benedict. *The Spectre of Comparisons: Nationalism, Southeast Asia, and the World.* Verso, 1998.

Anderson, Margaret. *Regeneration Through Empire: French Pronatalists and Colonial Settlement in the Third Republic*. University of Nebraska Press, 2015.

André, Charles. "Hydraulique: les compteurs d'eau." *Le génie civil* 3 (1883): 229–31.

André, Eugène Frédéric. *Note sur les variations de la circulation dans les rues de Paris de 1872 à 1887*. Paris: Imprimerie Nouvelle, 1888.

Annuaire de l'Observatoire de Montsouris pour l'an 1884: météorologie, agriculture, hygiene. Paris: Gauthier-Villars, 1884.

Anonymous. *Deux métropolitains*. Florence: Imprimerie coopérative, 1882.

Appadurai, Arjun. "Foreword." In Graham and McFarlane, *Infrastructural Lives*, xii–xiii.

Appelbaum, Stanley, ed. *French Satirical Drawings from "L'Assiette au Beurre."* Dover, 1978.

Armegnaud, Jules. *Nettoyons Paris*. Librarie Moderne Maurice Bauche, 1907.

Arnold, David. *Everyday Technology: Machines and the Making of India's Modernity*. University of Chicago Press, 2013.

"Art et hygiene: La Bièvre." *Paris médical: la semaine du clinician* 1912, no. 8: 317–19. J.-B. Baillière et fils, 1912.

Ashton, Rosemary. *One Hot Summer: Dickens, Darwin, Disraeli, and the Great Stink of 1858*. Yale University Press, 2017.

Aucoc, Léon. *La dérivation des sources pour l'alimentation des villes*. Paris: Alphonse Picard, 1889.

Augé, Marc. *In the Metro*. Translated by Tom Conley. University of Minnesota Press, 2002.

Auslander, Leora. *Taste and Power: Furnishing Modern France*. University of California Press, 1996.

Backer, Hector de. *Tramways: la traction par chevaux et la traction par machines sur les tramways: aperçu comparative*. Paris: Auguste Ghio, 1877.

Backouche, Isabelle. *La trace du fleuve: la Seine et Paris (1750–1850)*. École des hautes études en sciences sociales, 2000.

Baedeker, Karl. *Paris and Its Environs: With Routes from London to Paris: Handbook for Travellers*, 6th ed. Liepzig: Karl Baedeker, 1878.

Baedeker, Karl. *Paris and Environs: With Routes from London to Paris: Handbook for Travellers*, 8th ed. Liepzig: Karl Baedeker, 1884.

Baedeker, Karl. *Paris and Environs: With Routes from London to Paris: Handbook for Travellers*, 9th ed. Liepzig: Karl Baedeker, 1888.

Baedeker, Karl. *Paris and Environs: With Routes from London to Paris: Handbook for Travellers*, 10th ed. Liepzig: Karl Baedeker, 1891.

Baedeker, Karl. *Paris and Environs: With Routes from London to Paris: Handbook for Travellers*, 12th ed. Liepzig: Karl Baedeker, 1896.

Baedeker, Karl. *Paris and Environs: With Routes from London to Paris: Handbook for Travellers*, 13th ed. Liepzig: Karl Baedeker, 1898.

Baedeker, Karl. *Paris and Environs: With Routes from London to Paris: Handbook for Travellers*, 16th ed. Karl Baedeker, 1907.

Baker, Moses. "The Sewage Farms of Paris." In *British Sewage Works and Notes on the Sewage Farms of Paris and on Two German Works*, 139–41. Engineering News Publishing Co., 1904.

Bakker, Karen. "Water: Political, Biopolitical, Material." *Social Studies of Science* 42, no. 4, (August 2012): 616–23

Barbillion, Louis-Charles, and G. J. Griffisch. *Traité pratique de traction électrique*. E. Bernard et Cie., 1903.

Barles, Sabine. *L'invention des déchets urbains: France 1790–1970*. Champ Vallon, 2005.

Barles, Sabine. *La ville délétère: médecins et ingénieurs dans l'espace urbain, XVIIIe–XIXe siècle*. Champ Vallon, 1999.

Barles, Sabine. "Urban Metabolism and River Systems: An Historical Perspective—Paris and the Seine, 1790–1970." *Hydrology and Earth Sciences Discussions* 4 (2007): 1845–78.

Barles, Sabine, and André Guillerme, "La rue Parisienne au XIXe siècle: standardisation et contrôle?" *Romantisme* 1, no. 171 (2016): 15–28.

Barnes, David S. *The Great Stink of Paris and the Nineteenth-Century Struggle Against Filth and Germs*. Johns Hopkins University Press, 2006.

Barnes, David S. *The Making of a Social Disease: Tuberculosis in Nineteenth-Century France*. University of California Press, 1995.

Barnes, Jessica, and Samer Alatout, eds. "Water Worlds." Special issue, *Social Studies of Science* 42, no. 4 (August 2012): 483–627.

Baron, Armand. *Le paupérisme: ses causes et ses remèdes*. 2nd ed. Paris: Sandoz & Thuillier, 1882.

Barrère, Albert. *Argot and Slang: A New French and English Dictionary* . . . London: Whittaker and Co., 1887.

Barua, Maan. "Infrastructure and Non-human Life: A Wider Ontology." *Progress in Human Geography* (2021): 1–23.

Baudel, M. J. *Un an à Alger: excursions et souvenirs*. Paris: Charles Delgrave, 1887.

Bechmann, Georges. "Rapport de l'ingenieur en chef du Service technique des eaux et de l'assainissement (8 October 1902)." In *Alimentation de Paris en eau potable: mesures préliminaires en vue des nouvelles adductions d'eau*, 13–40. Chaix, 1902.

Bechmann, Georges. *Salubrité urbaine, distributions d'eau et assainissement*. 2 vols., 2nd ed. Paris: Encyclopédie des travaux publics, 1899.

Beck, Ulrich. *Risk Society: Towards a New Modernity*. Sage, 1992.

Bélanger, Pierre. "Landscape as Infrastructure." *Landscape Journal* 28, no. 1 (2009): 79–95.

Belenky, Masha. *Engine of Modernity: The Omnibus and Urban Culture in Nineteenth-Century Paris*. Manchester University Press, 2019.

Belgrand, Eugène. *Les travaux souterrain de Paris*. Vol. 5. Paris: 1887.

Beltran, Alain. "Du luxe au coeur du système: électricité et société dans la région Parisienne (1800–1939)." *Annales* 44, no. 5 (1989): 1113–36.

Beltran, Alain. "La difficile conquête d'une capitale: l'énergie électrique à Paris entre 1878 et 1907." *Histoire, économie et société* 4, no. 4–3 (1985): 369–95.

Beltran, Alain. "Les débuts du réseau électrique dans les villes: 1880–1920." *Bulletin d'histoire de l'électricité*, no. 7 (June 1986): 81–92.

Beltran, Alain, and Patrice A. Carré, *La fée et la servante: la société francaise face à l'électricité XIXe–XXe siècle*. Belin, 1991.

Benjamin, Walter. *The Arcades Project*. Translated by Howard Elland and Kevin McLaughlin. Harvard University Press, 2002.

Benjamin, Walter. "Paris, Capital of the Nineteenth Century." In *Arcades Project*, 14–26.

Berlanstein, Lenard. *Big Business and Industrial Conflict in Nineteenth-Century France: A Social History of the Parisian Gas Company*. University of California Press, 1991.

Berlant, Lauren. *Cruel Optimism*. Duke University Press, 2011.

Berman, Marshall. *All That Is Solid Melts into Air: The Experience of Modernity*. Penguin, 1988.

Bertillon, Jacques. "La crise des loyers." *Je sais tout* (1912): 697–704.

Berton, Claude, and Alexandre Ossadzow. *Fulgence Bienvenüe et la construction du métropolitain de Paris*. Presses Ponts et Chaussées, 2007.

Béry, Anatole. *La Bièvre autrefois et aujourd'hui*. G. Ficker, 1911.

Bess, Michael. "Artificialization and Its Discontents." *Environmental History* 10, no. 1 (January 2005): 31–33.

"Bibliographie." *Le génie civil* 9, no. 14 (31 July 1886): 224.

Bicknell, Ethel E. *Paris and Her Treasures*. Scribner's, 1912.

Biette, Louis. "Chemins de fer: le métropolitain de Paris." *Le génie civil* 58, no. 19 (11 March 1911): 385–97.

Biette, Louis. *Le métropolitain de Paris*. Chaix, 1906.

Biget, Henri. *Le logement de l'ouvrier: étude de législation des habitations à bon marché en France et à l'étranger.* Jouve et Cie, 1913.

Biggs, David. *Quagmire: Nation-Building and Nature in the Mekong Delta*. University of Washington Press/Weyerhauser, 2012.

Bijker, Wiebe. *Of Bicycles, Bakelites and Bulbs: Toward a Theory of Sociotechnical Change*. MIT Press, 1995.

Black, Brian. *Crude Reality: Petroleum in World History*. Roman and Littlefield, 2012.

Blanchot, Maurice. *The Writing of the Disaster.* Translated by Ann Smock. University of Nebraska Press, 1995.

Blau, Eve. "The City as Protagonist." In *Shaping the Great City: Modern Architecture in Central Europe, 1890–1937*, edited by Eve Blau and Monika Platzer, 11–24. Prestel Verlag, 1999.

Blerzy, H. "Études sur les travaux publics: routes, chemins et tramways." *La revue des deux mondes* 27 (1878): 628–58.

Block, Greet De, and Janet Polasky. "Light Railways and the Rural–Urban Continuum: Technology, Space and Society in Late Nineteenth-Century Belgium." *Journal of Historical Geography* 37, no. 3 (July 2011): 312–28.

Blondin, J. "La traction électrique urbaine et suburbaine." *Revue d'électricité* (26 July 1902): 120–33.

Bobrick, Benson. *Labyrinths of Iron: Subways in History, Myth, Art, Technology, and War.* Quill, 1986.

Bocquet, Denis, Konstantinos Chatzis, and Agnes Sander. "From Free Good to Commodity: Universalizing the Provision of Water in Paris (1830–1940)." *Geoforum* 39, no. 6 (2008): 1821–32.

Bogaert, R. "Les documents d'un conflit." *L'association Catholique*, year 33, vol. 66 (15 July 1908): 7–18.

Bogaert, R. "Les documents d'un conflit: le lock-out du bâtiment (fin)." *L'association Catholique*, year 33, vol. 66 (15 August 1908): 112–30.

Bon, Gustave le. *La psychologie des foules*. Felix Alcan, [1895] 1905.

Bonnet, Henri. "La carte des pauvres à Paris." *La revue des deux mondes*, fifth period, vol. 35, no. 2 (15 September 1906): 381–420.

Bonnin, R. "Les travaux du métropolitain sur la place de l'Opéra." *La nature*, no. 1555 (14 March 1903): 225–27.

Bonnin, R. "The Work of Superposing Three Lines of the Metropolitan Railway of Paris, at the Place de L'Opera." *Engineering News* 55, no. 5 (1 February 1906): 124–26.

Boucard, Max. *La vie de Paris*. Paris: Ollendorff, 1892.

Bouchet, Ghislaine. *Le cheval à Paris 1850–1914*. Droz, 1993.

Boudenoot, L. "Le métropolitain de Paris." *Le génie civil* 10, no. 10 (8 January 1887): 156–58.

Boudouard, O. "Recherches sur les odeurs de Paris." *La revue scientifique*, year 50, no. 20 (18 May 1912): 614–21.

Bouman, Mark. "Luxury and Control: The Urbanity of Street Lighting in Nineteenth-Century Cities." *Journal of Urban History* 14, no. 1 (November 1987): 7–37.

Bournon, Fernand. *La voie publique et son décor: colonnes, tours, portes, obélisques, fontaines, statues, etc.* Laurens, 1909.

Boutin, Aimé. *City of Noise: Sound and Nineteenth-Century Paris.* University of Illinois Press, 2015.

Bowie, Karen, ed. *La modernité avant Haussmann: formes de l'éspace urbain à Paris 1801–1853.* Editions Recherches, 2001.

Bowie, Karen. "Paris, 'plaque tournante' du réseau ferroviaire." In Bowie, *La modernité avant Haussmann*, 251–64.

Bowker, Geoffrey, and Susan Leigh Star. *Sorting Things Out: Classification and Its Consequences.* MIT Press, 1999.

Brantz, Dorothee, and Sonja Dümpelman, eds. *Greening the City: Urban Landscapes in the Twentieth Century.* University of Virginia Press, 2011.

Bray, Francesca. "Technics and Civilization in Late Imperial China: An Essay in the Cultural History of Technology." *Osiris*, 2nd ser., vol. 13 (1998): 11–33.

Brelay, Ernest, "Affaires municipales." *L'économiste français* 20, no. 2 (3 September 1892): 299–301.

Bressani, Martin. "Light into Darkness: Gaslight in Nineteenth-Century Paris." In *Cities of Light: Two Centuries of Urban Illumination*, edited by Sandy Isenstadt, Margaret Maile Petty, and Dietrich Neumann, 28–36. Routledge, 2015.

Brigstocke, Julian. *The Life of the City: Space, Humour, and the Experience of Truth in Fin-de-siècle Montmartre.* Routledge, 2014.

Brouardel, Paul. *La lutte contre la tuberculose.* Paris: Bailliere, 1901.

Brouardel, Paul. Preface to *L'étude et les progrès de l'hygiène en France de 1878 à 1882.* Paris: G. Masson, 1882.

Brown, Bill. "Thing Theory." *Critical Inquiry* 28, no. 1 (Autumn 2001): 1–22.

Bruant, Aristide, and Léon de Bercy. *L'argot au XXe siècle: dictionnaire français-argot.* Ernest Flammarion, 1901.

Brunfaut, Jules. *Les odeurs de Paris: assainissement.* Paris: J. Baudry, 1880.

Bullock, Nicholas, and James Read. *The Movement for Housing Reform in Germany and France, 1840–1914.* Cambridge University Press, 1985.

Burgraff, G. De. "Tramway funiculaire de Belleville." *Le magasin pittoresque*, year 58, series 2, vol. 8 (1890): 318–22.

Burke, Edmund. "Foreign and Colonial History." *The Annual Register* (1894), 217–39. Longmans, Green, and Co., 1895.

"The Busiest Underground Corner in Paris." *Popular Mechanics* 14, no. 1 (July 1910): 48.

Cadoux, Gaston. "Les eaux à Londres et à Paris." *La revue des deux mondes* (1902): 602–22.

Cain, Georges. *Nouvelles promenades dans Paris.* Paris: Flammarion, 1898.

Callet, Albert. *L'agonie du vieux Paris.* Paris: H. Daragon, 1911.

Callon, Michel. "Society in the Making: The Study of Technology as a Tool for Social Analysis." In Bijker et al., *Social Construction of Technological Systems*, 83–103.

Camp, Maxime du. *Paris, ses organes, ses fonctions et sa vie dans la second moitié du XIXe siècle.* 6 vols. Paris: Hachette, 1869–1875.

Canguilhem, Georges. *Knowledge of Life.* Translated by Stefanos Geroulanos and Daniela Ginsburg, introduction by Paola Marrati and Todd Meyers. Fordham University Press, 2008.

Cannon, James. *The Paris Zone: A Cultural History*. Routledge, 2015.

Capitan, Jeanne. *Notules sur la Bièvre, son présent, son passé*. H. Champion, 1908.

Cardoso da Silva, José Maria, and Emily Wheeler. "Ecosystems as Infrastructure." *Perspectives in Ecology and Conservation* 15 (2017): 32–35.

Carré, C. "L'enquête sur le tramway tubulaire souterrain à traction électrique." *La lumière électrique* 42, no. 41 (1891): 72–80.

Carré, Catherine, and Laurence Lestel, eds. *Les rivières urbaines et leur pollution*. Éditions Quae, 2021.

"Carrefour des écrasés." In *L'intermédiaire des chercheurs et curieux*, no. 1164 (10 October 1907): 655.

Carroll, Patrick. "Water and Technoscientific State Formation in California." *Social Studies of Science* 42, no. 2 (2012): 489–516.

Carse, Ashley. "Keyword: Infrastructure: How a Humble French Engineering Term Shaped the Modern World." In Penny et al., *Infrastructures and Social Complexity*, 27–39.

Carse, Ashley. "Nature as Infrastructure: Making and Managing the Panama Canal Watershed." *Social Studies of Science* 42, no. 4 (2012): 539–63.

Cartwright, Lisa. *Screening the Body: Tracing Medicine's Visual Culture*. University of Minnesota Press, 1995.

Castonguay, Stephane, and Matthew Evenden, eds. *Urban Rivers: Remaking Rivers, Cities, and Space in Europe and North America*. University of Pittsburgh Press, 2012.

Cate, Philip Dennis. "The Spirit of Montmartre." In Cate and Shaw, *Spirit of Montmartre*, 1–94.

Cate, Phillip Dennis, and Mary Shaw, eds. *The Spirit of Montmartre: Cabarets, Humor, and the Avant-Garde, 1875–1905*. Rutgers University Press, 1996.

Çelik, Zeynep. *Empire, Architecture, and the City: French-Ottoman Encounters, 1830–1914*. University of Washington Press, 2008.

Certeau, Michel de. *The Practice of Everyday Life*. Translated by Steven Rendell. University of California Press, 1998.

Chakrabarty, Dipesh. *Provincializing Europe: Postcolonial Thought and Historical Difference*. Princeton University Press, 2007.

Champsaur, Félicien. "Ballade du tramway." In *Entrée de clowns*, 89–102.

Champsaur, Félicien. *Entrée de clowns*. Paris: Jules Lévy, 1885.

Chardon, Eugène. Letter to Charles Garnier, 30 April 1887. Bibliothèque Nationale de France, 8-V Pièce-6342.

"Charivaria." *Le charivari*, 11 December 1883, 2.

Charpentier, Octave. *À travers Montmartre*. Le Croquis, 1913.

Charvet, Marie. *Les fortifications de Paris: de l'hygiènisme à l'urbanisme, 1880–1919*. Presses universitaires de Rennes, 2005.

Chateaublanc. "Le crue de la Seine." *Le cosmos*, no. 215 (9 March 1889): 397–400.

Chatzis, Konstantinos. "Brève histoire des compteurs d'eau à Paris, 1880–1930." *Terrains & Travaux* 2, no. 11 (2006): 159–78.

Chatzis, Konstantinos, and Olivier Coutard. "Water and Gas: Early Developments in the Utility Networks of Paris." *Journal of Urban Technology* 12, no. 3 (2005): 1–17.

Chauveau, Gustave. "Les tramways à traction mécanique." *Le chauffeur*, year 63, no. 103 (11 April 1901), 126–31.

"Chemin de fer atmosphérique." *Le magasin pittoresque* 47 (1879): 168.

Chevalier, Louis. *Laboring Classes and Dangerous Classes in Paris During the First Half of the Nineteenth Century*. Translated by Frank Jellinek. Howard Fertig, 1973.

Chevallier, Fabienne. *Le Paris moderne: histoire des politiques d'hygiène, 1855–1898*. Presses universitaires de Rennes, 2010.

Cheysson, Émile. "La circulation sur les routes nationales d'après les comptages de 1882." *Journal de la Société de statistique de Paris* 25 (1884): 136–59.

Choay, Françoise. *The Modern City: Planning in the 19th Century*. Translated by Marguerite Hugo and George R. Collins. George Braziller, 1969.

Chrétien, J. *Les odeurs de Paris: étude analytique des causes qui concourent à l'insalubrité de la ville et des moyens de les combattre*. Paris: J. Baudry, 1881.

Christianson, Rupert. *Paris Babylon: Grandeur, Decadence, and Revolution 1869–1875*. Pimlico, 2003.

"Chronique et revue de la presse industrielle." *La lumière électrique*, year 11, vol. 33, no. 32 (10 August 1889): 276–79.

"Chronique Parisienne." *Bibliothèque universelle et revue Suisse*, year 102, vol. 6, no. 16 (May 1897): 380–81.

"Chronique Parisienne." *Bibliothèque universelle et revue Suisse*, year 105, vol. 19, no. 55 (September 1900): 611–13.

"Chronique politique." *La quinzaine: revue littéraire, artistique, et scientifique* 24 (1898): 574–77.

"Chronique: le choléra." *La nature*, no. 15 (13 September 1873): 206.

Chrysale, Le Bonhomme. "Notes de la semaine: Les embarras de Paris." *Les annales politiques et littéraires* 25, no. 1278 (22 December 1907): 589–90.

Claretie, Jules. *La vie à Paris: 1880–1910*. Paris: G. Charpentier et E. Fasquelle, 1881–1911.

Clark, Catherine. *Paris and the Cliché of History: The City and Photographs, 1860–1970*. Oxford University Press, 2018.

Clark, Thomas Curtis. "Rapid Transit in Cities: I. The Problem." *Scribner's Magazine* (May/June 1892): 567–78.

Claux, Dr. V. du. "Le pavage en bois." *Les annales d'hygiène publique et de médecine légale*, 3rd series, vol. 11, no. 4 (1884): 305–17.

Clayson, Hollis. *Illuminated Paris: Essays on Art and Lighting in the Belle Époque*. University of Chicago Press, 2019.

Clayson, Hollis, and Andre Dombrowski, eds. *Is Paris Still the Capital of the Nineteenth Century? Essays on Art and Modernity, 1850–1900*. Routledge, 2016.

Cobb, Richard. *Paris and Elsewhere*. New York Review of Books Classics, 2004.

Cochut, André. "De l'enchérissement des marchandises et des services." *La revue des deux mondes*, 3rd period, vol. 60, no. 3 (1 December 1883): 512–51.

Cohen, Margaret. "Modernity on the Waterfront: The Case of Haussmann's Paris." In *Urban Imaginaries: Locating the Modern City*, edited by Alev Cindar and Thomas Bender, 55–76. University of Minnesota Press, 2007.

Cohen, Margaret. "The Waterways of Modernity: The Case of Haussmann's Paris." In *Visions/Revisions: Essays on Nineteenth-Century French Culture*, edited by Nigel Harkness et al., 15–36. Peter Lang, 2003.

Cohen, William A., and Ryan Johnson, eds. *Filth: Dirt, Disgust, and Modern Life*. University of Minnesota Press, 2005.

Cole, Joshua. *The Power of Large Numbers: Population, Politics, and Gender in Nineteenth-Century France*. University of California Press, 2000.

Colin, Léon. *Paris, sa topographie, son hygiene, ses maladies.* Paris: Masson, 1885.

Collection officielle des ordonnances de police depuis 1862 jusqu'à 1874. Paris: Boucquin, 1874.

Collin, Henri. *L'eau potable: ses dangers au point de vue sanitaire; moyens de purification. filtres.* Lille: L. Danel, 1894.

Commission d'enquête sur l'avant-projet d'un canal d'irrigation à l'aide des eaux d'égout entre Clichy et la forêt de Saint-Germain. *Assainissement de la Seine: épuration et utilisation des eaux d'égout.* Paris: Gauthier-Villars, 1876.

Commission d'extension de Paris. Vol. 1, *Aperçu historique.* Chaix, 1913.

Commission d'extension de Paris. Vol. 2, *Considerations techniques préliminaires.* Chaix, 1913.

Commission des inondations. *L'évacuation et la destruction des gadoues à Paris,* edited by Boreux and Tur. Imprimerie Nationale, 1910.

Commission des inondations. *Le chemin de fer métropolitain de Paris,* edited by Bienvenüe and Hétier. Imprimerie Nationale, 1910.

Commission des inondations. *Les communes suburbaines,* edited by Alexandre. Imprimerie Nationale, 1910.

Commission des inondations. *Rapport général de M. Alfred Picard.* Imprimerie Nationale, 1910.

Commission des logements insalubres. *Rapport général sur les travaux de la commission pendant les années 1877 à 1883.* Paris: Imprimeries Réunies, Etablissement D, 1884.

Commission des logements insalubres. *Rapport général sur les travaux de la commission pendant les années 1870 à 1876.* Paris: Charles de Mourgues frères, 1878.

Commission du Vieux Paris. *Procès-verbaux 1903.* Imprimerie Municipal, 1904.

Commission scientifique de perfectionnement de l'observatoire municipal de Montsouris. *Travaux de l'année 1902 sur les eaux d'alimentation et les eaux d'égouts de la ville de Paris.* Chaix, 1903.

"Compagnie du tramway électrique de Paris à Romainville." *L'industrie électrique* (1896): 429–32.

Congrès international de la tuberculose, tenu à Paris, du 2 au 7 Octobre 1905. Vol. 3, *Conférences; Fêtes, visites et excursions; Exposition.* Edited by Masson and Company, Library of the Académie de Médicine, 1906.

Conklin, Alice. *A Mission to Civilize: The Republican Idea of Empire in France and West Africa, 1895–1930.* Stanford University Press, 1997.

Conseil général du département de la Seine, Première session de 1892. *Mémoires de M. le Préfet de la Seine & de M. le Préfet de Police et procès-verbaux des délibérations.* Paris: Imprimerie Municipale, 1892.

Conseil municipal de Paris. *Note présenté par M. Maurice Binder, au sujet des divers accidents imputables aux sociétés d'électricité ou à la Compagnie du gaz et survenus au courts des années 1889, 1890 et du premier trimester 1891,* no. 35, 10 June 1891.

Conseil municipal de Paris. *Note sur la question des eaux, les acquisitions et le captage des sources et sur les légitimes recriminations du public en ce qui concerne le manque d'eau d'alimentation, au cours des mois de juillet et août 1911,* no. 80, 30 August 1911.

Conseil municipal de Paris. *Procès-verbaux.* Année 1889. Premier semestre. Imprimerie municipale, 1889.

Conseil municipal de Paris. *Procès-verbaux.* Année 1901. Deuxième semestre. Imprimerie municipale, 1902.

Conseil municipal de Paris. *Procès-verbaux.* Année 1906. Premier semestre. Imprimerie municipale, 1906.

Conseil municipal de Paris. *Procès-verbaux.* Année 1907. Deuxième semestre. Imprimerie municipal, 1908.

Conseil municipal de Paris. *Procès-verbaux.* Année 1911. Première partie. Imprimerie municipale, 1911.

Conseil municipal de Paris. *Rapport au nom de la 1re Commission, sur la concession du service des omnibus*, no. 60, 25 July 1907.

Conseil municipal de Paris. *Rapport au nom de la 2e Commission, sur la circulation générale des voitures et des piétons à Paris*, no. 17, 30 May 1910.

Conseil municipal de Paris. *Rapport au nom de la 2e Sous-Commission du budget et du contrôle, sur le budget des dépenses et de recettes de la Préfecture de Police pour l'exercise 1902*, no. 129, 21 December 1901.

Conseil municipal de Paris. *Rapport au nom de la 6e Commission, sur la mise à l'enquête, préalable à la déclaration d'utilité publique, de l'opération de la suppression de la Bièvre dans Paris*, no. 37, 12 June 1907.

Conseil municipal de Paris. *Rapport présenté par M. Benon, au nom de la 2e Commission, sur la navigation dans Paris (Seine et canaux) et sur la réforme de l'ordonnance du 25 octobre 1840 sur la police des fleuves et rivières*, no. 5, 4 February 1890.

Conseil municipal de Paris. *Rapport présenté par M. Paul Viguier, au nom de la 6e Sous-commission du Comité du budget et du contrôle, sur le chap. XXI, art. 5 des Dépenses du projet du budget de 1894 (Observatoire municipal de Montsouris)*, no. 99, 9 December 1893.

Coole, Diana, and Samantha Frost, eds. *New Materialisms: Ontology, Agency, and Politics.* Duke University Press, 2010.

Corbin, Alain. *Women for Hire: Prostitution and Sexuality in France after 1850.* Translated by Alan Sheridan. Harvard University Press, 1990.

Corday, Michel. "Science et moeurs: la ville." *La revue de Paris*, year 7, no. 4 (July–August 1900): 766–79.

Cordulack, Shelley Wood. "A Franco-American Battle of Beams: Electricity and the Selling of Modernity." *Journal of Design History* 18, no. 2 (Summer 2005): 147–66.

"Correspondance: Les chemins de fer à voie étroite." *L'économiste français*, vol. 6, part 1 (27 April 1878): 532.

"Correspondance: Les tramways dans les promenades publiques." *L'économiste français*, vol. 6, part 1 (13 April 1878): 465–66.

"Correspondance: Les tramways et la circulation dans les grandes villes." *L'économiste français*, vol. 6, part 1 (27 April 1878): 531–32.

Cottereau, Alain. "Les batailles du métropolitain: la compagnie du chemin de fer du Nord et les choix d'urbanisation." In Hallsted-Baumert et al., *Métro-Cité*, 75–84.

Cottereau, Alain. "Les batailles pour la création du Métro: un choix de mode de vie, un succès historique pour la démocratie locale." *Revue d'histoire du XIXe siècle* 29 (2004): 89–151.

Couche, Edouard. *Le service des eaux en juillet 1881.* Paris: Chaix, 1882.

Couche, Edouard. *Les eaux de Paris en 1884.* Paris: Chaix, 1884.

"Courrier de Paris." *Le monde illustré*, 13 February 1897, 98.

"Courrier de Paris." *Le monde illustré*, 27 July 1895, 50.

Coutard, Olivier, Benjamin Bothereau, and Joel Tarr. "History (and Stories) of Off-Grid Technologies: A Reappraisal." *Flux* 1, no. 131 (2023): 1–14.

Coutard, Olivier, and Jonathan Rutherford, eds. *Beyond the Networked City: Infrastructure Reconfigurations and Urban Change in the North and South.* Routledge, 2015.

Crew, David. *Germans on Welfare from Weimar to Hitler.* Oxford University Press, 1998.

Cronon, William. *Nature's Metropolis: Chicago and the Great West.* W. W. Norton, 1992.

Crue de la Seine: Paris inondé. Éditions ELD, 1910.

Cubitt, Geoffrey. "Martyrs of Charity, Heroes of Solidarity: Catholic and Republican Responses to the Fire at the Bazar de la Charité, Paris, 1897." *French History* 21 (2007): 331–52.

Culbert, John. "Nowhere in Paris: A Tour of the Bièvre." *Qui Parle* 8, no. 1 (Fall/Winter 1994): 1–28.

Cultural Anthropology 26, no. 4 (November 2011). Special section, "Water," 511–619.

Curtin, Philip. *Disease and Empire: The Health of European Troops in the Conquest of Africa.* Cambridge University Press, 1998.

Cushman, Carrie, and Nicholas Risteen. "The Efficient Ruin of Infrastructure." *Verge: Studies in Global Asias* 6, no. 2 (2020): 1–26.

Cutcliffe, Stephen H., and Martin Reuss, eds. *The Illusory Boundary: Environment and Technology in History.* University of Virginia Press, 2010.

d'Avenel, Georges. *Le mécanisme de la vie moderne.* 5 vols. Paris: Armand Colin, 1896–1905.

d'Haussonville, Othenin. "La misère à Paris." *La revue des deux mondes* 45 (1881): 812–49.

d'Haussonville, Othenin. *La vie et les salaires à Paris.* Paris: A. Quantin, 1883.

Datta, Venita. *Heroes and Legends of Fin-de-Siècle France: Gender, Politics, and National Identity.* Cambridge University Press, 2011.

Davies, Francis H. *Electric Power and Traction.* Archibald Constable and Co., Ltd., 1907.

Davis, Donald F., and Barbara Lorenzkowski. "A Platform for Gender Tensions: Women Working and Riding on Canadian Urban Public Transit in the 1940s." *Canadian Historical Review* 79, no. 3 (1998): 431–65.

De Block, Greet, and Janet Polasky. "Light Railways and the Rural–Urban Continuum: Technology, Space and Society in Late Nineteenth-Century Belgium." *Journal of Historical Geography* 37, no. 3 (July 2011): 312–28.

Defrance, Eugène. *Histoire de l'éclairage des rues de Paris.* Imprimerie nationale, 1904.

DeJean, Joan. *How Paris Became Paris: The Invention of the Modern City.* Bloomsbury, 2014.

Delattre, Simone. *Les douzes heures noires: la nuit à Paris au XIXe siècle.* Albin Michel, 2014.

Deleuze, Gilles, and Félix Guattari, *Franz Kafka: Toward a Minor Literature.* Translated by Dana Polan. University of Minnesota Press, 1986.

Delmer, Louis. *Les chemins de fer: petite encyclopédie populaire illustrée.* Paris: Schleicher Frères, 1899.

Deluermoz, Quentin. "Rouler à la baguette? Les agents du service des voitures et la difficile codification des circulations à Paris (1892–1914)." *Crime, Histoire & Sociétés/Crime, History & Societies* 14, no. 1 (2010): 5–27.

Delvau, Alfred. *Au bord de la Bièvre: impressions et souvenirs*, new edition. Paris: René Picebourd, 1873.

Denning, Andrew. "Infrastructural Propaganda: The Visual Culture of Colonial Roads and the Domestication of Nature in Italian East Africa." *Environmental History* 24, no. 2 (2019): 352–69.

Denning, Andrew. "Mobilizing Empire: The Citroën Central Africa Expedition and the Interwar Civilizing Mission." *Technology and Culture* 61, no. 1 (2020): 42–70.

Derville, Georges. "Montmartre pittoresque: Le Maquis." *Les annales politiques et littéraires*, no. 1269 (1907): 372.

Desabres, Pascal. "Decision Network: Who Decided What in the Building of the Paris Métropolitain, 1898–1920." *Business and Economic History On-Line* 2 (2004): 1–8.

Desabres, Pascal. "The Parisian Subway, 1880–1900: A Local or a National Interest Line? On the Concept of Globalization." *Business and Economic History On-Line* 1 (2003): 1–16.

Dickens, Charles. "At Bordeaux." *All the Year Round: A Weekly Journal* 43 (1879): 395–400.

Dickens, Charles. *Dickens's Dictionary of Paris.* London: Macmillan and Co., 1883.

Dictionnaire national des contemporains: contenant les notices des membres de l'Institut de France, du gouvernement et du parlement français, de l'Académie de médecine... sous la dir. de C.-E. Curinier. Vol. 5. Paris: Office général d'édition de librairie et d'imprimerie, 1899–1919.

Direction du travail. *Statistique des grèves et des recours à la conciliation et à l'arbitrage.* Imprimerie nationale, 1908.

Disco, Cornelis. "Taming the Rhine: Economic Connection and Urban Competition." In Hård and Misa, *Urban Machinery*, 23–48.

Doniol, A. *La règlementation des chemins de fer d'intérêt local des tramways et des automobiles.* Librarie polytechnique Ch. Béranger, 1900.

Douglas, Mary. *Purity and Danger: An Analysis of Concepts of Pollution and Taboo.* 2nd ed. Routledge, 2000.

Dousteyssier-Khoze, Catherine. "Fumisme: le rire jaune du Chat Noir." In *(Ab)normalities*, edited by Catherine Dousteyssier-Khoze and Paul Scott, 151–61. Durham University Press, 2001.

Drumont, Georges *Automobiles sur rails.* Paris: Gauthier-Villars, 1898.

Dubois, Félix. *Le péril anarchiste.* Paris: E. Flammarion, 1894.

Ducray, Camille. "Omnibus d'autrefois." *Touche à tout* (June 1912): 439–44.

Dumarest, "Quelques notes recueillies dans un service de chirurgie de l'Hôtel-Dieu de Paris." *Lyon médical* (11 May 1873): 276.

Dumas, A. "Le tramway électrique de Paris à Romainville." *Le génie civil* 29, no. 19 (5 September 1896): 289–94.

Dumas, Alexandre. "Le Speronare." *La revue de Paris* (November 1841): 29.

Dumas, Jean-Philippe. *Paris la rue: un autre 1900.* Archives de Paris/Paris Musées, 2000.

Du Mesnil, Octave. *L'hygiène à Paris: l'habitation du pauvre.* Paris: Bailliere et fils, 1890.

Dupain, S. *La Bièvre: nouvelles recherches historiques sur cette rivière et ses affluents, depuis les temps les plus reculés jusqu'a nos jours.* Paris: H. Champion, 1886.

Duprey, Le Mansois. *L'oeuvre sociale de la municipalité Parisienne 1871–1891.* Paris: Imprimerie municipal, 1892.

Dupuy, Paul. *La traction électrique: tramways, locomotives et métropolitains électriques.* Paris: Librairie de sciences générales, 1897.

Edgerton, David. *The Shock of the Old: Technology and Global History Since 1900.* Oxford University Press, 2007.

Edwards, Paul N. "Infrastructure and Modernity: Force, Time, and Social Organization in the History of Sociotechnical Systems." In Misa et al., *Modernity and Technology*, 185–225.

Edwards, Paul N. "The Mechanics of Invisibility: On Habit and Routine as Elements of Infrastructure." In Ruby and Ruby, *Infrastructure Space*, 327–37.

Edwards, Paul N. *A Vast Machine: Computer Models, Climate Data, and the Politics of Global Warming.* MIT Press, 2010.

E. F. "Les inondations du bassin de la Garonne." *Journal de la Société de statistique de Paris*, no. 10 (October 1875): 265–67.

Eglash, Ron, Jennifer Croissant, Giovanna Di Chiro, and Rayvon Fouché, eds. *Appropriating Technology: Vernacular Science and Social Power.* University of Minnesota Press, 2004.

Egyedi, Tineke M., and Donna Mehos, eds. *Inverse Infrastructures: Disrupting Networks from Below.* Edward Elgar, 2012.

"Electric Tramway Collision in Paris." *Manchester Guardian*, 18 December 1903, 5.

Eley, Geoff. "Labor History, Social History, Alltagsgeschichte: Experience, Culture, and the Politics of the Everyday: A New Direction for German Social History?" *Journal of Modern History* 61 (June 1989): 297–344.

Elias, Norbert. *The Civilizing Process.* Translated by Edmund Jephcott. Blackwell, 2000.

Ellis, Jack. *The Physician-Legislators of France: Medicine and Politics in the Early Third Republic, 1870–1914.* Cambridge University Press, 1990.

Elwitt, Sanford. *The Third Republic Defended: Bourgeois Reform in France, 1880–1914.* Louisiana State University Press, 1986.

Emanuel, Frank L. *The Illustrators of Montmartre.* A. Siegle, 1904.

Engels, Friedrich. *The Housing Question.* Progress, 1970.

Engels, Jens Ivo. "Rhythm Analysis: A Heuristic Tool for Historical Infrastructure Research." *Technology and Culture* 63, no. 3 (2022): 830–52.

"Engineering: The Paris Wood Pavements." *Kansas City Review of Science and Industry* 8 (1885): 245–47.

Esbester, Mike, and Jameson Wetmore, eds. "(Auto)mobility, Accidents, and Danger." *Technology and Culture*, special issue 56, no. 2 (April 2015).

"Études sur les travaux publics: Routes, chemins et tramways." *La revue des deux mondes* 27 (1 June 1878): 628–58.

Evenden, Matthew. "Beyond the Organic Machine? New Approaches in River Historiography." *Environmental History* 23, no. 4 (2018): 698–720.

Evenson, Norma. *Paris: A Century of Change, 1878–1978.* Yale University Press, 1979.

"The Extension of Paris." *American Architect* 119, no. 2352 (19 January 1921): 64–65.

Fallan, Kjetil. "De-scribing Design: Appropriating Script Analysis to Design History." *Design Issues* 24, no. 4 (Autumn 2008): 61–75.

Faure, Alain. "Les invasions barbares ou le tramway criminel (Paris, 1872–1914)." *Le movement social*, no. 266 (January–March 2019): 87–110.

Ferrand, Stanislas. "Les tramways de la Seine." *Le bâtiment* (10 July 1898).

Ferreira da Silva, Álvaro, and M. Luísa Sousa, "The 'Script' of a New Urban Layout: Mobility, Environment, and Embellishment in Lisbon's Streets (1850–1910)." *Technology and Culture* 60, no. 1 (January 2019): 65–97.

Ferry, Jules. *Comptes fantastiques d'Haussmann.* Paris: Armand le Chevalier, 1867.

Fevre, Henry. *La locomotive: poésies.* Paris: Flammarion, 1883.

Figuier, Louis, ed. *L'année scientifique et industrielle*, year 34 (1890) Paris: Hachette, 1891.

Figuier, Louis, ed. *L'année scientifique et industrielle*, year 36 (1892) Paris: Hachette, 1893.

Figuier, Louis, ed. *L'année scientifique et industrielle*, year 37 (1893) Paris: Hachette, 1894.

Figuier, Louis, ed. *L'année scientifique et industrielle*, year 38 (1894) Paris: Hachette, 1895.

Figuier, Louis. "La fièvre typhoïde et l'eau de Seine, à Paris." *La science illustrée* (30 November 1889): 306–7.

Figuier, Louis. "Le chemin de fer métropolitain de Paris." *L'année scientifique et industrielle* 30 (1 January 1886): 215–22.

Figuier, Louis. *Le téléphone, son histoire, sa description, ses usages.* Paris: Librairie illustrée, Ollendorff, 1885.

Figuier, Louis. "Les dangers de l'électricité," *L'année scientifique et industrielle* 34 (1890): 155–59.

Figuier, Louis. *Les merveilles de la science ou description populaire des inventions modernes.* 6 vols. Paris: Librairie Furne, Editeurs Jouvet et Cie., 1867–1891.

Figuier, Louis. *Les nouvelles conquêtes de la science.* Vol. 2, *Grand tunnels et railways métropolitains.* Paris: Girard and Boitte, 1884.

Figuier, Louis, and Émile Gautier, eds. *L'année scientifique et industrielle* 16 (1873) Paris: Hachette, 1874.

Fijalkow, Yankel. *La construction des îlots insalubres: Paris 1850–1945.* L'Harmattan, 1998.

Fillassier, A. *Les casiers sanitaires des villes et les oeuvres d'assistance.* Jules Rousset, 1906.

Flammarion, Camille. "L'été de 1911." *Société astronomique* 25 (October 1911): 427–34.

Flammarion, Camille. "Le fleau qui rampe." *Je sais tout*, vol. 62, year 6 (March 15, 1910): 172–82.

"Flâneries et curiosités." *Gazette anecdotique, littéraire, artistique et bibliographique* 1, no. 7 (15 April 1897): 210–17.

Flonneau, Jean-Marie. "Crise de vie chère et mouvement syndical 1910–1914." *Le mouvement social*, no. 72 (July–September 1970): 49–81.

Flonneau, Mathieu. "Du profil de la rue à la forme de la ville: l'intégration Parisienne du 18e arrondissement (1860–1940)." *Recherches contemporaines*, no. 4 (1997): 49–73.

Flonneau, Mathieu. *L'automobile à la conquête de Paris: chroniques illustrées.* Presse des Ponts et Chausées, 2003.

Flonneau, Mathieu. *Paris et l'automobile: un siècle de passions.* Hachette, 2005.

Floriane. "La vie mondaine: les feuillets de madame. Les voyageurs." *La revue hebdomadaire* 15, no. 7 (1906): n.p.

Folkers, Andreas. "Existential Provisions: The Technopolitics of Public Infrastructure." *Environment and Planning D: Society and Space* 35, no. 5 (October 2017): 855–74.

Ford, Caroline. *Natural Interests: The Contest over Environment in Modern France.* Harvard University Press, 2016.

Foucault, Michel. "Of Other Spaces." Translated by Jay Miskowiec. *Diacritics* 16, no. 1 (Spring, 1986): 22–27.

Foucault, Michel. *"Society Must Be Defended": Lectures at the College de France, 1975–76.* Picador, 2003.

Foucault, Michel. "What Is Enlightenment?" In *The Foucault Reader*, edited by Paul Rabinow, 32–50. Pantheon, 1984.

Foucher, Paul. *Les coulisses du passé.* Paris: E. Dentu, 1873.

Fox, Robert. *The Savant and the State: Science and Cultural Politics in Nineteenth-Century France.* Johns Hopkins University Press, 2012.

France, Anatole. *Histoire contemporaine: Monsieur Bergeret à Paris.* Calman Levy, 1901.

France, Assemblée nationale. *Annales du Sénat et de la Chambre des deputes*, session 13, December 1879. Paris: A. Wittersheim, 1879.

France, Direction du travail. *Statistique des grèves et des recours à la conciliation et à l'arbitrage.* Imprimerie nationale, 1907.

France, Direction du travail. *Statistique des grèves et des recours à la conciliation et à l'arbitrage.* Imprimerie nationale, 1909.

France, Ministère des travaux publics. *Le développement des travaux publics en France sous la IIIe République.* Chaix, 1915.

Francq, Léon. *Chemin de fer métropolitain: recueil des articles publiés dans le journal le Métropolitain à propos de la traction du métropolitain Parisien.* Paris: E. Bernard et Cie., 1892.

Fraser, Nancy. "From Redistribution to Recognition? Dilemmas of Justice in a 'Post-Socialist' Age." *New Left Review* 212 (July/August 1995): 68–93.

Fraser, Nancy. "Rethinking Recognition." *New Left Review* 3 (May/June 2000): 107–20.

Freed, Libby. "Networks of (Colonial) Power: Roads in French Central Africa After World War I." *History and Technology* 26, no. 3 (2010): 203–23.

Fressoz, Jean-Baptiste. "Beck Back in the 19th Century: Towards a Genealogy of Risk Society." *History and Technology* 23, no. 4 (2007): 333–50.

Fressoz, Jean-Baptiste. "The Gas Lighting Controversy: Technological Risk, Expertise, and Regulation in Nineteenth-Century Paris and London." *Journal of Urban History* 33, no. 5 (July 2007): 729–55.

Fressoz, Jean-Baptiste. *L'apocalypse joyeuse: une histoire du risque technologique.* Seuil, 2012.

Friedel, Robert. *A Culture of Improvement: Technology and the Western Millenium.* MIT Press, 2007.

Friedel, Robert. *Zipper: An Exploration in Novelty.* W.W. Norton, 1996.

Galignani, A. and W. and Co. *Galignani's Illustrated Paris Guide for 1889.* London: Galignani, 1889.

Galland, Gabriel. *Vingt mois au pays de la soif: la mission saharienne Foureau-Lamy, octobre 1898-mai 1900 par Gabriel Galland.* Librairie nationale d'éducation et de récréation, 1906.

Gandhi, Mohandas. *The Collected Works of Mahatma Gandhi.* Vol. 3. Publications Division, Ministry of Information and Broadcasting, Government of India, 1960.

Gandy, Matthew. *The Fabric of Space: Water, Modernity, and the Urban Imagination.* MIT Press, 2014.

Gandy, Matthew. "The Paris Sewers and the Rationalization of Urban Space." *Transactions of the Institute of British Geographers* 24 (1999): 23–44.

Gandy, Matthew. "Rethinking Urban Metabolism: Water, Space, and the Modern City." *City* 8 (2004): 363–79.

Garner, Kenneth and Richard Abel. "Regulating a Risky Business: Film, Censorship, and Public Safety in Prewar France, 1909–1914." *Yale French Studies*, no. 122 (2012): 160–85.

Garnier, Jules. *Avant-projet d'un chemin de fer aérien à voies superposées.* Paris: Imprimerie Chaix, 1884.

Geddes, Patrick. "Man and the Environment: A Study from the Paris Exposition." *International Monthly* 2 (July–December 1900): 169–95.

Geppert, Alexander. *Fleeting Cities: Imperial Expositions in Fin-de-Siècle Europe.* Palgrave Macmillan, 2010.

Gibbons, Helen Davenport. *Paris Vistas.* Century, 1919.

Giedion, Sigfried. *Mechanization Takes Command: A Contribution to Anonymous History.* Oxford University Press, 1948.

Giedion, Sigfried. *Space, Time, and Architecture: The Growth of a New Tradition.* 5th ed. Harvard University Press, 2009.

Gieryn, Thomas. "What Buildings Do." *Theory and Society* 31 (2002): 35–74.

Gondrecourt, Aristide-Henri-Ange de. *Le pays de la soif.* Paris: L. de Polter, 1864.

Goodman, David C., ed. *The European Cities and Technology Reader.* Routledge/Open University, 1999.

Gosser, H. Mark. "The Bazar de la Charité Fire: The Reality, the Aftermath and the Telling." *Film History* 10, no. 1 (1998): 70–89.

Goubert, Jean-Pierre. *The Conquest of Water: The Advent of Health in the Industrial Age.* Translated by Andrew Wilson. Polity Press, 1989.

Gould, Roger. *Insurgent Identities: Class Community and Protest in Paris from 1848 to the Commune.* University of Chicago Press, 1995.

Graffigny, Henry de. *L'électricité pour tous.* E. Bernard, 1905.

Graham, Stephen, ed. *Disrupted Cities: When Infrastructures Fail.* Routledge, 2010.

Graham, Stephen, and Simon Marvin, *Splintering Urbanism: Networked Infrastructures, Technological Mobilities and the Urban Condition.* Routledge, 2001.

Graham, Stephen, and Colin McFarlane, eds. *Infrastructural Lives: Urban Infrastructure in Context.* Routledge, 2014.

Graham, Stephen, and Colin McFarlane. "Introduction." In *Infrastructural Lives*, 1–14.

Grand-Carteret, John. *XIXe siècle (en France): classes, moeurs, usages, costumes, inventions, ouvrage illustré . . . d'après les principaux artistes du siècle et à l'aide des procédés modernes.* Paris: Firmin-Didot, 1893.

Graux, Lucien. *La tuberculose et l'habitation urbaine.* Jules Rousset, 1905.

Gray, Samuel Merrit. *Proposed Plan for a Sewerage System, and for the Disposal of the Sewage of the City of Providence.* Providence Press Company, 1884.

Green, Judith. *Risk and Misfortune: The Social Construction of Accidents.* Routledge, 2013.

Greene, Ann Norton. *Horses at Work: Harnessing Power in Industrial America.* Harvard University Press, 2008.

Grojnowski, Daniel. "Hydropathes and Company." In Cate and Shaw, *Spirit of Montmartre*, 95–110.

Grubbs, Caroline. "Terminus 1900: The Métro and the Universal Expositions in Fin-de-Siècle Paris." *Dix-Neuf* 24, nos. 2–3 (July 2020): 203–20.

Guérard, Albert L. *L'avenir de Paris.* Payot, 1929.

Guerrand, Roger-Henri. "La bataille du tout-a-l'égout." *L'histoire* 53 (February 1983): 66–74.

Guibourt, Nicolas Jean-Baptiste G. *Histoire naturelle des drogues simples: ou cours d'histoire naturelle.* Vol. 4, 4th ed. J. B. Baillière, 1851.

Guillerme, André. "Autophiles et autophobes." *Canadian Center for Architecture*, 6 April 2006. https://www.cca.qc.ca/en/events/55454/andre-guillerme-autophiles-and-autophobes.

Guizot, Francois-Pierre-Guillaume. *Cours d'histoire moderne: histoire générale de la civilisation en Europe depuis la chute de l'empire romain jusqu'à la Révolution française.* Paris: Pichon et Didier, 1828.

Guldi, Jo. *Roads to Power: Britain Invents the Infrastructure State.* Harvard University Press, 2012.

Guyot, Yves. *Trois ans aux Ministère des travaux publics: expériences et conclusions.* Paris: Léon Chailley, 1896.

Haag, Paul. "Le chemin de fer métropolitain de Paris." *Association française pour l'avancement des sciences, Comte rendu de la 12e session* 12 (1884): 253–57.

Haag, Paul. *Le métropolitain de Paris et l'élargissement de la rue Montmartre.* Paris: A. Lemerre, 1883.

Halliday, Stephen. *The Great Stink of London: Sir Joseph Bazalgette and the Cleansing of the Victorian Metropolis.* Sutton, 1999.

Hallsted-Baumert, Sheila, François Gasnault, and Henri Zuber, eds. *Métro-cité: le chemin de fer métropolitain à la conquête de Paris, 1871–1945.* Paris musées, 1997.

Hamel, Frédéric. *Histoire illustrée du second empire: l'empire libéral.* Paris: Librarie de L'Écho de la Sorbonne, 1874.

Haraway, Donna J. *The Companion Species Manifesto: Dogs, People, and Significant Otherness.* Prickly Paradigm Press, 2003.

Hård, Mikael, and Andrew Jamison, eds. *The Intellectual Appropriation of Technology: Discourses on Modernity, 1900–1939.* MIT Press, 1998.

Hård, Mikael, and Andreas Knie, "The Grammar of Technology: German and French Diesel Engineering, 1920–1940." *Technology and Culture* 40, no. 1 (1999): 26–46.

Hård, Mikael, and Thomas J. Misa, eds. *Urban Machinery: Inside Modern European Cities*. MIT Press, 2008.

Haroudin, Robert. *La vie des abeilles solitaires*. Gallimard, 1948.

Harrison, Frederic. "Paris in 1851 and in 1907." In *Realities and Ideals: Social, Political, Literary and Artistic*, 435–52. Macmillan, 1908.

Harvey, David. *Paris, Capital of Modernity*. Routledge, 2003.

Harvey, David. "The Right to the City." *New Left Review* 53 (September–October 2008): 23–40.

Harvey, Penny, Caspar Bruun Jensen, and Atsuro Morita, eds. *Infrastructures and Social Complexity: A Companion*. Routledge, 2017.

Harvey, Penny, Caspar Bruun Jensen, and Atsuro Morita. "Introduction." In *Infrastructures and Social Complexity*, ed. Harvey et al., 1–22.

Hausser, Elisabeth. *Paris au jour le jour: les événements vus par la presse, 1900–1919*. Les editions de minuit, 1968.

Haussmann, Georges-Eugène. *Mémoires du baron Haussmann*. Vol. 1, 3rd ed. Paris: Victor-Havard, 1890.

Haussmann, Georges-Eugène. *Mémoires du baron Haussmann*. Vol. 2, 3rd ed. Paris: Victor-Havard, 1890.

Haussmann, Georges-Eugène. *Mémoires du baron Haussmann*. Vol. 3. Paris: Victor-Havard, 1893.

"Haussmanniser." *L'Intermédiaire des chercheurs et curieux*, no. 681 (1895), 167.

Hautière, E. de la. *La constitution et les institutions*. Paris: Garnier frères, 1895.

Havard, Louis. *La maison salubre et la maison insalubre a l'exposition universelle de 1889: étude sur l'exposition du service de l'assainissement*. Paris: Imprimerie Charles Noblet et fils, 1890.

Hazan, Eric. *The Invention of Paris: A History in Footsteps*. Verso, 2011.

Heathcott, Joseph Jonathan Soffer, and Rae Zimmerman, eds. *Urban Infrastructure: Historical and Social Dimensions of an Interconnected World*. University of Pittsburgh Press, 2022.

Hecht, Gabrielle. *The Radiance of France: Nuclear Power and National Identity After World War II*. MIT Press, 1998.

Hénard, Eugène. *Études sur les transformations de Paris*. Fascicule 5. Librairies-imprimeries réunis, 1904.

Hénard, Eugène. *Études sur les transformations de Paris et autre écrits sur l'urbanisme*. L'equerre, 1982 [1903–1909].

Henriet, L.-H. "Au fil de la Bièvre." *Le magasin pittoresque* (1903): 565–71.

Henriot. "La question des eaux! . . ." *Journal amusant*, no. 59 (1900): 10.

Herf, Jeffrey. *Reactionary Modernism: Technology, Culture, and Politics in Weimar and the Third Reich*. Cambridge University Press, 1984.

Hervieu, Jules. *Le chemin de fer métropolitain municipal de Paris*. Vol. 1. Béranger, 1903.

Hervieu, Jules. *Le chemin de fer métropolitain municipal de Paris*. Vol. 2. Béranger, 1908.

Heuzé, Louis. *Chemin de fer transversal à air libre dans une rue Spéciale: passage couvert pour piétons*. Paris: A. Lévy, 1878.

Hewitt, Nicholas. *Montmartre: A Cultural History*. Liverpool University Press, 2017.

Heynen, Nicholas, Maria Kaika, and Erik Swyngedouw, eds. *In the Nature of Cities: Urban Political Ecology and the Politics of Urban Metabolism*. Taylor and Francis, 2005.

Higonnet, Patrice. *Paris Capital of the World*. Translated by Arthur Goldhammer. Belknap Press of Harvard University Press, 2005.

Hobsbawm, Eric. *The Age of Empire: 1875–1914*. Vintage, 1989.

Hommels, Anique. *Unbuilding Cities: Obduracy in Urban Sociotechnical Change*. MIT Press, 2008.

Hopkins, Richard S. *Planning the Greenspaces of Nineteenth-Century Paris*. Louisiana State University Press, 2015.

Horne, Janet. *A Social Laboratory for Modern France: The Musée Social and the Rise of the Welfare State*. Duke University Press, 2002.

Hovey, Tamara. *Paris Underground*. Orchard Books, 1991.

Howe, Cymene et al. "Paradoxical Infrastructures: Ruins, Retrofit, and Risk." *Science, Technology and Human Values* 41, no. 3 (2016): 547–65.

Hubault, Edgard. *Omnibus, tramways, métropolitain*. Imprimerie municipale, 1902.

Huet, E. *Chemin de fer d'intérêt local du Département de la Seine: Réseau urbain. Rapport de l'ingénieur en chef des chemins de fer municipaux*. Paris: Chaix, 1876.

Huet, E. *Les chemins de fer métropolitains de Londres*. Paris: Dunod, 1878.

Hughes, Thomas. *American Genesis: A Century of Invention and Technological Enthusiasm 1870–1970*. University of Chicago Press, 1989.

Hughes, Thomas. *Human-Built World: How to Think About Technology and Culture*. University of Chicago Press, 2004.

Hughes, Thomas. *Networks of Power: Electrification in Western Society, 1880–1930*. Johns Hopkins University Press, 1983.

Hugo, Victor. *Histoire d'un crime*. 4th ed. Paris: Calman Lévy, 1877.

Hugo, Victor. *Oeuvres complètes: Philosophie*. 2 vols. Ollendorff, 1934.

Hussey, Andrew. *Paris: The Secret History*. Bloomsbury, 2008.

Huysmans, J.-K., *La Bièvre, Les Gobelins, Saint-Séverin*. Société de propagation des livres d'art, 1901.

"Hygiène publique." *Comptes rendus hebdomaires des séances de l'Académie des sciences* 72 (1871): 89–92.

"Hygiène publique: choléra." *Gazette hebdomadaire de médecine et de chirurgie*, series 2, vol. 11, no. 2 (9 January 1874): 32.

Illich, Ivan. *H_2O and the Waters of Forgetfulness*. Dallas Institute of Humanities and Culture, 1985.

Jackson, Jeffrey H. "Artistic Community and Urban Development in 1920s Montmartre." *French Politics, Culture & Society* 24, no. 2 (Summer 2006): 1–25.

Jackson, Jeffrey H. "Solidarism in the City Streets: La Société protectrice contre les excès de l'automobilisme and the Problem of Traffic in Early Twentieth-Century Paris." *French Cultural Studies* 20 no. 3 (August 2009): 237–56.

Jackson, Jeffrey H. *Paris Under Water: How the City of Light Survived the Great Flood of 1910*. Palgrave Macmillan, 2010.

Jacquemet, Gérard. "Urbanisme Parisien: la bataille du tout-a-l'égout à la fin du XIXe siècle." *Revue d'histoire moderne et contemporaine* 26 (October 1979): 505–48.

Jacquemet, Gérard. "Voirie, transports et equipement urbain à Belleville de 1860 à 1914." *Paris et Ile-de-France: mémoires publiés par la Fédération des Sociétés historiques et archéologiques de Paris et de l'Ile-de-France* 33 (1982): 225–53.

Jacquin, Charles. "Le tramway électrique de la place de la République à Romainville." *L'éclairage électrique* 7, no. 18 (2 May 1896): 222–26.

Jalas, Mikko, Jenny Rinkinen, and Antti Silvast. "The Rhythms of Infrastructure." *Anthropology Today* 32, no. 4 (2016): 17–20.

James, Edmund. "Conditions Relating to the Treatment of Employees and Laborers Imposed by the City of Paris upon the Company to Which the Metropolitan Road Was Leased." *American Journal of Sociology* 5, no. 6 (May 1900): 826–28.

Jasanoff, Sheila, and Sang-Hyun Kim, eds. *Dreamscapes of Modernity: Sociotechnical Imaginaries and the Fabrication of Power.* University of Chicago Press, 2015.

Jenks, Andrew. *Perils of Progress: Environmental Disasters in the 20th Century.* Pearson, 2010.

Jerram, Leif. *Streetlife: The Untold History of Europe's Twentieth Century.* Oxford University Press, 2011.

Jerrold, William Blanchard. *Imperial Paris, Including New Scenes for Old Visitors.* London: Bradbury and Evans, 1855.

Johnson, Steven. *The Ghost Map: The Story of London's Most Terrifying Epidemic—and How It Changed Science, Cities and the Modern World.* Riverhead Books, 2006.

Jones, Christopher. *Routes of Power: Energy and Modern America.* Harvard University Press, 2016.

Jones, Colin. *Paris: The Biography of a City.* Viking, 2004.

Jones-Imhotep, Edward. *The Unreliable Nation: Hostile Nature and Technological Failure in the Cold War.* MIT Press, 2017.

Jonnes, Jill. *Eiffel's Tower: And the World's Fair Where Buffalo Bill Beguiled Paris, the Artists Quarreled, and Thomas Edison Became a Count.* Viking, 2009.

Jordan, David. "Haussmann and Haussmannization: The Legacy for Paris." *French Historical Studies* 27, no. 1 (Winter 2004): 87–113.

Jordan, David. *Transforming Paris: The Life and Labors of Baron Haussmann.* University of Chicago Press, 1996.

Josephson, Paul. *Fish Sticks, Sports Bras, and Aluminum Cans: The Politics of Everyday Technologies.* Johns Hopkins University Press, 2015.

Josephson, Paul. *Traffic.* Bloomsbury, 2017.

Jouy, Jules. "À la cloche de bois." *Le père peinard*, year 9, no. 74, 20–27 March 1898, 5.

Jouy, Jules. "Tout à l'égout!" In *Chansons de bataille*, 17–18. Paris: Marpon & Flammarion, 1889.

Joyce, Patrick. *The Rule of Freedom: Liberalism and the Modern City.* Verso, 2003.

Juuti, Petri S., and Tapio S. Katko, eds. *Water, Time and European Cities: History Matters for the Futures.* www.watertime.net.

Kafka, Franz. *The Diaries of Franz Kafka 1914–1923.* Edited by Max Brod. Schocken Books, 1968.

Kaika, Maria. *City of Flows: Modernity, Nature, and the City.* Routledge, 2005.

Kamoun, Patrick. *V'là Cochon qui démenage: prélude au droit au logement.* Ivan Davy, 2000.

Kaufmann, Richard. *Paris of To-day.* Translated by Olga Flinch. New York: Cassell, 1891.

Keller, Richard. *Fatal Isolation: The Devastating Paris Heat Wave of 2003.* University of Chicago Press, 2015.

Kenny, Nicolas. *The Feel of the City: Experiences of Urban Transformation.* University of Toronto Press, 2014.

Kenny, Nicolas. "Je Cherche Fortune: Identity, Counterculture, and Profit in Fin-de-Siècle Montmartre." *Urban History Review/Revue d'histoire urbaine* 32, no. 2 (2004): 21–32.

Killen, Andreas. *Berlin Electropolis: Shock, Nerves and German Modernity.* University of California Press, 2006.

Kinkela, David, Teresa Meade, and Enrique Ochoa, eds. "Water: History, Power, Crisis." Special issue, *Radical History Review* 116 (Spring 2013).

Klinenberg, Eric. *Heat Wave: A Social Autopsy of Disaster in Chicago.* University of Chicago Press, 2002.

Klinenberg, Eric. *Palaces for the People: How Social Infrastructure Can Help Fight Inequality, Polarization, and the Decline of Civic Life.* Crown, 2019.

Knoll, Martin, Uwe Lübken, and Dieter Schott, eds. *Rivers Lost, Rivers Regained: Rethinking City-River Relations.* University of Pittsburgh Press, 2017.

Knowles, Scott. "Learning from Disaster? The History of Technology and the Future of Disaster Research." *Technology and Culture* 55, no. 4 (October 2014): 773–84.

Koos, Leonard R. "A River Runs Through It: La Bièvre, Huysmans, and Nineteenth-Century Paris." In *French Literature In/and the City*, edited by Buford Norman, 95–109. Editions Rodopi, 1997.

Kostof, Spiro. "His Majesty the Pick: The Aesthetics of Demolition." In *Streets: Critical Perspectives on Public Space*, edited by Zeynep Çelik, Diane Favro, Richard Ingersoll, and Spiro Kostof, 9–22. University of California Press, 1994.

Krantz, Jean-Baptiste. *Ministère de l'agriculture et du commerce. Exposition universelle de 1878, à Paris. Rapport sur l'installation et la mise en mouvement des machines à l'Exposition universelle de 1878.* Paris: Imprimerie de Tolmer et I. Joseph, n.d.

Kudlick, Catherine. *Cholera in Post-Revolutionary Paris: A Cultural History.* University of California Press, 1996.

L. R., M. "Une visite à Paris en 1900." *Journal des économistes* (July–October 1900), 332.

"L'hygiène publique: les odeurs de Paris." *L'année scientifique* 24 (1881): 364–76.

L'inondation à Paris et dans la Banlieue Janvier 1910: 60 vues. Oeuvre des orphelins-apprentis d'Auteuil, 1910.

La Berge, Ann. *Mission and Method: The Early Nineteenth-Century French Public Health Movement.* Cambridge University Press, 1992.

"La brigade fluviale de Paris." *Revue municipale* (21 June 1902): 3866–68.

"La Brouette." *L'indicateur de la Savoie*, year 34, no. 1673 (26 August 1911): 1.

La clé des omnibus et tramways. Paris: Administration d'affichage, 1876.

"La gare de l'Opéra." *Revue universelle* 4, no. 122 (1904): 636.

"La lutte contre la tuberculose." *La réforme sociale* 5, series 5, year 23 (January–June 1903): 389–401.

"La question des loyers." *La baionnette*, 13 July 1916, 439.

La question du métropolitain: Conférence a l'École des hautes études commerciales (extrait de *la Revue scientifique*). Paris: Administration des deux revues, 1889.

"La traction mécanique dans Paris: discussion." *Bulletin de la Société internationale des électriciens* 14 (1897): 14–65.

Ladd, Brian. *Autophobia: Love and Hate in the Automobile Age.* University of Chicago Press, 2008.

Ladd, Brian. *The Ghosts of Berlin: Confronting German History in the Urban Landscape.* University of Chicago Press, 1997.

Ladd, Brian. *The Streets of Europe: The Sights, Sounds, and Smells That Shaped Its Great Cities.* University of Chicago Press, 2020.

Laere, L. de. "La question des tramways." *Le courrier bleu de Neuilly-Boulogne*, 17 October 1897.

Lafarelle, Francois Felix de. *Du progrès social au profit des classes populaires non indigentes ou études philosophiques et économiques sur l'amélioration matérielle et morale du plus grand nombre.* Paris: Maison, successeur de M. Audin, 1839.

Lafargue, Paul. "L'idéalisme et le matérialisme dans la conception de l'histoire" (1895). http://www.marxists.org/francais/lafargue/works/1895/00/idealisme.htm.
Lafargue, Paul. "La communisme et les services publics." *L'égalité*, 25 June and 2 July 1882.
Lafargue, Paul. *Monsieur Vautour et la réduction des loyers.* Parti socialiste—SFIO, 1909.
Lajoulet, E. *Amélioration morale et matérielle de la condition du travailleur.* Toulouse, 1848.
Lanfrey, Pierre. *Chroniques politiques, 1860–1865.* Vol. 2. Paris: Charpentier, 1883.
Lansdale, Maria Hornor. *Paris: Its Sites, Monuments and History.* Henry Coates, 1898.
Lanthier, Pierre. "The Relationship Between State and Private Electric Industry, France 1880–1920." In *Law and the Formation of the Big Enterprises in the 19th and Early 20th Centuries*, edited by Norbert Horn and Jürgen Kocka, 590–603. Vandenhoeck & Ruprecht, 1979.
Lapouge, Georges Vacher de. "Lois de la vie et la mort des nations." *Revue internationale de sociologie* 2, no. 1 (January 1894): 421–36.
Lapparent, Albert-Auguste de. "À propos du métropolitain: le projet Haag." *Le correspondant* (25 December 1890): 1066–77.
Lapparent, Albert-Auguste de. "La question du métropolitain" (part 1). *Le correspondant* (25 October 1890): 189–208.
Lapparent, Albert-Auguste de. "La question du métropolitain" (part 2). *Le correspondant* (10 November 1890): 397–422.
Larchey, Lorédan. *Dictionnaire historique d'argot.* Paris: E. Dentu, 1878.
Larivière, Guy. "Le climat, question de mode?" *Études rurales*, no. 118/119 (April–September 1990): 103–12.
Larkin, Brian. "The Politics and Poetics of Infrastructure." *Annual Review of Anthropology* 42 (2013): 327–43.
Larroque, Dominique. "Un aspect résiduel des conflits autour du métropolitain au XIXe siècle: l'impossible connexion." In *Métropolitain: l'autre dimension de la ville*, 65–73. Hôtel de Lamignon, 1988.
Latour, Bruno. *The Pasteurization of France.* Translated by Alan Sheridan and John Law. Harvard University Press, 1988.
Latour, Bruno. *We Have Never Been Modern.* Translated by Catherine Porter. Harvard University Press, 1993.
Launay, F. *Les champs d'épandage de la ville de Paris et le parc agricole d'Achères.* Paris: Masson et Cie, 1897.
Lavigne, Charles. "La brigade fluviale de l'exposition." In *Le livre d'or de l'exposition de 1900*, 88–90. Cornély, 1900.
Law, John. "Technology and Heterogeneous Engineering: The Case of Portuguese Expansion." In Bijker et al., *Social Construction of Technological Systems*, 111–34.
"Le caisson du métropolitain." *Je sais tout* 7 (15 August 1905): 338.
"Le carrefour des écrasés." *L'intermédiaire des chercheurs et curieux*, no. 1153 (10 July 1907): 8.
"Le kiosque-signal de carrefour Montmartre." *Comoedia*, 9 April 1912, 4.
Le livre d'or des salons, year 20. Vve F. Bender, 1905.
Le Roy Ladurie, Emmanuel. *Le réchauffement de 1860 à nos jours: histoire humaine et comparée du climat.* Vol. 3. Fayard, 2009.
Le Roux, Thomas. "Une rivière industrielle avant l'industrialisation: la Bièvre et le fardeau de la prédestination, 1670–1830." *Géocarrefour* 85, no. 3 (2010): 192–207.
Lechalas, Médéric Clément. *Hydraulique fluviale.* Paris: Baudry et Cie., 1884.
Lee, Min Kyung. *The Tyranny of the Straight Line: Mapping Modern Paris.* Yale University Press, 2024.

Lefebvre, Henri. *Rhythmanalysis: Space, Time and Everyday Life*. Translated by Stuart Elden and Gerald Moore. Continuum, 2004.

Lefebvre, Henri. "The Right to the City." In *Writings on Cities*, 147–59. Blackwell, 1996.

Lehaguez. *Le nouveau Paris [en 1878]: guide de l'étranger*. Paris: Bernardin-Béchet, 1878.

Le Hir, M. L. *Réseau des voies ferrées sous Paris*. Paris: Mémorial du commerce et de l'industrie, 1872.

Leman-Stefanovic, Ingrid, and Stephen B. Scharper, eds. *The Natural City: Re-Envisioning the Built Environment*. University of Toronto Press, 2012.

Lemoine, Em. *Étude sur la formation et l'emploi des capitaux engagés dans les tramways-nord & sud de Paris*. Paris: Imprimerie de la publicité, 1881.

Lemoine, G., and A. de Preaudeau, "Étude sur les crues de l'hiver 1882–1883 dans le bassin de la Seine." *Les annales des ponts et chaussées* 2, no. 53, pt.1 (1883): 314–46. Mémoires et documents, no. 50. Paris: A. Dumas, 1883.

Lenotre, Georges. "La Bièvre." *Le monde illustré*, 8 February 1896, 108–10.

Léon, Paul. *Eaux et fontaines de Paris*. Par Mont ct par Vaux, 1946.

Léon, Robert. "Paris et son alimentation en eau." *La revue de Paris* (15 December 1913): 849–69.

Leroy-Beaulieu, Paul. "Les chemins de fer d'intérêt local et les tramways en France." *L'économiste français* 2, year 26, no. 32 (6 August 1898): 177–79.

"Les chiens plongeurs." *Revue municipale* (28 December 1901): 3475.

"Les inondations." *Je sais tout*, vol. 62, year 6 (15 February 1910): 160.

"Les loyers à Paris." *Le magasin pittoresque* 50 (1882): 113–14.

"Les miettes de la vie." *La revue hebdomadaire* 11, year 11 (October 1902): 25–30.

"Les oeuvres et les hommes." *Le correspondant* 198 (1900): 416.

"Les passerelles." *La construction moderne* (14 December 1901): 128–29.

"Les prognostics de juillet." *La Vallée du Rhône: Journal illustré des stations du Valais*, year 11, no. 3 (20 July 1912): 27.

"Les tramways de Paris." *Le magasin pittoresque* 47 (1879): 113.

Levallois, Ernest. *Paris propre!* Edouard Cornély et Cie, 1910.

Levin, Miriam. "Bringing the Future to Earth in Paris, 1851–1914." In Levin et al., *Urban Modernity*, 13–75.

Levin, Miriam. "Dynamic Triad: City, Exposition, and Museum in Industrial Society." In Levin et al., *Urban Modernity*, 1–12.

Levin, Miriam. "The Eiffel Tower Revisited: Community Building Through Industrial Design in Late 19th Century France." *French Review* (Spring 1989): 1052–65.

Levin, Miriam. *Republican Art and Ideology in Late Nineteenth-Century France*. UMI Research Press, 1986.

Levin, Miriam. *When the Eiffel Tower Was New: French Visions of Progress at the Centennial of the Revolution*. University of Massachusetts Press, 1989.

Levin, Miriam, Sophie Forgan, Martina Hessler, Robert H. Kargon, and Morris Low. *Urban Modernity: Cultural Innovation in the Second Industrial Revolution*. MIT Press, 2010.

Lewis, Arnold. *An Early Encounter with Tomorrow: Europeans, Chicago's Loop, and the World's Columbian Exhibition*. University of Illinois Press, 1997.

Lindemann, Albert S. *Anti-Semitism Before the Holocaust*. Routledge, 2014.

Lions, Frédéric. "Essai de bibliographie des projets imprimés de métropolitain." In *Métropolitain: l'autre dimension de la ville*, edited by Jean Tricoire, 281–94. Hotel de Lamoignon, 1988.

Little, Richard G. "Managing the Risk of Cascading Failure in Complex Urban Infrastructures." In Graham, *Disrupted Cities*, 27–40.

"Locomotive sans fumée et sans feu." *La nature* 2, no. 53 (6 June 1874): 314–15.

Lopez, Alberte Martinez. "Belgian Investment in Tramways and Light Railways: An International Approach, 1892–1935." *Journal of Transport History* 24, no. 1 (March 2003): 59–77.

Lopéz Galviz, Carlos Andrés. *Cites, Railways, Modernities: London, Paris, and the Nineteenth Century*. Routledge, 2019.

Lopéz Galviz, Carlos Andrés, and Mustafa Dikec, "The Modern Atlas: Compressed Air and Cities, c. 1850–1930." *Journal of Historical Geography* 53 (2016): 11–27.

Lorang, Coline. "Les fontaines Wallace (1872–2012): hygiène, esthétique et patrimoine." *Sociétés & représentations*, no. 34 (Fall 2012): 213–27.

Loyer, Francois. *Paris XIXe siècle: l'immeuble et la rue*. Hazan, 1987.

Loyrette, Henri. *Gutsave Eiffel*. Rizzoli, 1985.

Loyselles, E. de. "Le métropolitain: l'ouvrage de superposition de la place de l'Opéra." *La science au XXe siècle* 3, no. 36 (15 December 1905): 363–68.

Ls., E. "Hygiène: l'eau potable à Paris." *Revue des cours scientifiques de la France et de l'étranger*, year 49, semester 2, no. 23 (2 December 1911): 724–25.

Lucas, Edward Verrall. *A Wanderer in Paris*. Macmillan, 1909–10/1922.

Lüdtke, Alf. *The History of Everyday Life*. Princeton University Press, 1995.

MacDonald, *French Dialogues: A Systematic Introduction to the Grammar and Idiom of Spoken French*. London: Macmillan, 1892.

Maisonneufve, V. F. "Les tanneries de la Bièvre." *La science illustrée* (1893): 245–46.

Mandell, Richard. *Paris 1900: The Great World's Fair*. University of Toronto Press, 1967.

Marchand, Bernard. *Paris, histoire d'une ville*. Editions du Seuil, 1993.

Marcus, Sharon. *Apartment Stories: City and Home in Nineteenth-Century Paris and London*. University of California Press, 1999.

Marcus, Sharon. "Haussmannization as Anti-Modernity: The Apartment House in Parisian Urban Discourse, 1850–1880." *Journal of Urban History* 27, no. 6 (September 2001): 7 23–45.

Maréchal, Henry. *Les tramways électriques*. Librarie Polytechnique Charles Beranger, 1902.

Margairaz, Michel. "Le réseau métropolitain et les pouvoirs publics: du compromis républicain à l'emprise technocratique." In Hallsted-Baumert et al., *Métro-Cité*, 165–76.

Marisy, Alexis Bailleux de. "Le déficit communal." *La revue des deux mondes*, 3rd period, vol. 54 (11 November 1882): 611–36.

Marsoulan. *Vérités nécessaires . . . !* Paris: Imprimerie Moderne, 1879.

Martial, René. "Le mouvement de population dans les villes." *Les annales d'hygiène publique, industrielle et sociale*, series 11, no. 10 (October 1933): 541–65.

Martin, Alexis. *Une visite à Paris, la ville et ses promenades vues en quinze jours*. 2nd ed. A. Hennuyer, 1909–1910.

Martin, Henry. *Production et distribution de l'énergie pour la traction électrique*. Librarie Polytechnique Charles Beranger, 1902.

Martin, Léon, ed. *Encyclopédie municipale de la ville de Paris*. Part 1. Georges Roustan, 1904.

Marvin, Carolyn. *When Old Technologies Were New: Thinking About Electric Communication in the Late Nineteenth Century*. Oxford University Press, 1988.

Massard-Guilbaud, Geneviève. "Introduction: The Urban Catastrophe." In *Cities and Catastrophes—Villes et catastrophes: Coping with Emergency in European History—Réactions face a l'urgence dans l'histoire européenne*, edited by Massard-Guilbaud, Harold L. Platt, and Dieter Scott, 9–41. Peter Lang, 2002.

Maubry, Victorien. "Le métropolitain." *Le magasin pittoresque* 6, series 2, year 66 (1898): 373.

Mavhunga, Clapperton Chakanetsa. "Cidades Esfumaçadas: Energy and the Rural-Urban Connection in Mozambique." *Public Culture* 25, no. 2 70 (2013): 261–71.

Mazoyer, C. M. *Les conditions du travail dans les chantiers de la ville de Paris*. Doctoral thesis, Law Faculty, University of Paris, 1899.

McClelland, Ellwood H. *Floods and Flood Protection: References to Books and Magazine Articles*. Carnegie Library of Pittsburgh, 1908.

McElligott, Anthony, ed. *The German Urban Experience 1900–1945: Modernity and Crisis*. Routledge, 2001.

McKay, John. *Tramways and Trolleys: The Rise of Urban Mass Transport in Europe*. Princeton University Press, 1976.

Mehta, Uday S. "Liberal Strategies of Exclusion." *Politics and Society* 18, no. 4 (December 1990): 427–54.

Melosi, Martin V. "Urban Infrastructure and the Cultural Turn." In *Concepts of Urban-Environmental History*, edited by Sebastian Haumann, Martin Knoll, and Detlev Mares, 145–54. transcript Verlag, 2020.

Mendelsohn, J. Andrew. "The Microscopist of Modern Life." *Osiris* 18, 2nd series, "Science and the City" (2003): 150–70.

Merriman, John. *The Dynamite Club: How a Bombing in Fin-de-Siècle Paris Ignited the Age of Modern Terror*. Houghton Mifflin Harcourt, 2009.

Merriman, John. "The French Revolutions 1830–1871." In *Revolutions and the Meaning of Freedom in the Nineteenth Century*, edited by Isser Woloch, 173–211. Stanford University Press, 1996.

Merriman, John. *The Margins of City Life: Explorations on the French Urban Frontier, 1815–1851*. Oxford University Press, 1991.

Meyer, Esther da Costa. *Dividing Paris: Urban Renewal and Social Inequality, 1852–1870*. Princeton University Press, 2022.

Michel, Albin, ed. *Paris dans l'eau: les inondations pendant les journées des 26–27–28–29–30 & 31 janvier 1910: cinquante vues photographiques inédites d'après nature*. Éditions de Luxe, 1910.

Michel, Georges. *Léon Say: sa vie, ses oeuvres*. Paris: Calmann Lévy, 1899.

Miles, Malcolm. *Urban Avant-Gardes: Art, Architecture and Change*. Routledge, 2004.

Miller, Michael. *The Bon Marché: Bourgeois Culture and the Department Store, 1869–1920*. Princeton University Press, 1981.

Misa, Thomas, Philip Brey, and Andrew Feenberg, eds. *Modernity and Technology*. MIT Press, 2003.

Mitchell, Allan. "Private Enterprise or Public Service? The Eastern Railway Company and the French State in the Nineteenth Century." *Journal of Modern History* 69, no. 1 (March 1997): 18–41.

Mitchell, Allan. *Rêves Parisiens: l'échec de projets de transport public en France au XIXe siècle*. Presses des Ponts, 2005.

Mitchell, Timothy. "Introduction: Life of Infrastructure." *Comparative Studies of South Asia, Africa and the Middle East* 34, no. 3 (2014): 437–39.

Mitchell, Timothy. *The Rule of Experts: Egypt, Techno-Politics, Modernity*. University of California Press, 2002.

Mithouard, Adrien. *La perdition de la Bièvre*. Bibliothèque de l'occident, 1906.

Mollier, Jean-Yves. *Le scandale de Panamá*. Fayard, 1991.

Monnory, Henri. "La locomotion électrique dans le monde." *À travers le monde*, new series, no. 1 (1895): 93–95.

Montorgeuil, Georges. *Les eaux et les fontaines de Paris*. Payot, 1928.

Mouchel, Jean-Marie et al., "Bathing Activities and Microbial River Water Quality in the Paris Area." In *The Seine River Basin*, edited by Nicolas Flipo, Pierre Labadie, and Laurence Lestel, 323–53. Springer, 2021.

Mouchelet, E. "Doléances du *Photographe* contre les excès de la civilisation." *Photo-gazette* (September 25, 1900): 201–5.

"Moyens de sécher et assainir les habitations humides." *Le magasin pittoresque* year 44 (1876): 358.

Mukerji, Chandra. *Impossible Engineering: Technology and Territoriality on the Canal du Midi*. Princeton University Press, 2009.

Muntz, A., and Ch. Girard. *Les engrais*. Vol. 1, 2nd ed. Paris: Librarie de Firmin-Didot et Cie, Imprimeurs de l'Institut, 1891.

Myllyntaus, Timo. "Prologue: Constructing Technology for Everyday Life." *Icon* 16, Special issue, "Technology in Everyday Life" (2010): 3–21.

Nansouty, Max de. "Causerie scientifique: Sciences appliquées: Le grand trou de la place de l'Opéra." *Le temps*, 31 July 1903.

Nansouty, Max de. "La question de l'eau en été." *Le génie civil* 21, year 12, no. 7 (18 June 1892): 104–6.

Nansouty, Max de. *La question du réseau métropolitain de Paris et le projet de la Cie des etablissements Eiffel*. Paris: Génie civil, 1891.

Nansouty, Max de. *La tour Eiffel de 300 mètres à l'Exposition universelle de 1889*. Paris: Tignol, 1889.

Nansouty, Max de. "Le réseau central métropolitain de Paris." *Le temps*, 22 July 1890, 3.

Nansouty, Max de. "Tout à l'égout?—Rien à l'égout?" *L'illustration* 100, no. 2590 (15 October 1892): 302–6.

Neuman, Michael. "Infiltrating Infrastructures: On the Nature of Networked Infrastructure." *Journal of Urban Technology* 13, no. 1 (2006): 3–31.

Ngalamulume, Kalala. "Coping with Disease in the French Empire: The Provision of Waterworks in Saint-Louis-du-Senegal, 1860–1914." In *Environmental History of Water*, edited by Petri S. Juuti, Tapio S. Katko, and Heikki S. Vuorinen, 147–64. IWA, 2007.

Niget, David, and Martin Pettitclerc. "Introduction: Le risque comme culture de la temporalité." In *Pour une histoire du risque: Québec, France, Belgique*, edited by Niget and Pettitclerc, 1–33. Presses universitaires de Rennes and Presses de l'Université du Québec, 2012.

"Nîmoiserie." *Nîmes-journal*, year 31, no. 1718 (5–12 July 1911): 3.

Nixon, Rob. *Slow Violence and the Environmentalism of the Poor*. Harvard University Press, 2013.

Noir, Louis. *L'Amour au pays de la soif*. Paris: A. Fayard, 1893.

Nolan, Mary. *Visions of Modernity: American Business and the Modernization of Germany*. Oxford University Press, 1994.

Nord, Philip. *The Politics of Resentment: Shopkeeper Protest in Nineteenth-Century Paris*. Princeton University Press, 1986.

Nord, Philip. *The Republican Moment: Struggles for Democracy in Nineteenth-Century France*. Harvard University Press, 1998.

Nord, Philip. "The Welfare State in France, 1870–1914." *French Historical Studies* 18, no. 3. (Spring 1994): 821–38.

Norton, Peter. *Fighting Traffic: The Dawn of the Motor Age in the American City*. MIT Press, 2011.

"Notes sur les hauteurs de la Seine en 1883." *Annuaire statistique de la ville de Paris* 1883 4 (1885): 51–55.

Notice sur le service des eaux et de l'assainissement de Paris. C. Béranger, 1900.

"Nouvelles du jour: l'inauguration du kiosque à signal." *Journal des débats politiques et littéraires*, 12 April 1912: 3.

Nye, David. *American Technological Sublime*. MIT Press, 1996.

Nye, David. *Electrifying America: Social Meanings of a New Technology*. MIT Press, 1992.

Nye, David. *When the Lights Went Out: A History of Blackouts in America*. MIT Press, 2010.

Nye, Robert. *Crime, Madness, and Politics in Modern France: The Medical Conception of National Decline*. Princeton University Press, 1984.

Observations des ingénieurs du service municipal au sujet des projets de rapport présentés par MM. A. Girard et Brouardel, M. Alfred Durand-Claye, rapporteur. Paris: Chaix, 1881.

Observations présentées au Conseil général de la Seine. Paris: Chaix, 1882.

Odelin, Joseph. *Métropolitain de pénétration centrale*. Paris: Chaix, 1890.

Olivier, Arsène. *Chemins de fer dans Paris et dans les grandes villes*. 3rd ed. Paris: Librarie Centrale des Arts et Manufactures (Auguste Lemoine), 1872.

Olson, Donald James. *The City as a Work of Art: London, Paris, Vienna*. Yale University Press, 1988.

Osborne, Michael. *The Emergence of Tropical Medicine in France*. University of Chicago Press, 2014.

Ossadzow, Alexandre. "Les pères du métropolitain: l'intervention des ingénieurs." In Hallsted-Baumert et al.., *Métro-Cité*, 57–72.

Otter, Chris. *The Victorian Eye: A Political History of Light and Vision in Britain, 1800–1910*. University of Chicago Press, 2008.

Otter, Chris. "The Vital City: Public Analysis, Dairies and Slaughterhouses in Nineteenth-Century Britain." *Cultural Geographies* 13 (2006): 517–37.

Oudshoorn, Nelly, and Trevor Pinch, eds. *How Users Matter*. MIT Press, 2003.

Ovenden, Mark. *Paris Underground: The Maps, Stations, and Design of the Metro*. Penguin, 2009.

P. L., Dr. "La maison salubre et la maison insalubre." In *L'exposition de 1889*. Paris: Librarie illustré, 1889.

Palus, Matthew. "Networked Infrastructure as the Material Culture of Liberal Government." In *Historical Archaeology and the Importance of Material Things*, edited by Julie Schablitsky and Mark Leone, 93–120. Society for Historical Archaeology, 2012.

Papayanis, Nicholas. "César Daly and the Emergence of Modern Urban Planning." *Planning Perspectives* 24, no. 1 (1 October 2006): 325–46.

Papayanis, Nicholas. "The Development of the Paris Cab Trade, 1855–1914." *Journal of Transport History* 8, no. 1 (1987): 52–65.

Papayanis, Nicholas. *Horse-Drawn Cabs and Omnibuses in Paris: The Idea of Circulation and the Business of Public Transit*. Louisiana State University Press, 1996.

Papayanis, Nicholas. *Planning Paris Before Haussmann*. Johns Hopkins University Press, 2004.

Parent-Duchâtelet, Alexandre-Jean-Baptiste. *Hygiène publique, ou, Mémoires sur les questions les plus importantes de l'hygiène appliquée aux professions et aux travaux d'utilité publique*. Paris: J.-B. Baillière, 1836.

Paris inconnu: l'inondation de 1910. Société Anonyme de l'Imprimerie de Vaugirard: Sté d'éditions artistiques de tourisme et de sports, 1910.

Paris inondé, 29 janvier 1910, 32 vues. Pierre Petit, 1910.

"The Paris Metropolitan Railway: The Place de l'Opera." *Engineering* 82, no. 2114 (6 July 1906): 25–26.

Paris Prefecture de Police. *Plan de Secours Spécialisé Inondations Zonal* 1 (1 January 2006). http://www.prefecture-police-paris.interieur.gouv.fr/prevention/innondation_janvier2006/sommaire.htm.

Park, Sun-Young. *Ideals of the Body: Architecture, Urbanism, and Hygiene in Postrevolutionary Paris.* University of Pittsburgh Press, 2018.

Parr, Joy. *Sensing Changes: Technologies, Environments, and the Everyday, 1953–2003.* UBC Press, 2010.

"Partie Financière." *L'économiste français* 2, year 7, no. 33 (16 August 1879): 209–14.

Parville, Henri de. *Les causeries scientifiques, découvertes et inventions.* Year 12. Paris: Rothschild, 1872.

Parville, Henri de. *Les causeries scientifiques, découvertes et inventions.* Year 13. Paris: Rothschild, 1873.

Parville, Henri de. *Les causeries scientifiques, découvertes et inventions.* Year 18. Paris: Rothschild, 1879.

Parville, Henri de. *Les tramways électriques.* 2nd ed. Béranger, 1902.

Passalacqua, Arnaud. "Reluctant Capitals: Transport Mobility and Tramways in London and Paris 1830–1950." *Town Planning Review* 85, no. 2 (2014): 203–16.

Passalacqua, Arnaud. "Resident Alien: France's Ambivalent Relationship to the Tramway Before 1914." In *Mobility in History: Themes in Transport, T2M Yearbook 2011*, edited by Gijs Mom et al., 71–86. Éditions Alphil, 2011.

Patrikios, Dr. "La lutte contre la tuberculose en Grèce." *Transactions of the Sixth International Congress on Tuberculosis, Washington, September 28 to October 12, 1908.* Vol. 4, part 1. William Fell, 1908.

Péchu, Cécile. "Entre résistance et contestation: la genèse du squat comme mode d'action." *Université de Lausanne Travaux de science politique/Political Science Working Paper Series*, no. 24 (2006). https://www.academia.edu/9635509/Entre_résistance_et_contestation_La_genèse_du_squat_comme_mode_daction.

Pelloutier, Fernand. "L'anarchisme et les syndicats ouvriers." *Bibliothèque libertaire* (1895). http://kropot.free.fr/Pelloutier-anarsynd.htm.

Pelloutier, Fernand. "La situation actuelle du socialisme." *Pelloutier.net* (1895). http://www.pelloutier.net/dossiers/dossiers.php?id_dossier=189.

Periton, Diana. "The 'Coupe Anatomique': Sections Through the Nineteenth Century Parisian Apartment Block." *Journal of Architecture* 9, no. 3 (2004): 289–304.

Perrot, Michelle, ed. *A History of Private Life.* Vol. 4, *From the Fires of Revolution to the Great War.* Translated by Arthur Goldhammer. Harvard University Press, 1994.

Perrow, Charles. *Normal Accidents: Living with High-Risk Technologies.* Princeton University Press, 1999.

Petitjean, Narcisse-Nicolas. *Les grandes travaux de Paris.* 3rd ed. Paris: L. Thouvenin, 1895.

Pick, Daniel. *Faces of Degeneration: A European Disorder, c. 1848–1918.* Cambridge University Press, 1989.

Picon, Antoine. "French Engineers and Social Thought, 18–20th Centuries: An Archeology of Technocratic Ideals." *History and Technology* 23, no. 3 (September 2007): 197–208.

Pike, David. *Subterranean Cities: The World Beneath Paris and London, 1800–1945.* Cornell University Press, 2005.

Pinch, Trevor, and Wiebe Bijker. "The Social Construction of Facts and Artifacts: Or How the Sociology of Science and the Sociology of Technology Might Benefit Each Other." *Social Studies of Science* 14, no. 3 (1984): 399–441.

Pinkney, David. "Napoleon III's Transformation of Paris: The Origins and Development of the Idea." *Journal of Modern History* 27, no. 2 (June 1955): 125–34.

Polasky, Janet. "Transplanting and Rooting Workers in London and Brussels: A Comparative History." *Journal of Modern History* 73 (September 2001): 528–60.

Pollet, E. "La nouvelle ligne métropolitaine." *Bulletin technologique*, no. 11 (November 1904): 1004–12.

Popelin, Pascal. *Le jour où l'eau reviendra: 100 ans apres la grande crue de 1910.* Jean-Claude Gawesewitch, 2009.

Porter, Theodore. *Trust in Numbers: The Pursuit of Objectivity in Science and Public Life.* Princeton University Press, 1995.

Poubelle, Eugène. "Mémoire au Conseil municipal" (8 December 1884). In *Nouvelle derivation d'eau de source.* Paris: Imprimerie municipale, 1884.

Pouget, Émile. "À la cloche de bois." *Le père peinard*, 2nd series, no. 14, 24–31 January 1897, 4.

Pouget, Émile. "À propos des chiottes." *Le père peinard*, 16 February 1890, 89–91.

Pouget, Émile. *Almanach du père peinard.* Paris: Père peinard, 1894.

Pouget, Émile. "Barbarie française." *Le père peinard*, 12 January 1890, 89.

Pouget, Émile. "Garce d'influenza." *Le père peinard*, year 1, no. 47, 26 January 1890, 1–3.

Pouget, Émile. "La fête aux proprios." *Le père peinard*, year 4, no. 149, 24–31 January 1892, 1–2.

Pouget, Émile. "La mistoufle." *Le père peinard*, year 4, no. 171, 26 June–3 July 1892, 4–5.

Pouget, Émile. "Le choléra." *Le père peinard*, year 4, no. 166, 22–29 May 1892, 2–3.

Pouget, Émile. *Le père peinard: Un journal "espatrouillant" (1889–1900).* Les Nuits Rouges, 2006.

Pouget, Émile. "Les bouffe-gallette socialos." *Le père peinard*, 22 December 1889, 81–83.

Pouget, Émile. "Les tramways de Marseille." *Le père peinard*, 12–19 November 1893, 2.

Pouget, Émile. "Mon drapeau rouge." *Le père peinard*, 12 May 1889, 40.

Pouget, Émile. "Socialo à la Manque." *Le père peinard*, 14–21 February 1892, 5.

Pouget, Émile. "Toujours la dèche." *Le père peinard*, 29 December 1889, 85–87.

Pouget, Émile. "Toujours les grinches." *Le père peinard*, no. 37, 10 November 1889, 10–11.

Pouget, Émile. "Tous kif-kif!" *Le père peinard*, year 4, no. 151, 7–14 February 1892, 3.

Pouget, Émile. "V'la le choléra." *Le père peinard*, year 4, no. 182, 11–18 September 1892, 1–2.

"Pour lire avant la sieste." *La voie*, year 2, no. 66 (21 December 1935): 1.

Preaudeau, A. de. *Manuel hydrologique du bassin de la Seine.* Paris: Imprimerie nationale, 1884.

Prefecture de la Seine. *Commission d'extension de Paris.* 2 vols. Chaix, 1913.

Prefecture de la Seine, Direction des affaires municipales, Service de la statistique municipale. *Annuaire statistique de la ville de Paris*, year 15 (1894). Masson et Cie, 1896.

Prefecture de la Seine, Direction des affaires municipales, Service de la statistique municipale. *Annuaire statistique de la ville de Paris*, year 21 (1900). Masson et Cie, 1902.

Prefecture de la Seine, Direction des affaires municipales, Service de la statistique municipale. *Annuaire statistique de la ville de Paris*, year 23 (1902). Masson et Cie, 1904.

Prefecture de la Seine, Direction des affaires municipales, Service de la statistique municipale. *Annuaire statistique de la ville de Paris*, year 32 (1911). Masson et Cie, 1913.

Prefecture de la Seine, Secrétariat générale, Service de la statistique municipale. *Annuaire statistique de la ville de Paris*, year 1 (1880). Paris: Imprimerie nationale, 1881.

Prefecture de la Seine, Secrétariat générale, Service de la statistique municipale. *Annuaire statistique de la ville de Paris*, year 2 (1881). Paris: Imprimerie nationale, 1882.

Pritchard, Sara B. *Confluence: The Nature of Technology and the Remaking of the Rhône*. Harvard University Press, 2011.

Pritchard, Sara B. "An Envirotechnical Disaster: Nature, Technology, and Politics at Fukushima." *Environmental History* 17, no. 2 (1 April 2012): 219–43.

Pritchard, Sara B. "From Hydroimperialism to Hydrocapitalism: 'French' Hydraulics in France, North Africa, and Beyond." *Social Studies of Science* 42, no. 4 (2012): 591–615.

Pritchard, Sara B., and Carl Zimring. *Technology and Environment in History*. Johns Hopkins University Press, 2020.

Prochasson, Christophe. *Les années électriques, 1880–1910*. La découverte, 1991.

Projet de chemin de fer métropolitain pour Paris dressé par M. le Chatelier, ingénieur des Ponts et Chaussées: notice par A. Flamant, ingénieur en chef des Ponts et Chaussées. Paris: Publications du journal Le génie civil, 1889.

"Questions scientifiques: Les tramways électriques." *Le dimanche: Bulletin des corporations* 14, no. 28 (12 July 1902): 457–64.

Rabinbach, Anson. *The Human Motor: Energy, Fatigue and the Origins of Modernity*. University of California Press, 1992.

Rabinow, Paul. *French Modern: Norms and Forms of the Social Environment*. University of Chicago Press, 1995.

Ragon, Michel, ed. *L'assiette au beurre (1901–1912): l'âge d'or de la caricature*. Les nuits rouges, 2007.

Rammell, T. W. *Tramway souterrain à propulsion atmosphérique ou Pneumatic-railway, système Rammell: note sur une demande de concession faite pour l'application de ce système à Paris*. Paris: Chaix, 1881.

Ramsay, M.W. "L'épuration des eaux d'égout." *Sixth International Congress on Applied Chemistry, Rome, April 24 to May 3, 1906*. Bureau de la revue et du répertoire, 1906.

Rapport au nom de la Commission du métropolitain sur l'accident du chemin de fer métropolitain du 10 août 1903. Imprimerie municipale no. 61, 1904.

Reclus, Elisée. *Correspondance*. Vol. 2. Schleicher Frères: A. Costes, 1911–1925.

Regnier, L. R. "La lutte contre la tuberculose." *Journal d'hygiène* (3 February 1898): 52.

Reid, Donald. *Paris Sewers and Sewermen: Realities and Representations*. Harvard University Press, 1991.

Reid, Donald. "Putting Social Reform into Practice: Labor Inspectors in France, 1892–1914." *Journal of Social History* 20, no. 1 (1 October 1986): 67–87.

"Renseignements divers." *Revue générale des chemins de fer et des tramways* 25 (1902): 70.

"Résultat de concours de poésie: (Paris future) Le dernier cocher." *Le pêle-mêle*, 8 January 1898, 8.

"Revue foncière et bulletin de mutations immobilières." *La vie Parisienne* (2 April 1892): 194.

Reynolds, Sian. *Paris-Edinburgh: Cultural Connections in the Belle Epoque*. Ashgate, 2007.

Rider, John Hall. *Electric Traction: A Practical Handbook on the Application of Electricity as a Locomotive Power*. Whittaker and Co., 1903.

Rieger, Bernhard. *The People's Car: A Global History of the Volkswagen Beetle*. Harvard University Press, 2013.

Rieger, Bernhard. *Technology and the Culture of Modernity in Britain and Germany 1870–1930*. Cambridge University Press, 2005.

Rigaud, Lucien. *Dictionnaire d'argot moderne.* Paris: Paul Ollendorff, 1881.
Rigaud, Lucien. *Dictionnaire du jargon Parisien: l'argot ancien et l'argot moderne.* Paris: F. Aureau, 1878.
Robbins, Bruce. "The Smell of Infrastructure: Notes Toward an Archive." *boundary 2* 34, no. 1 (2007): 25–33.
Robert, Jean. *Les tramways Parisiens.* 3rd ed. RATP, 1992.
Robida, Albert. "L'embellissement de Paris par le métropolitain." *La caricature: journal hebdomadaire*, no. 338 (19 June 1886), cover.
Robida, Albert. *Le vingtième siècle: la vie électrique.* Paris: La librarie illustré, 1890.
Robin, Albert. "La lutte contre la tuberculose." *La revue de Paris* 10, no. 14 (15 July 1903): 269–91.
"Robinet intermittent." *La nature*, no. 745 (10 September 1887): 227.
Rochard, Jules Eugène. *Traité d'hygiène publique et privée.* Paris: Delahaye et Lecrosnier, 1888.
Rochegude, Félix Marquis de. *Guide pratique à travers le vieux Paris: maisons historiques ou curieuses anciens hotels pouvant être visités en trente-trois itinéraires détaillés.* Hachette, 1905.
Rochegude, Félix Marquis de. *Promenades dans toutes les rues de Paris par arrondissements.* Hachette, 1910.
Rollet, Catherine. "La canicule de 1911: obsérvations démographiques et médicales et réactions politiques." *Annales de démographie historique*, no. 120 (2010): 105–30.
Ross, Kristin. *Communal Luxury: The Political Imaginary of the Paris Commune.* Verso, 2015.
Ross, Kristin. *The Emergence of Social Space: Rimbaud and the Paris Commune.* University of Minnesota Press, 1988.
Rousseau, Daniel. "Les temperatures mensuelles en région Parisienne de 1676 à 2008." *La météorologie*, no. 67 (November 2009): 43–55.
Rousselle. "Compte rendu du Sécretariat: système Berlier, pour la réception et l'élimination des matières de vidange." *Journal d'hygiène* 8, year 9, no. 350 (7 June 1883): 282–83.
Ruby, Andreas, and Ilka Ruby, eds. *Infrastructure Space.* Ruby Press, 2017.
Ruby, Andreas, and Ilka Ruby. "Preface." In Ruby and Ruby, *Infrastructure Space*, 5–6.
Russell, Andrew, and Lee Vinsel. "After Innovation, Turn to Maintenance." *Technology and Culture* 59, no. 1 (January 2018): 1–25.
Rybczynski, Witold. *City Life.* Touchstone, 1995.
Saalman, Howard. *Haussmann: Paris Transformed.* G. Braziller, 1971.
Salet, Dr. *Mémoire sur l'avant-projet de dérivation des eaux d'égout de la ville de Paris.* St. Germain en Laye: T. Lancelin, 1876.
Sancha. "La plus Parisienne—celle qui ne quitte jamais Paris." *L'assiette au beurre*, no. 73, 23 August 1902, 1208.
Sante, Luc. *The Other Paris.* Farrar, Straus and Giroux, 2015.
Saraiva, Tiago, and Ana Cardoso De Matos. "Technological Nocturne: The Lisbon Industrial Institute and Romantic Engineering (1849–1888)." *Technology and Culture* 58, no. 2 (April 2017): 422–58.
Sarcey, Francisque. *Les odeurs de Paris: assainissement de la Seine.* Paris: Gauthier-Villars, 1882.
Savary, Robert, and Dr. Collet. "La lutte contre la tuberculose en France." *Les annales des sciences politiques* (1904): 490–94.
Say, Jean Baptiste. *Traité d'économie politique.* 8th ed. Paris: Guillaumin et Cie., 1876.
Schatzberg, Eric. "Culture and Technology in the City: Opposition to Mechanized Street Transportation in Late-Nineteenth Century America." In *Technology and History: Essays*

in Honor of Thomas Parke Hughes and Agatha Chipley Hughes, edited by Gabrielle Hecht and Michael Allen, 57–94. MIT Press, 2001.

Schatzberg, Eric. "The Mechanization of Urban Transit in the United States: Electricity and Its Competitors." In *Technological Competitiveness: Contemporary and Historical Perspectives on Electrical, Electronics, and Computer Industries*, edited by William Aspray, 225–42. IEEE Press, 1993.

Schipper, Frank, and Johan Schot. "Infrastructural Europeanism, or the Project of Building Europe on Infrastructures: An Introduction." *History and Technology* 27, no. 3 (2011): 245–64.

Schivelbusch, Wolfgang. *Disenchanted Night: The Industrialization of Light in the Nineteenth Century*. Translated by Angela Davies. University of California Press, 1995.

Schivelbusch, Wolfgang. *The Railway Journey: The Industrialization of Time and Space in the Nineteenth Century*. University of California Press, 1986.

Schmucki, Barbara. "The Machine in the City: Public Appropriation of the Tramway in Britain and Germany, 1870–1915." *Journal of Urban History* 38, no. 6 (2012): 1060–93.

Schmucki, Barbara. "On the Trams: Women, Men, and Urban Public Transport in Germany." *Journal of Transport History* 23, no. 1 (2002): 60–72.

Schmucki, Barbara. "Toward New Routes in Tramway History." In *Mobility in History: Themes in Transport*, edited by Gijs Mom, 127–33. Éditions Alphil, 2010.

Schneider, Daniel. *Hybrid Nature: Sewage Treatment and the Contradictions of the Industrial Ecosystem*. MIT Press, 2011.

Schneider, William. *An Empire for the Masses: The French Popular Image of Africa, 1870–1900*. Greenwood Press, 1982.

Schoeller, Adolphe. *Les chemins de fer: les tramways, les chemins de fer éléctriques*. Librarie J. B. Baillière et Fils, 1902.

Schott, Dieter, Bill Luckin, and Geneviève Massard-Guilbaud, eds. *Resources of the City: Contributions to an Environmental History of Modern Europe*. Routledge, 2005.

Schroeder-Gudehus, Brigitte, and Anne Rasmussen. *Les fastes du progrès: le guide des expositions universelles*. Flammarion, 1992.

Schwartz, Vanessa. *Spectacular Realities: Early Mass Culture in Fin-de-Siècle Paris*. University of California Press, 1998.

Scott, James. *Seeing Like a State: How Certain Schemes to Improve the Human Condition Have Failed*. Yale University Press, 1998.

Sedgwick, Eve Kosofsky. *Touching Feeling: Affect, Pedagogy, Performativity*. Duke University Press, 2003.

Seigel, Jerrold. *Bohemian Paris: Culture, Politics, and the Boundaries of Bourgeois Life, 1830–1930*. Johns Hopkins University Press, 1986.

Sérafon, Félix. *Étude sur les chemins de fer, les tramways et les moyens de transport en commun à Paris et à Londres*. Paris: Dunod, 1872.

Sérafon, Félix. *La verité sur les tramways-nord & sud de Paris*. Paris: Imprimerie de la publicité, 1882.

Shapiro, Ann-Louise. *Housing the Poor of Paris 1850–1902*. University of Wisconsin Press, 1985.

Shapiro, Michael. *The Politics of Representation: Writing Practices in Biography, Photography, and Political Analysis*. University of Wisconsin Press, 1988.

Shattuck, Roger. *The Banquet Years: The Origins of the Avant-Garde in France, 1885 to World War I*. Revised edition. Vintage Books, 1968.

Shaw, Albert. *Municipal Government in Continental Europe*. New York: Century Co., 1895.

Shaya, Gregory. "The Flâneur, the Badaud, and the Making of a Mass Public in France, Circa 1860–1910." *American Historical Review* 109, no. 1 (February 2004): 41–77.

Shorter, Edward, and Charles Tilly. *Strikes in France 1830–1968*. Cambridge University Press, 1974.

Shukin, Nicole. *Animal Capital: Rendering Life in Biopolitical Times*. University of Minnesota Press, 2009.

Sibey, Charles. "Jurisprudence: Tramways électriques et cochers imprudents." *L'électricien* 10, series 2, no. 255 (16 November 1895): 314–16.

Simon, Jules. "Les comptes fantastiques d'Haussmann." *Le Gaulois*, 3 May 1882, 1.

Simon, Linda. *Dark Light: Electricity and Anxiety from the Telegraph to the X-Ray*. Harcourt Trade, 2005.

Simond, Charles. *La vie Parisienne à travers le XIXe siècle: Paris de 1800 à 1900 d'après les estampes et les mémoires du temps*. Vol. 2. Plon, 1900.

Simond, Charles. *La vie Parisienne à travers le XIXe siècle: Paris de 1800 à 1900 d'après les estampes et les mémoires du temps*. Vol. 3. Plon, 1901.

Sizeranne, Robert de La. "Les prisons de l'art." *La revue des deux mondes*, fourth period, vol. 156 (1 November 1899): 114–38.

Smith, Cecil O. Jr. "The Longest Run: Public Engineers and Planning in France." *American Historical Review* 95 (1990): 657–92.

Smith, Robert Henry. *Electric Traction*. Harper, 1905.

Société des ingénieurs civils de France. *Discussion sur l'emploi de l'air comprimé pour la locomotion mécanique par la procédés L. Mékarski*. Paris: Imprimerie Viéville et Capiomont, 1876.

Société des ingénieurs civils de France. *Mémoires et compte rendu des travaux*. Vol. 25. Paris: Librarie scientifique, industrielle et agricole, 1872.

Société des ingénieurs civils de France. *Mémoires et compte rendu des travaux*. Vol. 38, 2nd semester. Paris, 1882.

Société des ingénieurs civils de France. *Mémoires et compte rendu des travaux*. Vol. 39, 1st semester. Paris, 1883.

Société des ingénieurs civils de France. *Mémoires et compte rendu des travaux*. Vol. 40, 2nd semester. Paris, 1883.

Société générale d'éclairage et de force motrice. *Tramways pneumatiques Popp-Conti*. Paris: Chaix, 1896.

"Société Médicale des Hôpitaux." *Gazette hebdomadaire de medecine et de chiurgie*, no. 3 (16 January 1874): 44.

Soffer, Jonathan. "Introduction." In Heathcott et al., *Urban Infrastructure*, 3–8.

Sonn, Richard. *Anarchism and Cultural Politics in Fin de Siècle France*. University of Nebraska Press, 1989.

Sonn, Richard. "Marginality and Transgression: Anarchy's Subversive Allure." In Weisberg, *Montmartre*, 120–41.

Soppelsa, Peter. "Finding Fragility in Paris: The Politics of Infrastructure After Haussmann." *Proceedings of the Western Society for French History* 37 (2009): 233–47.

Soppelsa, Peter. "How Haussmann Haunted the Early Third Republic." In Clayton and Dombrowski, *Is Paris Still the Capital?* 35–51.

Soppelsa, Peter. "The Instrumentalisation of Horses in Nineteenth-Century Paris." In *Anthropocentrism: Humans, Animals, Environments*, edited by Rob Boddice, 245–64. Brill, 2011.

Soppelsa, Peter. "Paris's 1900 Universal Exposition and the Politics of Urban Disaster." *French Historical Studies* 36, no. 2 (Spring 2013): 271–98.

Soppelsa, Peter. "Reworking Appropriation: The Language of Paris Railways." *Transfers: Interdisciplinary Journal of Mobility Studies* 4, no. 2 (Summer 2014): 104–23.

Soppelsa, Peter. "Theorizing Infrastructure and Affect." In Heathcott et al., *Urban Infrastructure*, 207–22.

Soppelsa, Peter. "Urban Railways, Industrial Infrastructure and the Paris Cityscape, 1870–1914." In *Trains, Culture and Mobility*, edited by Benjamin Fraser and Steven D. Spalding, 117–44. Lexington Books, 2011.

Soppelsa, Peter. "Visualizing Viaducts in 1880s Paris." *History and Technology* 27 no. 3 (September 2011): 371–77.

Soppelsa, Peter. "Water and Power in Modern France." *French Politics, Culture and Society* 31, no. 2 (Summer 2013): 117–32.

Soppelsa, Peter, and Amy S. Rodgers. "Origins of the Flyswatter." *Technology and Culture* 60, no. 3 (July 2019): 886–95.

"Souvenirs de l'exposition d'électricité." *Le magasin pittoresque* 50 (1882): 59–62.

Spector, Scott. *Prague Territories: National Conflict and Cultural Innovation in Franz Kafka's Fin de Siecle*. University of California, 2002.

Spencer, Michael. "A Fresh Look at Rimbaud's 'Métropolitain.'" *Modern Language Review* 63 (October 1968): 849–53.

Starosielski, Nicole. *The Undersea Network*. Duke University Press, 2015.

Starosielski, Nicole. "'Warning: Do Not Dig': Negotiating the Visibility of Critical Infrastructures." *Journal of Visual Culture* 11, no. 1 (April 2012): 38–57.

Steinberg, Ted. *Acts of God: The Unnatural History of Natural Disaster in America*. Oxford University Press, 2000.

Stone, Judith. *The Search for Social Peace: Reform Legislation in France, 1890–1914*. State University of New York Press, 1985.

Stovall, Tyler. "The Consumers' War: Paris, 1914–1918." *French Historical Studies* 31, no. 2 (2008): 293–325.

Stovall, Tyler. *The Rise of the Paris Red Belt*. University of California Press, 1990.

Strauss, Jonathan. *Human Remains: Medicine, Death, and Desire in Nineteenth-Century Paris*. Fordham University Press, 2012.

Strauss, Paul. "La disette d'eau à Paris." *La revue bleue* 6, ser. 4, no. 4 (25 July 1896): 130–34.

Strauss, Paul. "La vie municipale: le budget et les grands travaux de Paris." *La revue bleue* 6, ser. 4, no. 11 (12 September 1896): 322–26.

Strauss, Paul. "Les eaux de Paris." *La revue politique et littéraire* (1893): 430–34.

Suominen, Jaakko, Antti Silvast, and J. Tuomas Harviainen. "Smelling Machine History." *Technology and Culture* 59, no. 2 (April 2018): 313–37.

"Sur la traction mécanique à Paris." *Bulletin de la Société internationale des électriciens* 13 (1896): 427–61.

Sutcliffe, Anthony. *The Autumn of Central Paris: The Defeat of Town Planning, 1850–1970*. McGill-Queen's University Press, 1971.

Sutcliffe, Anthony. *Paris: An Architectural History*. Yale University Press, 1993.

Swyngedouw, Erik. "Circulations and Metabolisms: (Hybrid) Natures and (Cyborg) Cities." In *Technonatures: Environments, Technologies, Spaces, and Places in the Twenty-First Century*, edited by Damian F. White and Chris Wilbert, 61–84. Wilfrid Laurier University Press, 2010.

Swyngedouw, Erik. "Metabolic Urbanization: The Making of Cyborg Cities." In Heynen et al., *In the Nature of Cities*, 21–40.

T., A. "Les odeurs de Paris en 1911." *La nature* 40, no. 2056 (19 October 1912): 349–50.

Tables du journal Le Temps. Vol. 3, 1871–1875. Éditions du Centre national de la recherche scientifique, 1968.

Tainturier, C. *La traction électrique: tramways, chemins de fer et petites voitures. Étude d'une ligne. Usines génératrices. Matériel roulant. Lois et règlements*. Paris: J. Fritsch, 1897.

Tarr, Joel. "The Metabolism of the Industrial City: The Case of Pittsburgh." *Journal of Urban History* 28, no. 5 (2002): 511–45.

Tarr, Joel, and Gabriel Dupuy, eds. *Technology and the Rise of the Networked City*. Temple University Press, 1988.

Tarr, Joel, and Clay McShane. *The Horse in the City: Living Machines in the Nineteenth Century*. Johns Hopkins University Press, 2007.

"Taxes on Transport." *Nineteenth Century* 31 (1892): 134–49.

Taylor, E. A. *Paris: Past and Present*. "The Studio" Ltd., 1915.

Ten Hoor, Meredith. "Architecture and Biopolitics at Les Halles." *French Politics, Culture & Society* 25, no. 2 (Summer 2007): 73–92.

Terdiman, Richard. "The Mnemonics of Dispossession: 'Le Cygne' in 1859." In *Home and Its Dislocations in Nineteenth-Century France*, edited by Suzanne Nash, 169–89. State University of New York Press, 1993.

Terrail, Mermeix Gabriel. *L'Angleterre: aspects inconnus*. Ollendorff, 1911.

Théry, J.A. *Les chemins métropolitains de Paris: la vérité sur l'exécution et la dépense des chemins métropolitains souterrains*. Paris: Lambert, 1882.

Thierry, J.B. *Étude sur le métropolitain de Paris: ses installations intérieures, ce qu'elles sont—ce qu'elles devraient* être. Librarie polytechnique Charles Beranger, 1907.

Thoinot, L. "La fièvre typhoïde à Paris de 1870 à 1899: role actuel des eaux de source." *Bulletins et mémoires de la Société médicale des hôpitaux de Paris* 16, 3rd series (Paris: Masson, 1899): 645–69.

Thoinot, L. "La vie et l'oeuvre de Paul Brouardel (1837–1906)." *Les annales d'hygiène publique et de médecine légale* 6, series 4, no. 3 (1906): 229.

Thompson, Victoria. "Review Essay: Boulevard Dreams: Paris and the Myths of Modernity." *Journal of Urban History* 33 (2007): 664–69.

Tissanie-Noir, Joëlle. "Le système de santé des sociétés de transports en commun (Paris, 1855–1948)." *Les cahiers du Centre de recherches historiques* 12 (February 1994). https://journals.openedition.org/ccrh/2736.

Tomory, Leslie. "London's Water Supply Before 1800 and the Roots of the Networked City." *Technology and Culture* 56, no. 3 (July 2015): 704–37.

Toulhouse, M. de. *Aux pays de la soif*. Paris: Vic et Amat, 1891.

Tourdonnet, Comte A. de. "De la crise des loyers." *La revue contemporaine* 30 (February–March 1857): 696–730.

"Tout à l'égout, rien à la Seine." *Journal d'hygiène* 9, no. 427 (27 November 1884): 576.

Tracy, Roger S. "The Problem of Municipal Nuisances." *Popular Science* (March 1881): 585–99.

Transactions of the British Congress on Tuberculosis for the Prevention of Consumption. Vol. 1. William Clowes and Sons, Ltd., 1902.

Tresca, Alphonse. "Arts mécaniques." *Bulletin de la Société d'encouragement pour l'industrie nationale* 90 (1891): 25–27.

Tresch, John. *The Romantic Machine: Utopian Science and Technology After Napoleon*. University of Chicago Press, 2012.

"Tribunal civil de la Seine (20 avril 1898)." In *Revue générale d'administration*, 71–72. Ministère de l'intérieur, 1898.

Tricoire, Jean. "L'exploitation du métropolitain: de l'improvisation à la méthode." In Hallsted-Baumert et al., *Métro-Cité*, 103–16.

Tricoire, Jean, ed. *Le Métro de Paris: 1899–1911 images de la construction*. Paris musées/RATP, 1999.

Tricoire, Jean, ed. *Métropolitain: l'autre dimension de la ville*. Hôtel de Lamoignon, 1988.

Trillat, A. *Commission des odeurs de Paris: rapport général*. Imprimerie générale, 1928.

Tucker, Kenneth. *French Revolutionary Syndicalism and the Public Sphere*. Cambridge University Press, 1996.

"Un chasse-neige pneumatique." *La nature*, no. 505 (5 February 1883), 145–46.

Union photographique française. *Chemin de fer métropolitain municipal de Paris*. Vol. 1, *Ligne circulaire nord 1900–1903*. École nationale des ponts et chaussées, 1900–1903.

Union photographique française. *Chemin de fer métropolitain municipal de Paris*. Vol. 2, *Ligne circulaire sud 1902–1905*. École nationale des ponts et chaussées, 1902–1905.

Union photographique française. *Chemin de fer métropolitain municipal de Paris*. Vol. 3, *Ligne du boulevard de Courcelles à Ménilmontant 1902–1904*. École nationale des ponts et chaussées, 1902–1904.

Union vélocipédique de France. "Faux départ." *Bulletin officiel*, year 9 no. 51 (December 10, 1899): 532–33.

Uzanne, Octave. *La locomotion à travers l'histoire et les moeurs*. Paris: Paul Ollendorff, 1900.

Van Zanten, David. "Paris Space: What Might Have Constituted Haussmannization." In *Manifestoes and Transformations in the Early Modernist City*, edited by Christian Hermansen Cordua, 179–210. Routledge, 2010.

Vanderslaghmolen, F. *De l'amélioration de la condition morale et matérielle des ouvriers en Belgique*. M. J. Poot, 1865.

Velu, H. "La guerre microbienne est-elle réalisable?" *Recueil de médecine vétérinaire* 111 (1935): 822–52.

Verne, Jules. *Journey Through the Impossible*. Translated by Edward Baxter. Prometheus Books, 2003.

Verne, Jules. *La tour du monde en quatre-vingt jours*. Paris: Hetzel, 1873.

Verne, Jules. *Robur-le-Conquérant*. Paris: Hetzel, 1886.

Vernon-Harcourt, Leveson Francis. *Rivers and Canals: The Flow, Control, and Improvement of Rivers and the Design, Construction, and Development of Canals both for Navigation and Irrigation, with Statistics of the Traffic on Inland Waterways*. Oxford: Clarendon, 1896.

Vernon-Harcourt, Leveson Francis. *The River Seine*. London: Institute of Civil Engineers, 1886.

Veron, Pierre. "Humble requête d'un barbillon d'Argenteuil à sa toute-puissance M. le Préfet de la Seine." *Le charivari*, 15 September 1865, 1–2.

Veuillot, Louis. *Les odeurs de Paris*. Paris: Palmé, 1867.

Vèze, G.-R. "Les embarras de Paris." *Le mois litteraire et pittoresque* (1905): 347.

Vibert, Charles. *La catastrophe du métropolitain* (Extrait des *Annales d'hygiène publique et de médicine légale*). Librarie J.-B. Baillière et fils, 1905.

Vibert, Paul. *La concurrence étrangère, les transports par terre et par mer*. 2 vols. Paris: Berger-Levrault, 1896–97.

Vignes, Émile. *La traction méchanique des tramways*. Paris: E. Bernard, 1894.

Villain, Paul. *Le Métro qu'on peut faire*. Paris: Grande Imprimerie, 1891.

Villain, Paul, and E. Mauger. *Un métropolitain qui ne coûte rien et ne trouble rien*. Paris: Grande Imprimerie, 1892.

Virmaitre, Charles. *Dictionnaire d'argot fin-de-siècle*. Paris: A Charles, 1894.

"Vital Statistics of Paris and Vienna." *Medical Times and Gazette* (3 December 1881): 666.

Wagenaar, Michiel. "Conquest of the Center or Flight to the Suburbs? Divergent Metropolitan Strategies in Europe, 1850–1914." *Journal of Urban History* 19, no. 1 (November 1992): 61–62.

Wagner, Matthieu. "Décembre 1879 à Paris: de neige, de glace . . . et de sel!" *Méchaniques du ciel* (blog), 26 February 2013. https://blog.francetvinfo.fr/mecaniques-du-ciel/2013/02/26/decembre-1879-a-paris-de-neige-de-glace-et-de-sel.html.

Wakeman, Rosemary. *The Heroic City: Paris, 1945–1958*. University of Chicago Press, 2009.

Wakeman, Rosemary. "Nostalgic Modernism and the Invention of Paris in the Twentieth Century." *French Historical Studies* 27, no. 1 (Winter 2004): 115–44.

Walton, William. *Paris, from the Earliest Period to the Present Day*. 9 vols. George Barrie and Son, 1900.

Walton, William. *Paris Known and Unknown*. Vol. 1. Philadelphia: George Barrie and Son, 1898.

Walz, Robin. *Pulp Surrealism: Insolent Popular Culture in Early Twentieth-Century Paris*. University of California Press, 2000.

Warner, Sam Bass. *Streetcar Suburbs: The Process of Growth in Boston, 1870–1900*. 2nd ed. Harvard University Press, 1978.

"Water." *Technology and Culture*. Special issue 49, no. 3 (July 2008).

Weber, Eugen. *France, Fin de Siècle*. Harvard University Press, 1986.

Weber, Eugen. *Peasants into Frenchmen: The Modernization of Rural France, 1870–1914*. Stanford University Press, 1976.

Weil, Kari. "Purebreds and Amazons: Saying Things with Horses in Late-Nineteenth-Century France." *differences: A Journal of Feminist Cultural Studies* 11, no. 1 (1999): 1–37.

Weisberg, Gabriel P., ed. *Montmartre and the Making of Mass Culture*. Rutgers University Press, 2001.

Weiss, Sean. "The 'Dravert Affair': Paperwork and the Administration of Negligent Street Maintenance in Modern Paris." *Journal of Urban History* 46, no. 2 (2020): 289–309.

Weiss, Sean. "Making Engineering Visible: Photography and the Politics of Drinking Water in Modern Paris." *Technology and Culture* 61, no. 3 (2020): 739–71.

Weisz, George. *The Medical Mandarins: The French Academy of Medicine in the Nineteenth and Early Twentieth Centuries*. Oxford University Press, 1995.

Whatmore, Sarah. *Hybrid Geographies: Natures, Cultures, Spaces*. Sage, 2002.

White, Richard. "From Wilderness to Hybrid Landscapes: The Cultural Turn in Environmental History." *Historian* (2004): 559–64.

White, Richard. *The Organic Machine*. Hill and Wang, 1995.

Wiebe Bijker, Thomas Hughes, and Trevor Pinch, eds. *The Social Construction of Technological* Systems. MIT Press, 1987.

Wileman, Donald G. "Not the Radical Republic: Liberal Ideology and Central Blandishment in France, 1901–1914." *Historical Journal* 37, no. 3 (September 1994): 593–614.

Willems, Phillipe. "A Stereoscopic Vision of the Future: Albert Robida's Twentieth Century." *Science Fiction Studies* 26, no. 3 (1999): 354–78.

Williams, Rosalind. *Dream Worlds: Mass Consumption in Late Nineteenth-Century France*. University of California Press, 1982.

Williams, Rosalind. *Notes on the Underground: An Essay on Technology, Society and the Imagination.* MIT Press, 1992.

Willms, Johannes. *Paris, Capital of Europe: From the Revolution to the Belle Epoque.* Holmes and Meier, 1997.

Withington, John. *Flood: Nature and Culture.* Reaktion Books, 2013.

Wolf, Peter M. *Eugène Hénard and the Beginning of Urbanism in Paris 1900–1914.* International Federation for Housing and Planning/Centre de recherche d'urbanisme (IHP-CRU), 1968.

Worms, Dr. Jules. *Rapport adressé a M. Le Préfet de la Seine sur l'épidémie de choléra qui a sévi a Paris pendant les mois de septembre, octobre et novembre 1873.* Paris: Charles de Mourgues frères, 1874.

Worster, Donald. *Rivers of Empire: Water, Aridity, and the Growth of the American West.* Oxford University Press, 1985.

Yates, Alexia. *Selling Paris: Property and Commercial Culture in the Fin-de-Siècle Capital.* Harvard University Press, 2015.

Zahra, Tara. "Behind the Storm." *Nation*, 12 November 2013. https://www.thenation.com/article/archive/behind-storm/.

Zaretsky, Rob. "The Sewers of Paris." *The Engines of Our Ingenuity*, radio program, University of Houston KUHF, episode 1966. http://www.uh.edu/engines/epi1966.htm.

Zeller, Thomas. "Aiming for Control, Haunted by Its Failure: Towards an Envirotechnical Understanding of Infrastructures." *Global Environment* 10, no. 1 (April 2017): 202–28.

Zola, Émile. *The "Assommoir": A Realistic Novel.* Translated by Vizetelly. London: Vizetelly, 1884.

Zola, Émile. *The Flood.* Warren Press, 1911.

Zola, Émile. *The Ladies' Paradise.* Translated by Brian Nelson. Oxford University Press, 1998.

Zylstra, Geoff D. "Whiteness, Freedom, and Technology: The Racial Struggle over Philadelphia's Streetcars, 1859–1867." *Technology and Culture* 52, no. 4 (2011): 678–702.

Periodicals

À travers le monde
American Journal of Sociology
Annual Register
Banque Industrielle & Coloniale
Bibliothèque universelle et revue Suisse
Bulletin de l'Académie nationale de médecine
Bulletin médical
Bulletin municipal officiel de la ville de Paris
Bulletin technologique
Bulletins et mémoires de la Société médicale des hôpitaux de Paris
Chicago Daily
Comoedia
Comptes rendus hebdomaires des séances de l'Académie des sciences
Cosmos
Engineering News
Gazette anecdotique, littéraire, artistique et bibliographique
Gil Blas
Je sais tout
Journal d'hygiène

Journal de la Société de statistique de Paris
Journal des débats politiques et littéraires
Journal officiel de la République française
L'action
L'actualité
L'anti-concièrge: organe officiel de la défense des locataires
L'assiette au beurre
L'aurore
L'autorité
L'avenir républicaine
L'éclair
L'éclairage électrique
L'économiste français
L'égalité
L'électricien: Revue internationale de l'électricité et de ses applications
L'événement
L'humanité
L'intransigeant
L'univers
La baionnette
La construction moderne
La croix
La justice
La lanterne
La liberté
La libre parole
La lumière électrique
La nature
La patrie
La petite presse
La petite république
La presse
La quinzaine: revue littéraire, artistique, et scientifique
La réforme sociale
La revue bleue
La revue contemporaine
La revue d'électricité
La revue de Paris
La revue des deux mondes
La revue mondiale
La revue scientifique
La revue universelle
La rive gauche
La science au XXe siècle: nouvelle revue illustrée des sciences
La vie Parisienne
Le bâtiment

Le charivari
Le correspondant
Le courrier bleu de Neuilly-Boulogne
Le dimanche: Bulletin des corporations
Le écho de Paris
Le Figaro
Le Gaulois
Le génie civil
Le journal
Le journal des transports
Le magasin pittoresque
Le matin
Le mois littéraire et pittoresque
Le monde illustré
Le moniteur de Paris
Le parti national
Le pêle-mêle
Le père peinard
Le petit journal
Le petit Parisien
Le radical
Le rappel
Le siècle
Le soleil
Le temps
Le vélo
Le XIXe siècle
Les annales d'hygiène publique et de médecine légale
Les annales d'hygiène publique, industrielle et sociale
Les annales des ponts et chaussées
Les annales des sciences politiques
Les annales politiques et littéraires
Les causeries scientifiques
Les merveilles de la science
Les nouvelles
Medical Times and Gazette
Mercure de France
Nineteenth Century
Paris ouest
Popular Mechanics
Popular Science Monthly
Punch
Recueil de médicine vétérinaire
Revue générale de chimie
Scribner's Magazine
Times (London)

Index

Note: Page numbers in *italics* indicate figures.

Printed and bound by CPI Group (UK) Ltd, Croydon, CR0 4YY
16/09/2026
14963482-0001